Great Sleds & Wagons!

The YOUTH'S COMPANION combined with American Boy
January 1934
Founded 1827

Joan Palicia

Schiffer Publishing Ltd
4880 Lower Valley Road Atglen, Pennsylvania 19310

In Memorium:

In memory of
Bessie & Francis Reise, my parents;
Dr. Gerald F. Reise, DDS, my brother;
Tillie & Joseph Palicia, my inlaws;
& Hollyberry 11, my Beloved English Springer Spaniel

Dedication

To my husband Mel and my wonderful family, Deborah, Glen, Sharon and Megan, affectionately known as "The Megster." Thank you for supporting my decision to do another book. This book was truly a labor of love and you made it all possible.

Library of Congress Control Number: 2008935907

Covers designed by Bruce Waters
Type set in Korinna BT

ISBN: 978-0-7643-3217-3
Printed in China

Published by Schiffer Publishing Ltd.
4880 Lower Valley Road
Atglen, PA 19310
Phone: (610) 593-1777; Fax: (610) 593-2002
E-mail: Info@schifferbooks.com

Contents

Acknowledgments

Writing a book is not an easy task, and can not be accomplished alone. Many individuals, historians, institutions, and collectors around the country have made this book possible. The decision to do another book was mainly due to all the collectors out there that continually reminded me that without another book some of their prize possessions would never be identified. That is not to say, I have been able to identify every Sled manufacturing company in this country. I'm sure there are hundreds, if not thousands, that will remain a mystery. Part of the allure of collecting is the mystery that surrounds that one item you can not identify.

To the collectors that shared their collections with me, to the historians for access to historical ephemera, trade literature, and other research material, I owe you a debt of gratitude, many thanks. Special thanks to Edie Raught, volunteer at the Mt. Jewett Pa. Public Library, for putting me in touch with Barbara Hagg, granddaughter of the founder of the Safety Sled Company. Barbara, many thanks to you for sharing your wonderful Salesman's Samples, and allowing us to photograph them and your Safety-Line Sled Catalog. To my old friends Carol & Lou Scudillo, for going above and beyond, allowing me to photograph your wonderful Sleds in the midst of your move out of state. I will miss having you a stones through away to compare notes, thank you. To former strangers and now new friends, I would to thank Paul Cote, Jim Escher, Sam Tressler, Carroll "Skip" Palmer, Jane Dwyer Garton, Ann Huenink Mc Intyre, and Carol & Jim Pauzé, for allowing us to photograph their fantastic collections of Sleds, Wagons, and Ephemera, without it this book would not be complete. To my husband Mel, words can not express my love and gratitude, for bailing me out and traveling to Maine to complete the photography for this book,"Super Job!" To my family and friends, for sharing in all my excitement and accompanying me on all the journeys to "Where?" It would have been impossible without your help many thanks, and thank God for the GPS! It goes without saying, many thanks to the entire Schiffer Publishing Staff. To Peter for convincing me to do another book and to Doug Condon Martin editor and Chief for keeping me focused and calming me down when I thought everything was lost in "Cyber space." Thanks, Doug. As the author I welcome any additional information or comments. You may write to Joan Palicia, 15 Canton Rd. Wayne, N.J. 07470. Please send self addressed stamped envelope for reply or e-mail jpsledssnofn@nac.net

Introduction

Several years have passed since my first book was published, but that's not to say thoughts of sleds and sledding were never more than a blink of an eye away. As I look around my office I am surrounded by constant reminders. On the walls are scenes like "Breakfast at Copperfields" by Charles Wysocki, with all of those great sleds, their decks brightly decorated, lined up against the fence in front of the Inn, waiting for their owners to claim them, before heading back to "The Hill" for another adventure. There are magazine covers, like Norman Rockwells "Deadman's Hill" and "Down Hill Racing," or "Grandma Takes a Ride" by an unknown artist...that one really hits home.

Instantly, my mind is flooded with memories and scenes of bygone years. The human mind is a wonderful thing. It allows us to revisit our past and imagine our future by projecting our thoughts and memories on that giant cinema screen we call our subconscious. For me, in my minds eye, the past was racing down Sandy Hill Road, on my vintage Flexible Flyer and being the first to reach the bottom of the hill to claim the title, in today's vernacular, "Queen of the Hill."

Fast forward, the scene changes, I find myself racing down the hill on my Flyer gathering up everything from advertising, vintage posters, company catalogs, membership pins, sleds, "odd ball" snow vehicles, and anything with runners. I realize I have created an avalanche. Digging through the mounds of information I'm faced with a whiteout. The tools for compiling antique research are mainly catalogs, company advertising, giveaways, and photographs. These items are what validate and make a specialized collection both interesting and valuable. As collectors we know how valuable it is to have an accurate resource to use as a guide, and I am grateful for your support and encouragement to do another sled book.

Once again the scene changes and I'm faced with that nagging question, "Where do I begin?" As with any collection we try to stay focused, on one subject, but how can that apply when we find something so incredible we just have to have it. I'm sure you have all been there at one time or another, and I am no exception. Along the way I stumbled onto small child-sized sleds, some even smaller, I was told many different versions of the story behind these miniatures. Some said that they were toys, others said salesman's samples, and, I found through research, that some are actually US Patent models. Several examples of these rare items are included in this book

As mentioned in my first book, research is painstakingly slow. Over the years I have been able to research other American companies not mentioned in my first book. The rules remain the same, pay close attention to the smallest details, even a slight change in style, design, or color can put your sled in a different decade, and increase or decrease its value.

As collectors we are all aware that sled manufacturing is a seasonal business. Like S.L. Allen, who realized that his agricultural business Planet Junior, makers of farm and garden equipment, needed something to keep his employees busy all year, the not-so-famous companies that jumped on the band wagon to get into the sled business soon realized that they needed to manufacture other things. Soon they were making such things as children's furniture, wagons, shoo-flys, and "odd ball" snow vehicles in order to compete.

In this book I have included, whenever possible, some of these items as they appeared in the company catalogs. As I said before it's hard to stay focused when you see, for instance, a neat wagon or patrol wagon, but in my case, I have resisted all temptation. The deterrent has not been for lack of interest, but for lack of time. I figured that if it took twenty plus years of collecting and researching sleds to do my first book and ten years plus, to do my second, I would not have enough time to research another subject. I will leave that to anyone willing to pick up where I have left off.

Leafing through these pages, I hope you find something that touches your subconscious and allows you also to revisit the past and imagine your future if only for a brief moment. Whatever your interest, sleds, wagons, or just nostalgic things, sit back, relax, close your eyes, and enjoy the views.

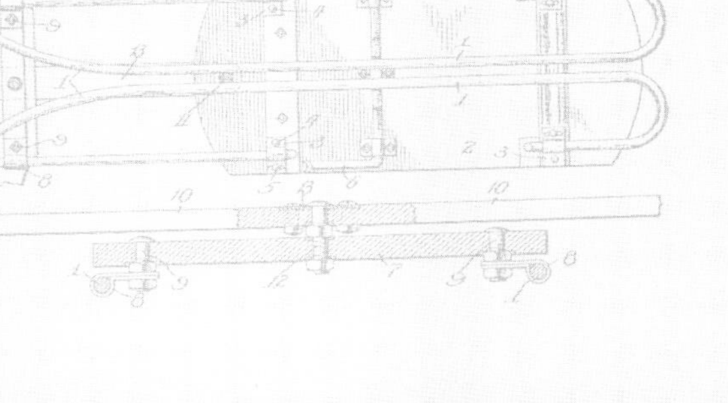

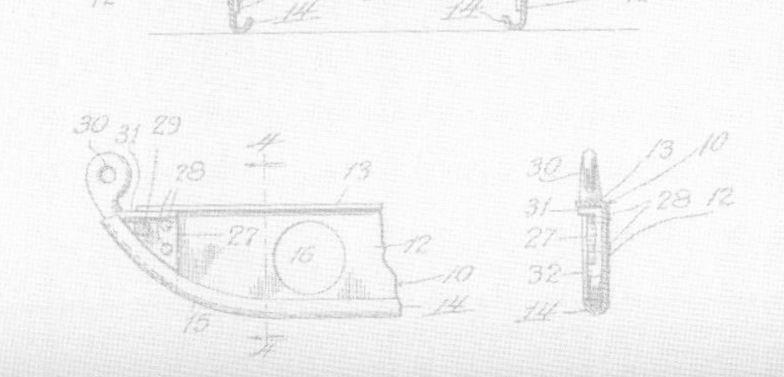

Common Terms

Artillery wheels. Small radiating bars (spokes) inserted into the hub of the wheel to support the rim.

Coaster. A small vehicle as a sled or wagon used for coasting

Daisy. A first rate person or thing. Top of the line

Dasher. A board or apron in front of the wagon, or removable sides of a cart to prevent mud from being splashed upon the interior of the vehicle.

Excelsior. Highest quality; trade name

Felloes. The exterior rim of the wheel supported by spokes. The curved pieces of wood which, when joined together, form the circular rim of the wheel.

Plush. A fabric with even pile longer and less dense than velvet

Satin Russé. Deep ruby brocade fabric characterized by raised designs.

Shoe-Fly. A child's rocker, having a seat built between supports representing an animal figure.

Tongue. The pole of a vehicle (handle)

Vermillion. Bright red/vivid orange. The pigment contains mercuric oxide

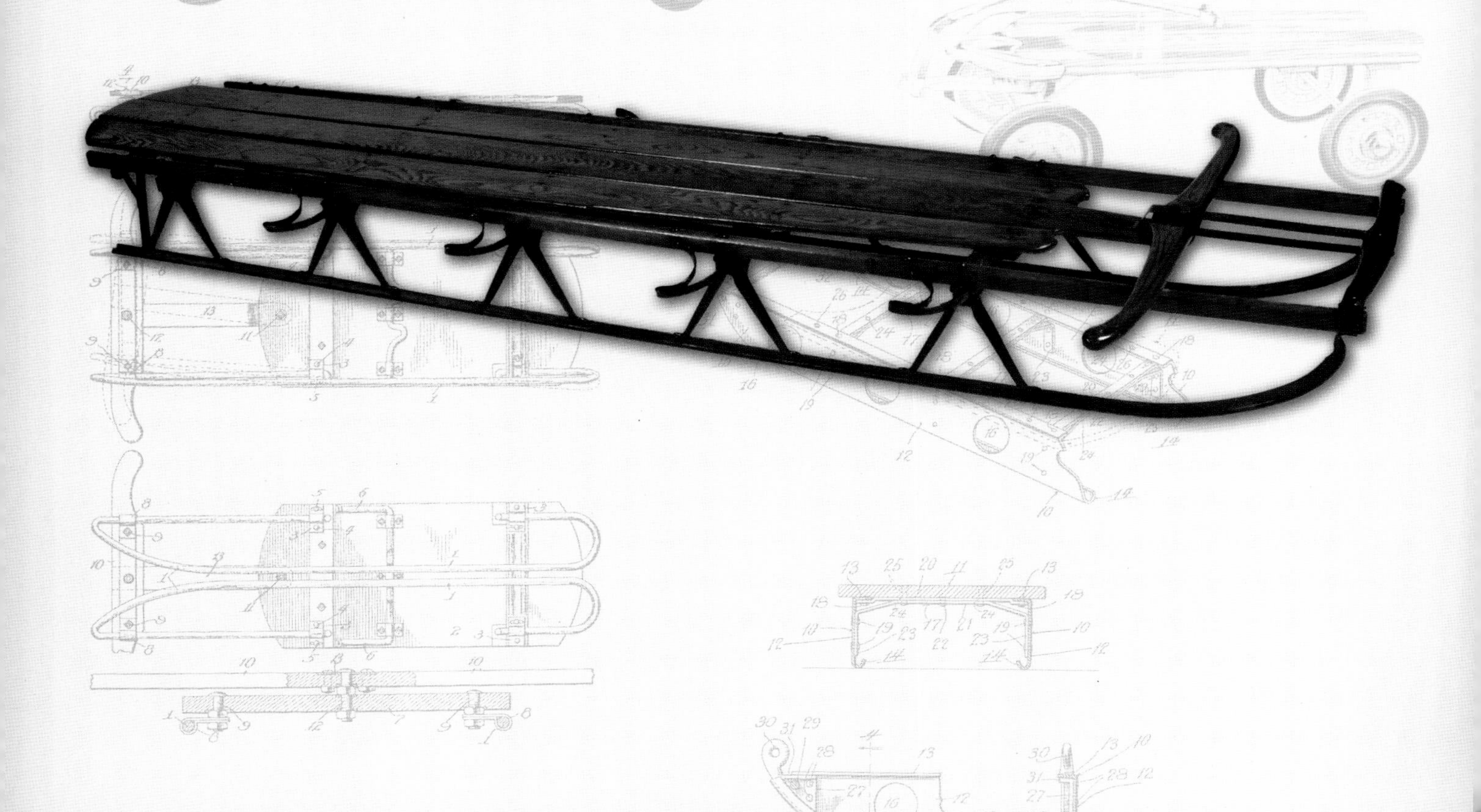

Scene 1

Company Up Dates

Flexible Flyer

We all know Samuel Leeds Allen was a master at marketing his famous sled. Without the catalogs, advertising covers, pinback buttons, and other promotionals, it would have been impossible to compile a company timeline. From 1900 on, there are numerous pieces of ephemera to validate and identify years and models of sleds. However, from 1889 to 1900, it is a little more difficult. The 1892 Chicago Worlds Fair postcard and brochure are the only known sources of pictures showing the early Flyers. The following updated information has been validated by this printed ephemera, company literature, and family interviews.

According to company records the # 6 sled was introduced somewhere around 1890. These early # 6's had green or black pinstriping and plain decks. The words Flexible Flyer were stenciled on the steel brace connecting the wooden bumper to the deck of the sled. The wooden bumpers and steering bars were decorated with a unique pattern of black lines, the standard pinstriping, as we know it, replaced the unique lines sometime between 1895 and 1899. The extra steel brace behind the fifth knee was also discontinued after the 1908 sledding season.

In my first book the Fire Fly "B" series was used as the pre-trademark example of Mr. Allen's new invention. Knowing there was an "A" series and not having one in mint condition at the time of that publication was extremely frustrating. As luck would have it, and I truly mean luck, I spotted one in an antique shop in Connecticut, and purchased it. Take note that the "A" Series is much more detailed, with two color lettering (red and black), and the deck is elaborately decorated with the holly motif. Mr. Allen's business sense and dedication to producing a great product for the public and maintaining a profit, was paramount. I can see why his "A" Series was not really cost effective, and much more labor intensive.

In 1908 the Tuxedo Racer made its debut in the catalog, remembering this sled was originally designed for the fashionable Tuxedo Club in Tuxedo, New York., it was the first Flexible Flyer to have an all-steel frame. It was not until 1915 that all-steel frames were introduced for all models. The runners are chrome-nickel steel, superior to ordinary steel. The deck is supported by four steel standards and benches, unique to the Tuxedo. No other Flexible Flyer has this feature. The front two benches are wood and steel and the back two benches are steel only, causing the back to slope down two and a half inches from front to back and giving it a "racy" appearance. All this steel came with a higher price and a weight of twenty-two and a half pounds. Advertising clearly states "Not recommended for children, but we do recommend it for those who want the strongest, fastest, and best sled that can be had."

Fast forward to 1935. Those of you who actually read all of the advertisements in my first book, must have realized that a very important sled had been left out, the Airline Eagle #151 also referred to as the "Flagship of the Flexible Flyer Fleet." This was the first major design change for the Flexible Flyer. The patent calls for a steering mechanism resembling handlebars, with rubber grippers, a change in standard knees and a more streamlined deck. This is the only model with half-moon knees and no side rails. The side rails are incorporated in the deck of the sled making it unique. This was the sled to own in 1936, as advertised in December 1935. According to company records, the Airline Pursuit was also offered with the handle bar steering mechanism later that year.

Updated Chronology of Flexible Flyer

1997 • Brunswick Corporation purchased the company and, in 1999, decided to stop manufacturing the steel runner sled.

2000 • Pacific Cycle purchased the Flexible Flyer trademark

2005 • December. Paricon Inc. acquired the exclusive use of the Flexible Flyer trademark in the U.S. and Canada for winter recreational products. Once again the world's most recognizable trademark is in familiar territory. Mr. Henry Morton, President of Paricon Inc. is the great grandson of the founder of the Paris Manufacturing Company, South Paris, Maine. He is committed to releasing a new steel runner sled, for the coming winter season 2006-2007, worthy of the name "Flexible Flyer"

Flexible Flyer postcard, c. 1895-1900, depicting the unique black line pattern on the deck with a center motif, as seen on the Flexible Flyer #6 on page 9. $15-25.

Fire Fly #12a, 1889. Note the wooden bumper, ornate holly motif, double wreath and two-tone lettering. L: 46"' W: 14"; H: 8". $300-500.

Flexible Flyer #6. Note unique pattern of black lines on wooden bumper and steering bar. The words Flexible Flyer are stenciled on the steel brace connecting the bumper to the deck. Also note the deck is plain with green pinstriping and an extra steel brace behind the fifth knee. Discontinued after the 1908 sledding season. $1500-2500.

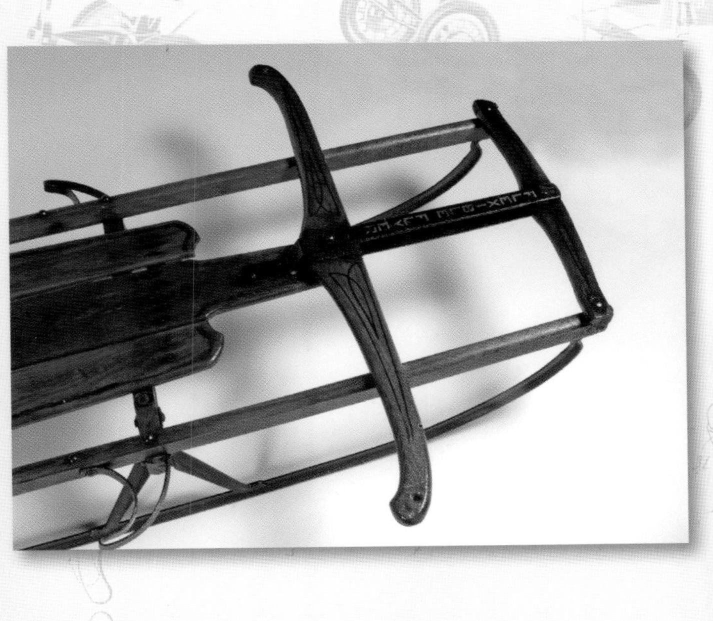

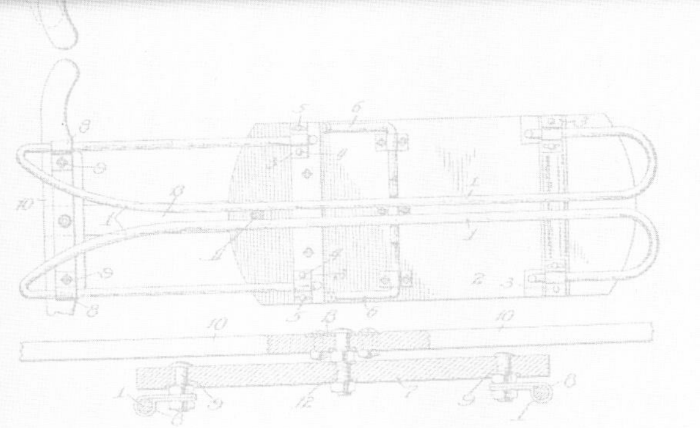

Flexible Flyer, Tuxedo Racer, 1908. This is the only Flexible flyer with 4 knees and 4 steel braces. $200-275. *Courtesy of Jim Pauzé.*

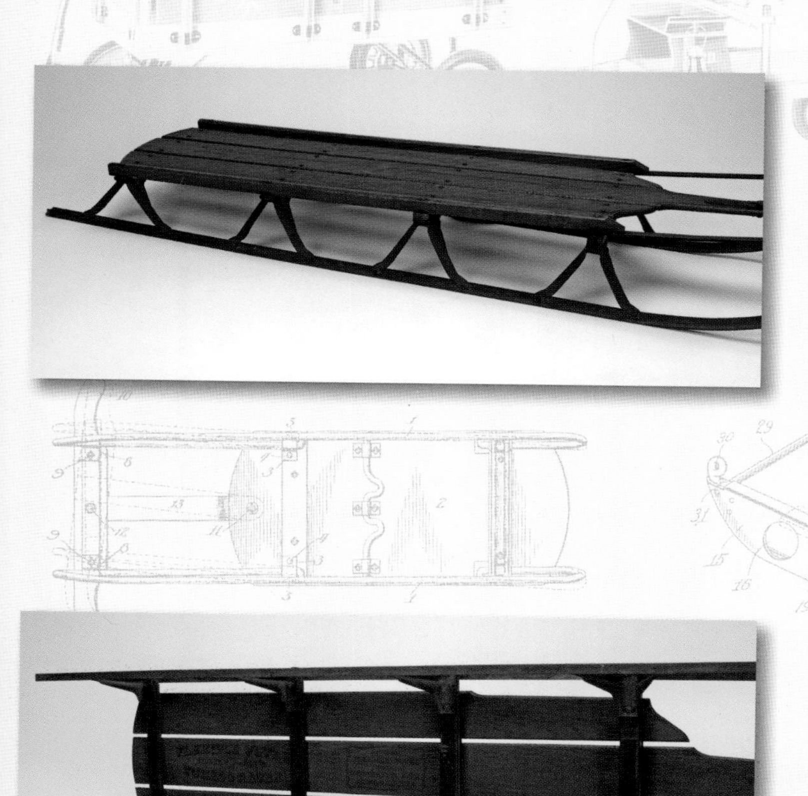

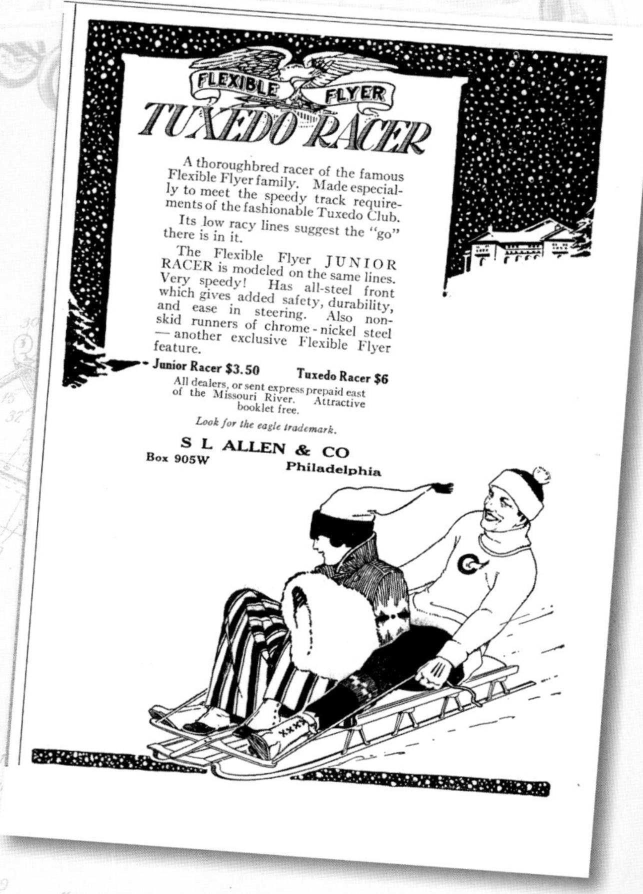

Flexible Flyer Tuxedo Racer advertisement, December 1915. $5-7.

Airline Eagle #151B, the flagship of the Flexible Flyer Fleet, December 1935. Note the steering mechanism with rubber grippers, the half-moon knees, and no side rails. L: 51"; W: 11"; H: 5 ½" in back, 6 ½" in front. $200-300.

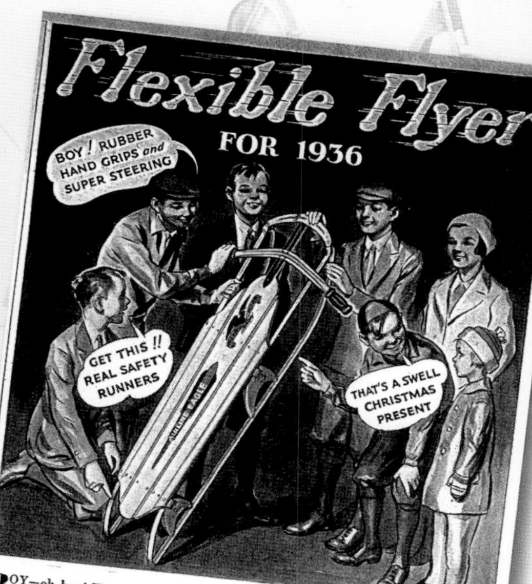

Flexible Flyer Airline Eagle advertisement, December 1935, *Popular Science*. $5-10.

Airline Pursuit #47, with handlebar grippers, 1936. L: 47";
W: 11 3/4"; H: 5 1/4" in back, 6 1/4" in front. $250-350.

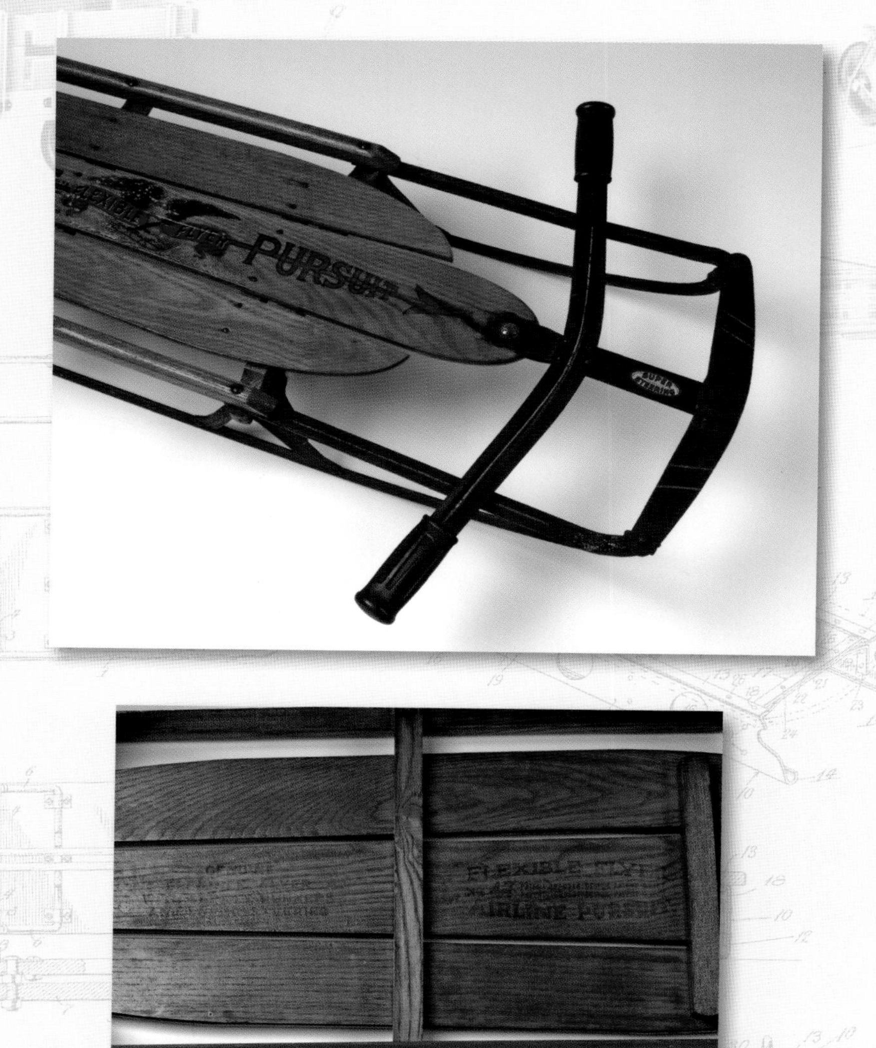

Auto Wheel Coaster Company

(Buffalo Sled Co., 1904-1921)

In addition to the familiar Fleetwing Racers, Coaster Kings, Cyclone Racers, and Fleetwing Flash (Spring-top) Sleds, the company manufactured three different models of the Fleetwing Bob. The first to arrive on the scene, in the early 1930s, was described as "the most exciting thing on the hill, goes like the wind, and steers like an auto." Note that the front resembles the hood of a car. As was advertised in *Boy's Life* in 1934. One could own this beauty for $10.00

As a purist, that is one who advocates not to restore, it is interesting that company literature describes color as: "top stained green, and varnished with orange lettering and striping....hood and Bobs enameled orange, with black steering wheel and gear." In my years of collecting, I cannot tell you how many of these "Bob's" I have passed up because they were advertised as "all original," and on seeing the Bob's, discovered they had been skillfully restored. That brings me to the numbers that appear on the hoods of the Bobs. The only original numbers are 66, 72, and 84. These numbers indicate overall length of each Bob.

Two new Bob's were introduced in the 1940 catalog. Instead of the auto-type hood they boasted a more streamlined look. In addition to the selected northern ash top, they advertised hard maple Bobs with 3/8 round steel runners and a sheet steel hood. New, also, was the powerful brake for coasting safety. The saw-toothed scraper blade is attached to the rear of the sled and activated by hand, causing the teeth to bite into the ice and snow. The Bob's of the thirties left braking to the imagination.

The Auto-Wheel Coaster Company, formerly known as the Buffalo Sled Company was famous for their Coasters, that is wagons, not sleds. It is not clear why their name indicated they were a sled company, when their main products were Coaster Wagons. Much like Flexible Flyer, early advertisements (1919-1920) offered free giveaways, such as pennants, shirts, and captain caps to anyone who wanted to belong to the Auto-wheel Coaster Club. They also offered a monthly publication called the "Auto- Wheel Spokes-man" boasting the attributes of their Coasters (wagons, that is) According to a publication of the Spokes-man in 1919 the Buffalo Sled company launched an essay contest for boys and girls under the age of fifteen. The best letter of 75 words on the "Uses of the Auto Coaster - Roadster" would win a Convertible Roadster, second prize, an Auto Wheel Coaster, and third prize, a Fleetwing Steerable Sled. Response was so overwhelming, that the judges decided that a number of letters were very good and needed some sort of recognition or award of merit. A small bank, made of metal, bearing a picture of an Auto Wheel Coaster was the award. The idea was to save up for a Coaster, putting away nickels and dimes, and in no time one would have one paid for. The bank held $10.00. The Auto Wheel Coaster Dealer in your area retained the key that would open the bank. In exchange for the bank your Coaster Wagon was waiting for you. The contests continued every few months over the next few years. In 1921 the Buffalo Sled Company became the Auto Wheel Coaster Company, and the promotions for their Coasters continued. By 1923 prizes for the story writing expanded to five top prizes all Coaster Wagons; not one Fleetwing Steerable Sled was offered. In all the years of collecting I have never seen, nor do I know of any promotions for their sleds. The company continued to sell their products until 1964, when they were forced to close their doors due to bankruptcy.

A Companion Your Boy Will Prize

All through boyhood—affording him endless pleasure in the great outdoors and aiding him to be useful—nothing equals the

Auto Wheel Coaster

King of Coaster Wagons—only one built like an auto; has roller bearing wheels, steel axles, dust-proof hub caps.

Goes down a hill like a streak of lightning—and stands up under heaviest load.

Get the genuine Auto Wheel Coaster—don't accept a substitute or imitation. For sale by hardware and department stores.

Buffalo Sled Co., No. Tonawanda, N. Y.

Buffalo Sled Company blotter, circa 1910-1915. *Courtesy of Jim Pauzé.*

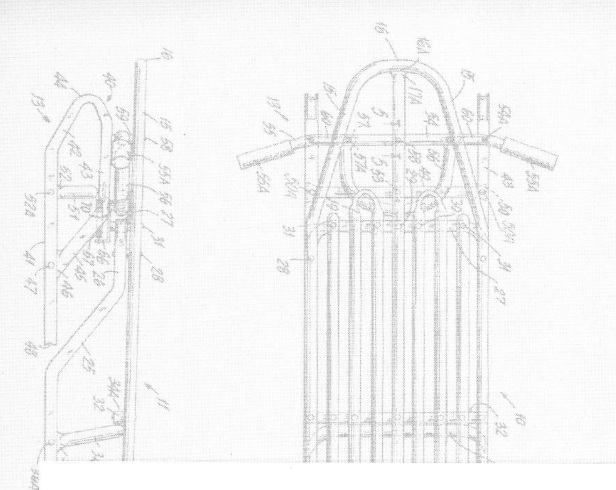

Buffalo Sled Co., Auto-Wheel Coaster advertisement from *The American Boy*, August 1919. $15-20.

RUNNERS FOR AUTO-WHEEL COASTER

Made to fit Wagons Nos. 1 to 4 and 11 to 44.
When ordering, give number of wagon for which runners are wanted

Price, per set complete to fit one wagon $1.50

THE BUFFALO SLED COMPANY
NORTH TONAWANDA, N. Y.

Buffalo Sled Co. runners for Auto Wheel Coaster, c. 1910. *Courtesy of Jim Pauzé.*

Buffalo Sled Co., Auto-Wheel Coaster advertisement from *The American Boy*, April 1920. $15-20.

Vol. 4 No. 22 Published by AUTO-WHEEL COASTER COMPANY, Inc., North Tonawanda, N. Y. May - June, 1923

STORY WRITING CONTEST BIG SUCCESS

Stories Received From Thirty States — Write A Story And Win A Prize

"AUTO-WHEEL" Builds Better Boys

The Short Story Contest which started in the last issue of the Spokes-man has proven itself amazingly popular with the large number of AUTO-WHEEL COASTER boosters, and if the flood of early entries is any criterion of the future this contest is bound to be the greatest ever held by the SPOKES-MAN.

1st Prize — No. 63 AUTO-WHEEL COASTER, Disc Wheels with new oversize rubber
2nd. Prize — No. 62 AUTO-WHEEL COASTER, Disc Wheels with new oversize rubber
3rd. Prize — No. 20 AUTO-WHEEL COASTER, Disc wheels with steel tires.
4th. Prize — No. 61 AUTO-WHEEL COASTER, Disc wheels with rubber tires.
5th. Prize — No 16 AUTO CART, Disc wheels with rubber tires.

HAVE YOU GOT THEM ALL CLEAR? YOU HAVE? ALRIGHT THEN LET'S GO.

Another Sample Story on Page 5

I'll be named by Next Month

FEBRUARY, 1920

SOME LUCKY BOY WILL GET A PRIZE FOR NAMING ME!

Published Monthly by THE BUFFALO SLED CO., North Tonawanda, N. Y.

In the Interests of the Boy Who Owns or Would Like to Own a Coaster

Mailed with our Compliments

EDITORIAL STAFF
Managing Editor J. J. SCHNEIDER
Business Manager J. E. BAUMLER

FEBRUARY 1920

Best Name Wins Prize!

How to Win a Coaster!

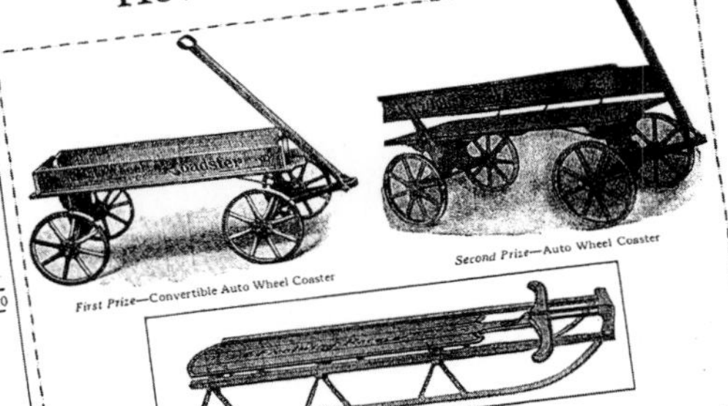

First Prize—Convertible Auto Wheel Coaster

Second Prize—Auto Wheel Coaster

Third Prize—Fleetwing Steering Sled

Don't delay—you've only 20 days to get in your suggestion for name for this magazine

EDITORIAL

PRIZES

First Prize: Auto Wheel Convertible Roadster-Coaster

Second Prize: Auto Wheel Roadster, with quick detachable box. Makes dandy coaster.

Third Prize: Fleetwing Steering Sled. Strong, roomy and handsome. Stands the hard knocks.

First Prize and title of Associate Editor to Winner.

Second and Third Prizes for two next best names.

IT'S A BIG HONOR

The Editor.

FEBRUARY, 1920

Someone Will Win a Prize by Naming Me

The Buffalo Sled Announces Winners!

Offered Prizes Best Articles

"Uses of Auto Coaster-Roadster"

Thousands Enter Contest—The Winners

First Prize

HENRY JACOBS, Brooklyn, N. Y. — JAMES OTTO PIERCE, W. Plains, Mo. — JAMES ALBERT BISHOP, Knoxville, Tenn. — ARTHUR RICK, Syracuse, N. Y. — CHAS. E. BRADY, Glenellyn, Ill.

You, too, may find yourself winner in the next contest announced in this issue—Go in and try hard for a prize!

MORE CONTESTS COMING

FOR THE SCOUT HIKE

1. To haul supplies.

Auto-Wheel "Spokesman" newsletter, February, 1920. Winners of Essay Contest announced and merit award banks issued to a select few.

Celluloid Merit Award bank, 1920. 2-1/2" diameter. $50-100.

December, 1919 31

The Day After Christmas

One gift stands out above all the others—the one that you can use at once for good, hard, rollicking play—

The Auto-Wheel

You can jump right on it on Christmas morning, and you'll *know* it's Christmas. And then the next day, and the next, and on through your boyhood, your Auto-Wheel is your most dependable friend. It's always ready for fun or work. Somehow, it coasts faster than the others, and it's easier to pull with a load.

It has real roller bearings, and the steel tires and axles make it strong and long-wearing.

There Are Two Types

There is the Auto-Wheel Coaster—the old reliable—a marvel for coasting and general use. And then there is the Auto-Wheel Roadster — two wagons in one. Open up the sides, and you have a box wagon. Fold them toward the center and there is the snappiest roadster you ever saw. And it takes but a moment to make the change.

Put the Auto-Wheel on your Christmas list. The other fellows want it, too. And when you all gather for the first coast—form your Auto-Wheel Coaster Club. We'll send you a cap for each member, and a special cap for the Captain, as soon as your club is organized

FELT PENNANT FREE

Send us the names of three local coaster dealers, telling which one handles the Auto-Wheel. We'll send you a handsome Felt Pennant and the Auto-Wheel Booklet, showing these wagons in detail.

Write us at once, so you can show the folks what your Christmas present ought to be.

BUFFALO SLED COMPANY
125 Schenck Street
N. Tonawanda,
N[ew Y]ork
Preston, Ont.

27

November, 1926

The Queerest Prescription a doctor ever wrote

Still, it accomplished more than medicines had. Three months after his visit this particular youngster was a perfect picture of health—ruddy cheeks, sturdy body, bright-eyed and full of animation.

He never had been puny—no, it wasn't that—but he was thin and somewhat smallish for his age. And sickness found him an easy prey. The doctor said "medicines and tonics undoubtedly will help, but what that child needs is outdoors and plenty of exercise." My boy was just the same—until I bought him an Auto-Wheel. Has this little lad an Auto-Wheel?" The mother's head shook. The doctor's pencil wrote a few words and he handed the slip to the mother.

"Get him an Auto-Wheel" it said.

Girls need Auto-Wheels, too. All children, from the day they begin to crawl, should have the health-promoting aid Auto-Wheel toys offer. There's an Auto-Wheel for each age and stage of childhood—and the benefits derived are precious assets for the years and struggles ahead.

To be sure, there are many toys similar to Auto-Wheel—but to a boy or a girl who has owned both kinds, there is all the difference in the world between an Auto-Wheel and "something just as good." To a mother, too, who permits her keen value-judgment its rightful exercise, there is a vast difference. The Auto-Wheel is a more economical purchase, for its life is measured by years—not mere months.

And so the Auto-Wheel toy you buy for your boy or girl *really* does build up the health and stamina you want that youngster to have. Really does build strength and sturdiness in that fine little body. For the Auto-Wheel can withstand the hard service it is given day in and day out without breaking faith with its owner.

For a great many years now we have been building these fine toys. We take considerable pride in our work, for we feel that through these fine products we are helping in no small way to build healthier, sturdier boys and girls. Our own youngsters have Auto-Wheels—we receive letters from children and grown folks all over the country telling us about *their* Auto-Wheels. And now—we would like to share the good fun and good health that the Auto-Wheel promotes—with your boy or girl.

If your hardware dealer or department store cannot supply you with the Auto-Wheel toy you want—accept no substitute but fill out and mail the attached coupon to us.

As a Baby-Walker — As a Stroller

The New AUTO-WALK-STROLLER
Teaches baby to walk. Strengthens little legs. Beautifully finished in blue and ivory. Handsomely shaped steel bolsters, extra wide to prevent tipping. All-metal back with perforated Blue Bird design; all-metal tray, free from sharp corners. Rubber-bound bumpers, front and rear. Adjustable height. Jiffy-attachable handle, folding type—no bolts or nuts needed; used either front or rear. Sturdy and long lasting.

The New, Improved AUTO-SPEEDER
12 features make this the most-wanted wagon in America. Trim, fleet and sturdy. Rides low and easy on its smooth-running roller-bearing wheels with their Firestone tires. Splendidly finished in every way, and built to give unusually fine service.

DeLuxe AUTO-SCOOT Coaster
Built like the modern steel automobile. Balanced steel frame. Non-spill steering post. Rubber suction mat. Non-skid, heavy rib tread type balloon tires. Real automobile external contracting type drum brake. New design, durable, speedy wheels. Striking colors. The first real "safety first" scooter. Also other styles and size Scooters.

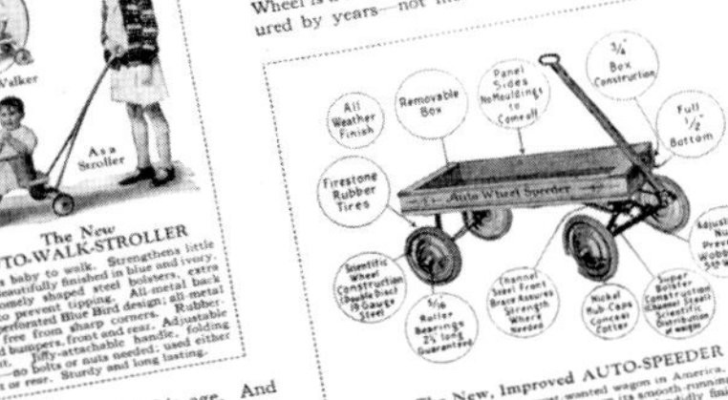

AUTO-PEDAL CAR
For tiny tots. Easy to steer. Quiet running. Splendid construction, attractively finished. Two sizes, with or without rubber pedals.

The famous **AUTO-WHEEL Coaster**
Most widely known coaster wagon in the world. Owned and loved by boys and girls everywhere. In popularity and advanced features "King of 'Em All."

Auto-Wheel
COASTERS · SPEEDERS · CARTS · SCOOTERS · PEDAL CARS
BABY WALKERS · FLEETWING SLEDS

CLIP AND MAIL TO
AUTO-WHEEL COASTER CO., Inc.,
North Tonawanda, N. Y.
Please send descriptive matter, prices and name of dealer who can supply me with AUTO-WHEEL SPEEDER ( ) AUTO-WHEEL COASTER ( ) AUTO-SCOOT-COASTER ( ) AUTO-PEDAL CAR ( ) AUTO-WALK-STROLLER ( ) (Check items you are interested in.)
Name
Address
My dealer's Name

AUTO-WHEEL COASTER CO., Inc. NORTH TONAWANDA NEW YORK

Buffalo Sled Co., Auto-Wheel Coaster advertisement from *The American Boy*, December 1919. Captain's cap offered as soon as your Auto-Wheel Coaster Club is organized. $15-20.

Auto-Wheel Coaster Company Inc. advertisement from *Junior Home Magazine*, November 1926. $5-10.

Auto Wheel Coaster Co., Fleetwing Bob Sled, no brake, 1930. L 66" W 14 ½". $1000-1500.

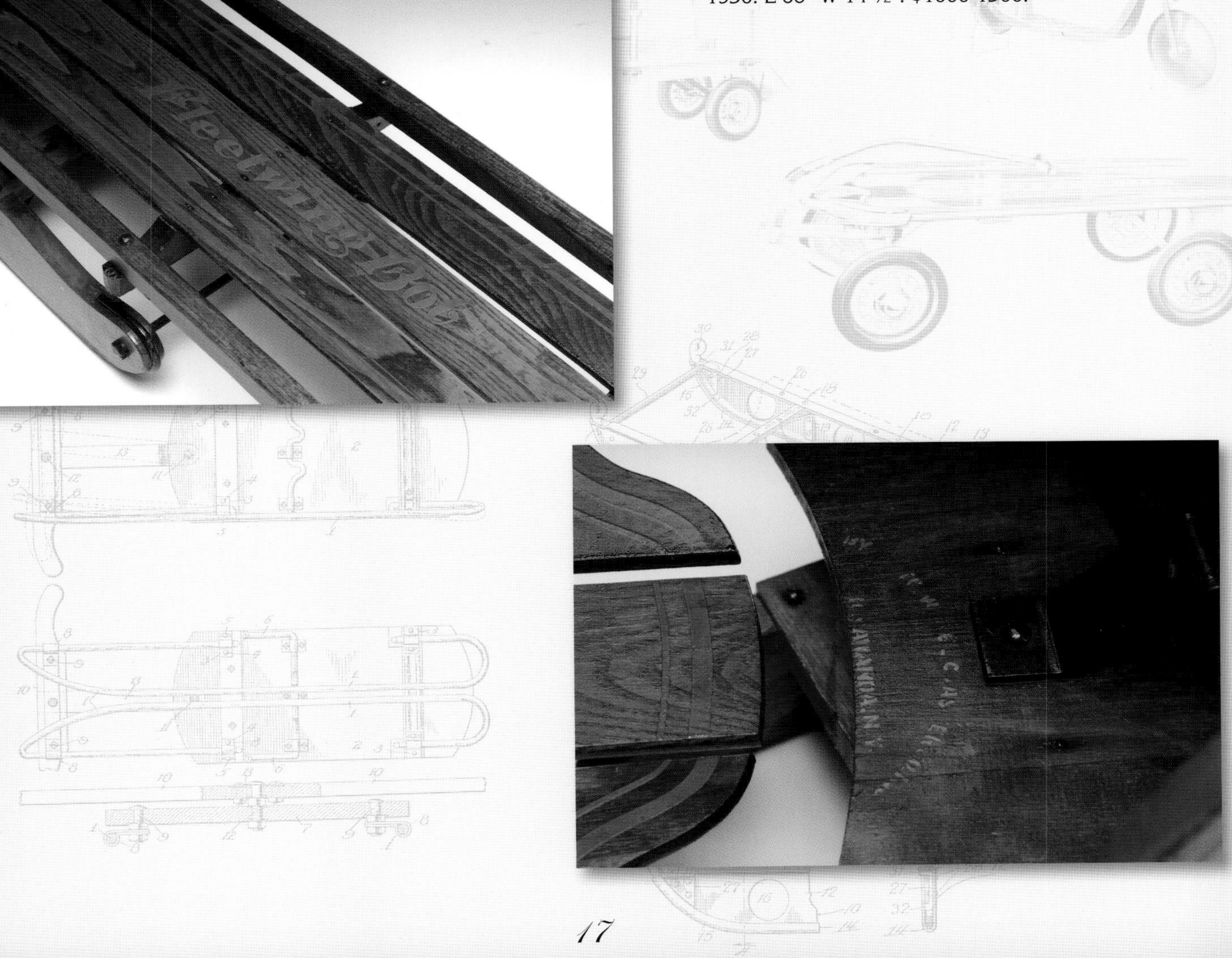

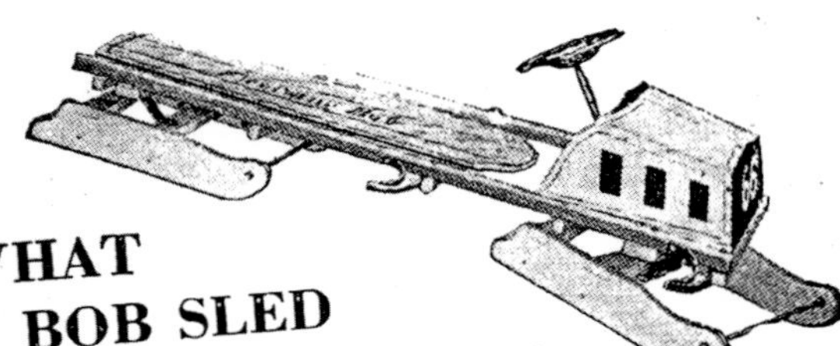

WHAT A BOB SLED

Here is something new in sliding down hill—steers like an auto, holds three kids or two grown ups, an goes like the wind. The most exciting thing on the hill, and boy can it break records! If you cannot get this Bob Sled from your Dealer, you can have it by writing to AUTO-WHEEL COASTER COMPANY, INC., North Tonawanda, N. Y., and sending $10.00. (If you live west of the Mississippi River, $12.00). If you want the thrill that comes once in a life time just ask Dad for the money, put it in an envelope and send it off to us as soon as possible, and the first thin know you will be on your way down the s hill in the neighborhood. Lickity-Split. $10.00. No Express to pay.

AUTO-WHEEL COASTER COMPANY, North Tonawanda, New York

Magazine advertisement, 1930, for the Fleetwing Bob Sled. Note the auto front with the number 66.

Auto-Wheel Coaster Co., Inc. 1940 catalog, *courtesy of the Historical Society of Tonawanda.*

AUTO WHEEL COASTER PRODUCTS 1

536 RED RACER

Heavy, kiln dried hard wood box.

Roller Bearing wheels.

Two piece front bolster.

Box varnished natural, gear black, wheels duo-tone red and black.

Model	Body Size	Wheels	Rubber Tires	Wt. Ea. Lbs.
536	16 x 36"	10" Airflow	¾"	34
536-B	16 x 36"	10" Airflow	1"	35
5366	16 x 36"	10" Disc	2"	36¼

No. 5366 has 2" semi-pneumatic tires.

All packed 1 per carton, K.D. flat.

Wood Spoke wheel, used on 736-S

Heavy, kiln dried, hard wood box, reinforced with moulding.

Roller Bearing wheels.

Two-piece front bolster.

Box varnished natural, gear black, wheels duo-tone red and black.

736 RED RACER

Model	Body Size	Wheels	Tires	Wt. Ea. Lbs.
732	14 x 32"	8" Double Disc	¾" Rubber	27
736	16 x 36"	10" Airflow	¾" Rubber	33
736-S	16 x 36"	10" Wood Spoke	¾" Steel	32

Packed 1 per carton K.D. flat.

2 AUTO WHEEL COASTER PRODUCTS

7360 FLEETWING "Spring Top" FLASH

Heavy, kiln dried, hardwood box, reinforced with moulding. Equipped with "Knee Action" springs, 2 in front and 2 in rear, which absorb the shocks.

Two-piece channel steel front and rear bolsters.

Roller Bearing, airflow wheels.

Box varnished natural, gear black, wheels duo tone red and black.

Model	Body Size	Wheels	Rubber Tires	Wt. Ea. Lbs.
7360	16 x 36"	10"	¾"	35½
7360-B	16 x 36"	10"	1"	36½

Packed 1 per carton K.D. flat.

HEALTH BUILDERS FOR CHILDREN 3

BODY — Heavy gauge steel, 14½ x 34"; Rolled edge top and bottom.
WHEELS — Roller Bearing, 10" Airflow.
TIRES — ¾" rubber.
COLOR — Body red with cream panel, gear black, wheels duo-tone red and black.
WEIGHT — 35 lbs. each. Packed 1 per carton, K.D. flat.

34 RED "ALL STEEL" RACER

234 AUTO-WHEEL COASTER DE LUXE

BODY — Heavy gauge steel, 14½ x 34". Rolled edge top and bottom. Two-piece front bolster. Steel sliding brake.
WHEELS — Roller bearing, 10" Airflow.
TIRES — 1" rubber.
COLOR — Body blue with cream bottom, gear black, wheels duo-tone blue and cream, handle silver painted with rubber handle protector.
WEIGHT — 35½ lbs. each. Packed 1 per carton, K.D. flat.

4 AUTO WHEEL COASTER PRODUCTS

3-S AUTO-WHEEL COASTER

MODEL 3-S

Heavy, kiln dried maple box, hardwood bottom, panelled sides.

Two-piece front bolster.

Roller bearing wheels.

Brake on 3-S only.

Box varnished natural, gear black, wheels red.

Model	Body Size	Wheels	Tires	Wt. Ea. Lbs.
3-S	16 x 36"	10" Wood Spoke	¾" Steel	34½
4-S	18 x 40"	10" Wood Spoke	¾" Steel	36

Packed 1 per carton K.D. flat.

MODEL 3600 AUTO-WHEEL COASTER "SPRING TOP"

Equipped with "Knee Action" springs, 2 in front and 2 in rear, similar to spring construction on page 2.
BOX —Heavy, kiln dried Maple, hardwood bottom, panelled sides. Size 16 x 36".
WHEELS —Roller bearing, 10" Airflow, 1" rubber tires.
COLOR —Box varnished natural, gear black, wheels duo-tone red and black.
WEIGHT —36½ lbs. each. Packed 1 per carton K.D. flat.

MODEL 464

Heavy kiln dried, Maple box, hardwood bottom, panelled sides, Hard Maple bolsters.

Roller Bearing wheels.

Wood parts varnished natural, wheels duo-tone red and black.

Model	Body Size	Wheels	Tires	Wt. Ea. Lbs.
464	18x40"	10" Airflow	1" Rubber	42
424	18x40"	10" Wood Spoke	¾" Steel	38½
4430	22x42⅜"	10" Wood Spoke	¾" Steel	50

Packed 1 per carton K.D. flat.

No. 4430 equipped with tubular steel handle; extra heavy wood bolsters. Axles attached to bolsters with U-Bolts.

464 AUTO-WHEEL COASTER

HEALTH BUILDERS FOR CHILDREN 5

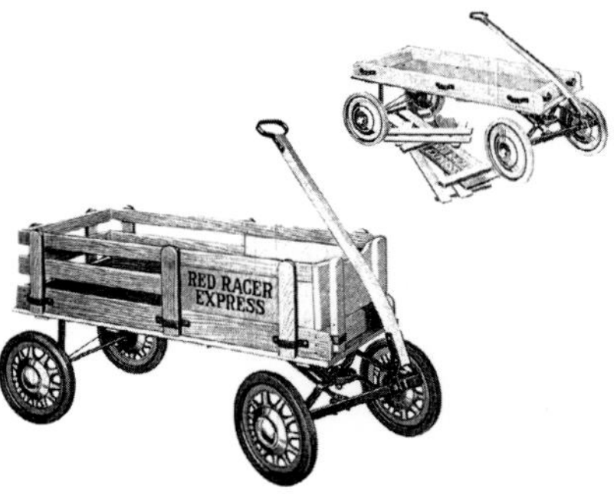

371 RED RACER EXPRESS

Heavy, kiln dried hard wood box and sides.

Roller Bearing, Airflow wheels.

Two-piece front bolster.

Box and stake sides varnished natural, panels and front red, gear black, wheels duo-tone red and black.

Model	Body Size	Wheels	Rubber Tires	Wt. Ea. Lbs.
370	16 x 36"	10"	¾"	40
371	16 x 36"	10"	1"	41
*371-X	16 x 36"	10"	1"	46½
471	16 x 40"	10"	1"	44
*471-X	16 x 40"	10"	1"	49

*Six wheels (dual wheels in rear)

Packed 1 per carton K.D. flat.

6 AUTO WHEEL COASTER PRODUCTS

613-615-616 AUTO-CARTS

Kiln Dried Hard Wood

Box varnished natural, lettered in black. Metal parts black.

Wheels: No. 613 all red. Nos. 615 and 616—red with white stripe.

Model	Body Size	Bolsters	Wheels	Tires	Wt. Per Carton
613	9 x 18"	Wood	4" Wood		21 lbs.
615	11 x 24"	Channel Steel	6" Disc	⅝" Rubber	28 lbs.
616	12 x 28"	Channel Steel	7" Disc	¾" Rubber	34 lbs.

Packing—No. 613 packed 6 per carton K.D. flat.
Nos. 615 and 616 packed 2 per carton K.D. flat.

HEALTH BUILDERS FOR CHILDREN 7

53 AUTO-WHEEL SPEEDER

TOP — Heavy, kiln dried hard wood, 13½ x 54". Sheet Steel Hood. Equipped with brake and folding footrests.
WHEELS — Roller bearing, 8" double disc, ¾" rubber tires.
BRAKE — Wood.
COLOR — Top and hood red with yellow ornamentation, side boards yellow, wheels red with yellow stripe, gear black.
WEIGHT — 34 lbs. each. Packed 1 in carton K.D. flat.

Model 533

Same as 53, with 8" disc wheels, 2" semi-pneumatic tires.
WEIGHT — 36½ lbs. each. Packed 1 per carton K.D. flat.

37 FLEETWING RACER

TOP — Heavy, kiln dried hard wood, 13 x 37". Two piece front bolster.
WHEELS — Roller bearing, 10" Airflow, ¾" rubber tires.
BRAKE — Wood.
COLOR — Top center varnished natural, side boards red, wheels duo-tone red and black, gear black.
WEIGHT — 29½ lbs. each. Packed 1 per carton K.D. flat.

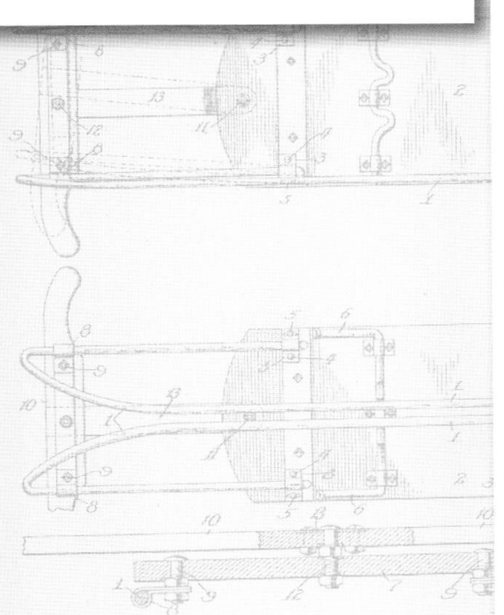

Auto Wheel Coaster Co., Fleetwing Bob Sled. With hand brake, 1940. L 84", W 16 ½". $1200-2000

Fleetwing Racer post-card, late 1940s or early 1950s. $5-10.

Auto Wheel Coaster Co., Fleetwing Bob Sled with hand brake, 1940. L72", W 14 ½". $750-1200.

The Garton Toy Company

Updating the Garton Toy Company, would have been impossible without the catalogs and assistance of Jane Dwyre Garton and Ann Huenink Mc Intyre. According to legend, Eusebius B Garton, placed four wheels on a cigar box and called it a wagon. Founding a small company three years later in 1879, he introduced the first completely wooden coaster wagon to the nation. Before long he was producing, express wagons, willow carriages, toy chairs, parlor swings, and sleds. Fifteen different wagon models were offered between 1889-1892. Early Garton wagons were unique. In addition to the stenciling and scroll work, some of the models offered hand painted scenic side panels, much like the sled decks of the era. Wisconsin, also nicknamed the Badger State, may have been the inspiration for Mr. Garton to introduce the Badger Line of coaster wagons. In the early teens, steel wagons were introduced. Names like "The Ironclad,"and "The Bulldog" adorned the side panels implying that they were as indestructible as a naval vessel or as strong as a bulldog. It was not long after that cutters and clippers began appearing in the catalogs. By 1916 an all-steel Badger Express wagon was introduced along with the first steerable sleds, "The Eskimo"and the "Dreadnaught." cutters and clippers were produced along with the new steerable sleds, but completely disappeared from the scene in 1920. Garton continued to dominate the market place with its sidewalk autos and wagons, The Teenie Weenie and Dreadnaught Coaster Wagons, appear in a 1924 catalog along with the Go-Dink, a scooter on runners. It is not clear when the Teenie Weenie and Dreadnaught Coaster wagons and sleds were discontinued but they do not appear in the 1930 catalogs. Clarence Garton, son of Eusebius served as President of the company from 1915 and in 1937 he applied for a patent for an ornamental design for a new sled, later to be known as the Streamliner, joining the already popular Eskimo, Royal Racer, and Silver Streak. It was also in 1937 that Garton made its first acquisition, the Globe Company, manufactures of Globe Biltwell Line vehicles. Acquiring the Pratt Mfg. Company in 1952 and Kalamazoo, their leading competitor in sled manufacturing sometime later, made Garton Toys an international leader in the toy industry. The company continued to flourish and in 1962, moved to their new offices and manufacturing facilities just north of Sheboygan, in a plant that was constructed entirely out of steel and concrete, making it less vulnerable to fire. Eleven years after moving into their new seven and a half acre plant the business was sold to Monitor Corporation, and "Toy" was dropped from the name. Business operations were suspended in 1975.

By creating toys for children, a business that spanned ninety six-years continues to bring pleasure to both young and old collectors. An appropriate company motto today, might read "Garton Toys – Pleasure is our Business."

Chronology of the Garton Toy Company

1889 • Company founded by Eusebius B.Garton in Sheboygan, Wisconsin

1887-1892 • Wagon Models – Pony Express, US. Express, American Express & Merchants Express Wagon

1910-1914 • The "Ironclad"& "Bulldog" Wagons appear in catalog

1915 • Cutters introduced, known for round fenders & half-oval shoes, Clippers, the Red Ryder & Black Beauty spring shod with full round spring shoes (runners)

1916 • The "Eskimo," "Dread (battleship graphic) Naught"& "Dreadnaught Special Torpedo Racer"steerable sleds & the "Badger Express" wagon constructed of sheet steel introduced

1924 • The Teenie Weenie Coaster Wagon with disc wheels appears in the catalog. The" Dreadnaught" Ball Bearing double disc Wheeled Coaster with rubber tires and the Badger Roller Bearing Coaster Wagon introduced. The "Go Dink," a scooter on runners, introduced

1929 • Battleship graphics removed from the decks of the "Dreadnaught" Line & the Super Dreadnaught Racer previews. All sleds advertised with grooved runners.

1930 • The Teenie Weenie & Dreadnaught Coaster Wagons & Sleds disappear from the market place.

1937 • Clarence E. Garton granted patent for an ornamental sled design. Garton acquired The Globe Company, manufactures of Globe Biltwell vehicles

1939 • The Streamliner Sled equipped with the new ornamental design and the Flying Star join the Garton Fleet.

1941-1946 • No steel toys produced during the war years

1952 • Garton acquires Pratt manufacturing Company

195? • Kalamazoo also acquired

1973 • Company sold to Monitor Corporation. "Toy" dropped from name

1975 • Business operations suspended

ILLUSTRATED CATALOGUE

AND PRICE LIST

GARTON TOY CO.

SHEBOYGAN, WISCONSIN.

MANUFACTURERS OF

EXPRESS ∴ WAGONS

WILLOW CARRIAGES, SLEDS, TOY CHAIRS

TOY CRADLES, PARLOR SWINGS ETC.

PONY EXPRESS.

NO. 00.

Body, 7 x 14 inches; wheels, 6 and 8 inches. Body painted red. Front wheels turn under body.

Price, per doz......................$4.00.

NO. 0.

Body, 9½ x 17 inches; wheels, 6 and 8 inches. Stenciled and varnished.

Price, per doz......................$5.00.

NO. 1.

Body, 10 x 19 inches; wheels, 8 and 10 inches, with iron tire. Stenciled and varnished. Front wheels turn under body.

Price, per doz......................$7.00.

Garton Toy Company Catalog, 1887. *Courtesy of Ann Huenink McIntyre.* (continued on following page)

EXPRESS AND CHILD'S WAGON COMBINED.

NO. 4.

Wood axles; body, 12 x 25 inches; wheels, 11 and 14 inches. Varnished and stenciled. Front wheels turn under the body. The above cut shows style and shape. It has a high back, and is being extensively used in place of more expensive children's carriages, combining a pleasing toy with a very useful article.

Price, per doz......................$15.00.

Painted and ornamented like cut..... 17.00.

With iron axle, $1.50 per doz. net extra.

No. 9.

Iron axles; body 15x29 inches; wheels 12x16 inches. Hard wood paneled body, stenciled, scrolled and nicely painted. Has heavy iron axles in thimble skeins, iron braces, malleable iron tongue fastening and fifth wheel and hub caps.

Price, per doz.......................................$22.00

Note.—Seats for any of the foregoing Wagons for $1.50 per doz, net extra.

11

No. 10.

Iron axles; body 14x28 inches; wheels 12 and 16 inches. Has heavy iron axles in iron thimble skeins, and malleable iron tongue fastening and fifth wheel. New style hardwood paneled body, high seat and dashboard, iron braced, hub caps and bent handle.

Price, per doz...$26.00

No. 11.

Iron axles; 14x28 inches; wheels 12 and 16 inches. Hardwood paneled body, landscape painting, scrolled and varnished, hub caps, high seat and dashboard. Iron braced, heavy iron axles in iron thimble skeins, oval tires welded and shrunk on. Same as cut.

Price, per doz...$32.00.

12

BOYS' HEAVY STAKE WAGON.

No. 12.

Iron axles; body 15x32 inches; wheels 14 and 20 inches. Has heavy square iron axles and is heavily braced and bolted, oval tires or heavy flat tires, welded and shrunk on. All hardwood, new style, paneled body, high seat, dashboard and stakes, landscape painting, stenciled, scrolled and nicely finished, making a very stylish, substantial and useful wagon. The handle is the strongest in the market and has the best attachments, being made entirely of malleable iron. (*See page 3*)

Price, per doz...$50.00.

No. 13.

Iron axle. Same as No. 12, only painted in fancy colors and ornamented.

Price, per doz...$56.00.

13

NO. 14.

Iron axles; body, 18 x 38 inches; wheels, 14 and 20 inches. Has heavy flat tire, welded and shrunk on, paneled body, seat and dashboard, nicely painted, striped and scrolled. The most substantial and largest wagon in the market. The handle is the strongest in the market, and has the best attachments, being made entirely of malleable iron.

Price, per doz.............................$72.00.

Can furnish shafts with irons, for dog or goat wagon, for 75c. each net.

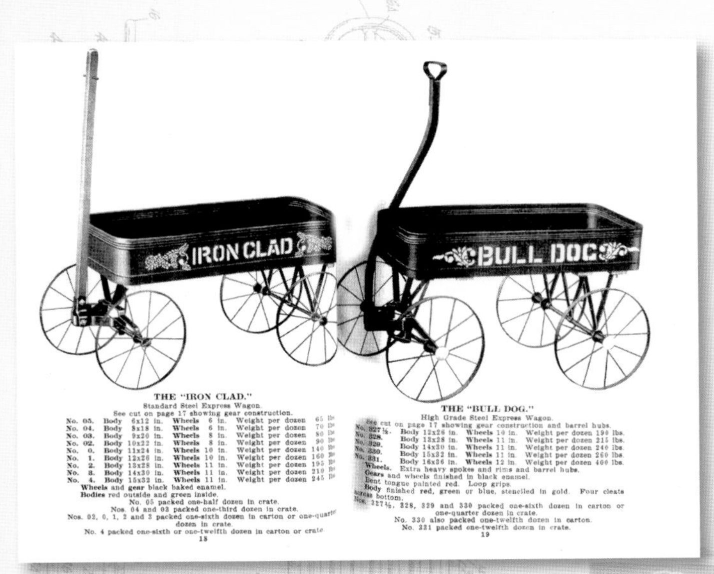

THE "IRON CLAD."
Standard Steel Express Wagon.
See cut on page 17 showing gear construction.

No. 05.	Body 6x12 in.	Wheels 6 in.	Weight per dozen 65 lbs
No. 04.	Body 8x18 in.	Wheels 6 in.	Weight per dozen 70 lbs
No. 03.	Body 9x20 in.	Wheels 8 in.	Weight per dozen 80 lbs
No. 02.	Body 10x22 in.	Wheels 8 in.	Weight per dozen 90 lbs
No. 0.	Body 11x24 in.	Wheels 10 in.	Weight per dozen 140 lbs
No. 1.	Body 12x26 in.	Wheels 10 in.	Weight per dozen 160 lbs
No. 2.	Body 13x28 in.	Wheels 11 in.	Weight per dozen 195 lbs
No. 3.	Body 14x30 in.	Wheels 11 in.	Weight per dozen 210 lbs
No. 4.	Body 15x32 in.	Wheels 11 in.	Weight per dozen 245 lbs

Wheels and **gear** black baked enamel.
Bodies red outside and green inside.
No. 05 packed one-half dozen in crate.
Nos. 04 and 03 packed one-third dozen in crate.
Nos. 02, 0, 1, 2 and 3 packed one-sixth dozen in carton or one-quarter dozen in crate.
No. 4 packed one-sixth or one-twelfth dozen in carton or crate.
18

THE "BULL DOG."
High Grade Steel Express Wagon.
See cut on page 17 showing gear construction and barrel hubs.

No. 327½.	Body 12x26 in.	Wheels 10 in.	Weight per dozen 190 lbs.
No. 328.	Body 13x28 in.	Wheels 11 in.	Weight per dozen 215 lbs.
No. 329.	Body 14x30 in.	Wheels 11 in.	Weight per dozen 240 lbs.
No. 330.	Body 15x32 in.	Wheels 11 in.	Weight per dozen 260 lbs.
No. 331.	Body 16x36 in.	Wheels 12 in.	Weight per dozen 400 lbs.

Wheels. Extra heavy spokes and rims and barrel hubs.
Gears and wheels finished in black enamel.
Bent tongue painted red. Loop grips.
Body finished red, green or blue, stenciled in gold. Four cleats across bottom.
Nos. 327½, 328, 329 and 330 packed one-sixth dozen in carton or one-quarter dozen in crate.
No. 330 also packed one-twelfth dozen in carton.
No. 331 packed one-twelfth dozen in crate.
19

Garton Toy Badger Line Catalog, 1916-1917. *Courtesy of Ann Huenink McIntyre.* (continued on following two pages)

Showing gear construction of Standard Steel Wagon.

The "Iron Clad."

Showing gear construction of High Grade Steel Wagon.

The "Bull Dog."

17

THE "BADGER EXPRESS."

BUILT FOR STRENGTH.

Body constructed of No. 24 sheet steel.

Wheels extra heavy, tires 11/16x3/32 in. oval steel, 16 spokes of No. 1[illegible] wire. Finished in green enamel.

Wood gear, same as used on our Badger Coaster Wagon. Strong steel braces. 9/16 in. round steel axles.

Body and **gear** painted red, green or blue, striped with gold; inside painted green.

Not the Cheapest, but the Best.

No. 227.	**Body** 12x24 in.	**Wheels** 11 and 14 in.	**Shipping weight** 270 lbs. per dozen.
No. 227½.	**Body** 13x26 in.	**Wheels** 11 and 14 in.	**Shipping weight** 287 lbs. per dozen.
No. 228.	**Body** 14x28 in.	**Wheels** 11 and 14 in.	**Shipping weight** 318 lbs. per dozen.
No. 229.	**Body** 15x30 in.	**Wheels** 11 and 14 in.	**Shipping weight** 325 lbs. per dozen.

This line also equipped with ½ inch rubber tire wheels at additional cost.

One-sixth dozen in crate.

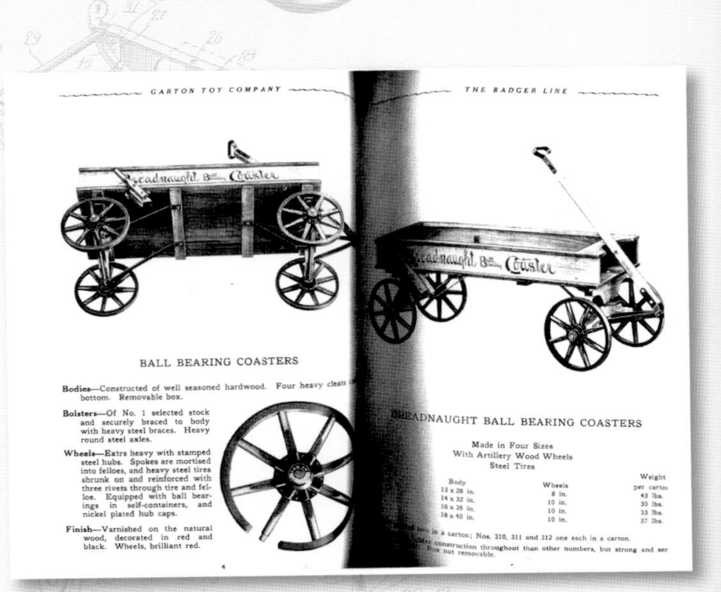

GARTON TOY COMPANY — THE BADGER LINE

BALL BEARING COASTERS

Bodies—Constructed of well seasoned hardwood. Four heavy cleats [illegible] bottom. Removable box.

Bolsters—Of No. 1 selected stock and securely braced to body with heavy steel braces. Heavy round steel axles.

Wheels—Extra heavy with stamped steel hubs. Spokes are mortised into felloes, and heavy steel tires shrunk on and reinforced with three rivets through tire and felloe. Equipped with ball bearings in self-containers, and nickel plated hub caps.

Finish—Varnished on the natural wood, decorated in red and black. Wheels, brilliant red.

4

DREADNAUGHT BALL BEARING COASTERS

Made in Four Sizes
With Artillery Wood Wheels
Steel Tires

Body	Wheels	Weight per carton
13 x 28 in.	8 in.	43 lbs.
14 x 32 in.	10 in.	30 lbs.
16 x 36 in.	10 in.	33 lbs.
18 x 40 in.	10 in.	37 lbs.

[illegible] two in a carton; Nos. 310, 311 and 312 one each in a carton.
[illegible] construction throughout than other numbers, but strong and ser[illegible]. Box not removable.

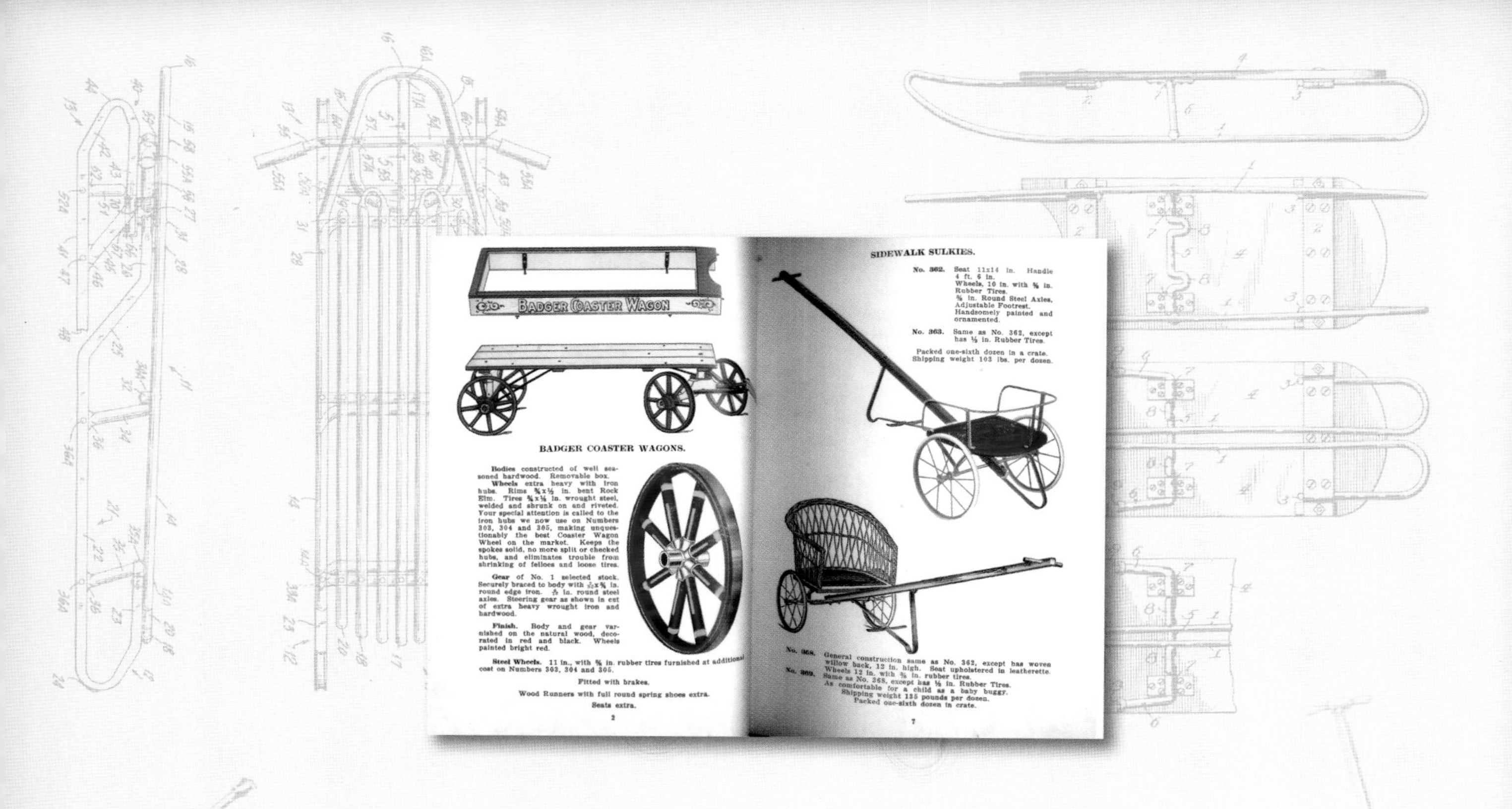

BADGER COASTER WAGONS.

Bodies constructed of well seasoned hardwood. Removable box.

Wheels extra heavy with iron hubs. Rims ⅝x½ in. bent Rock Elm. Tires ⅝x⅛ in. wrought steel, welded and shrunk on and riveted. Your special attention is called to the iron hubs we now use on Numbers 303, 304 and 305, making unquestionably the best Coaster Wagon Wheel on the market. Keeps the spokes solid, no more split or checked hubs, and eliminates trouble from shrinking of felloes and loose tires.

Gear of No. 1 selected stock. Securely braced to body with 3/32x⅝ in. round edge iron. 7/16 in. round steel axles. Steering gear as shown in cut of extra heavy wrought iron and hardwood.

Finish. Body and gear varnished on the natural wood, decorated in red and black. Wheels painted bright red.

Steel Wheels. 11 in., with ⅝ in. rubber tires furnished at additional cost on Numbers 303, 304 and 305.

Fitted with brakes.

Wood Runners with full round spring shoes extra.

Seats extra.

2

SIDEWALK SULKIES.

No. 362. Seat 11x14 in. Handle 4 ft. 6 in. Wheels, 10 in. with ⅜ in. Rubber Tires. ⅝ in. Round Steel Axles, Adjustable Footrest. Handsomely painted and ornamented.

No. 363. Same as No. 362, except has ½ in. Rubber Tires.

Packed one-sixth dozen in a crate.
Shipping weight 103 lbs. per dozen.

No. 368. General construction same as No. 362, except has woven willow back, 12 in. high. Seat upholstered in leatherette. Wheels 12 in. with ⅜ in. rubber tires.

No. 369. Same as No. 368, except has ½ in. Rubber Tires. As comfortable for a child as a baby buggy.

Shipping weight 135 pounds per dozen.
Packed one-sixth dozen in crate.

7

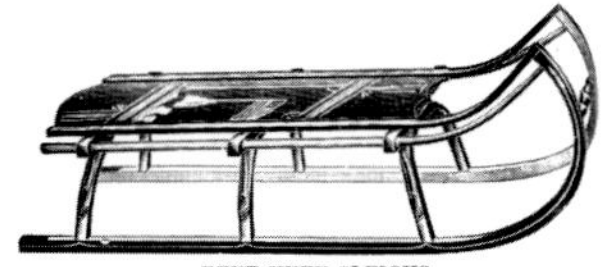

BENT KNEE SLEIGHS.

No. 40. Size 15x33 in. Varnished and striped, top board painted and ornamented, round side fenders, two knees, iron braces, flat shoes. Shipping weight 42 lbs. per dozen.

No. 41. Same as No. 40, with half-oval shoes. Shipping weight 5 lbs. per dozen.

No. 42. Size 15x36 in. Varnished and striped, top painted in handsome colors, round rail, three knees, iron braces, flat shoes. Shipping weight 60 lbs. per dozen.

No. 43. Same as No. 42, with half-oval shoes. Shipping weight 6 lbs. per dozen.

Packed one-sixth dozen in a bundle.

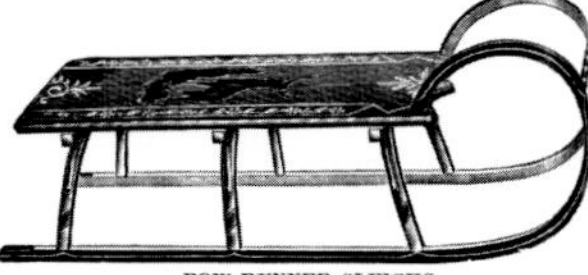

BOW RUNNER SLEIGHS.

No. 44. Size 12x34 in. Varnished on the wood, fancy painted top board, two knees, iron braces, flat shoes. Shipping weight 51 lbs. per dozen.

No. 45. Same as No. 44, with half-oval shoes. Shipping weight 6 lbs. per dozen.

No. 46. Same as cut. Size 12x37 in. Varnished on the wood, top board finely painted, three knees, iron braces, flat shoes. Shipping weight 66 lbs. per dozen.

No. 47. Same as No. 46, with half-oval shoes. Shipping weight 72 lbs. per dozen.

Packed one-sixth dozen in a bundle.

68

BENT KNEE SLEIGHS.

Size 15x33 in. Varnished and stenciled, top board painted and ornamented in bright colors, round side fenders, two knees, iron braces, dragon heads, flat shoes. Shipping weight 50 lbs. per dozen.

Same as No. 52, with half-oval shoes. Shipping weight 56 lbs. per dozen.

Packed one-sixth dozen in a bundle.

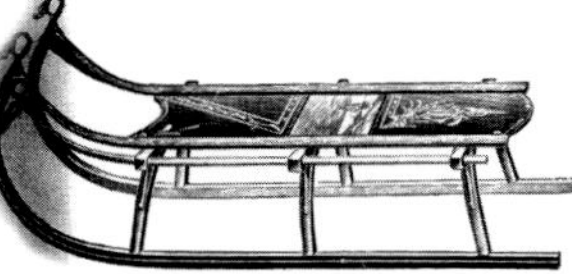

BENT KNEE SLEIGHS.

Size 15x36 in. Varnished on the wood, fancy painted top, round side fenders, three knees, iron braces, dragon heads, flat shoes. Shipping weight 63 lbs. per dozen.

Same as No. 54, with half-oval shoes. Shipping weight 72 lbs. per dozen.

Packed one-sixth dozen in a bundle.

69

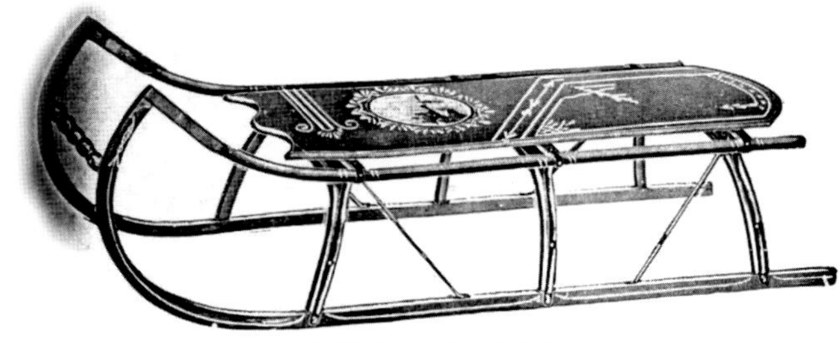

ROCK ELM SLEIGHS.

No. 48. Size 13x36 in. Made of selected rock elm, finished on natural wood. Top board highly painted and ornamented, three bent knees, round side fenders with brightly plated braces extending to runners, half-oval shoes. It is light, yet very strong and durable.

No. 49. Same as No. 48, except is beautifully painted and ornamented throughout.

Shipping weight 78 lbs. per dozen.

Packed one-sixth dozen in a bundle.

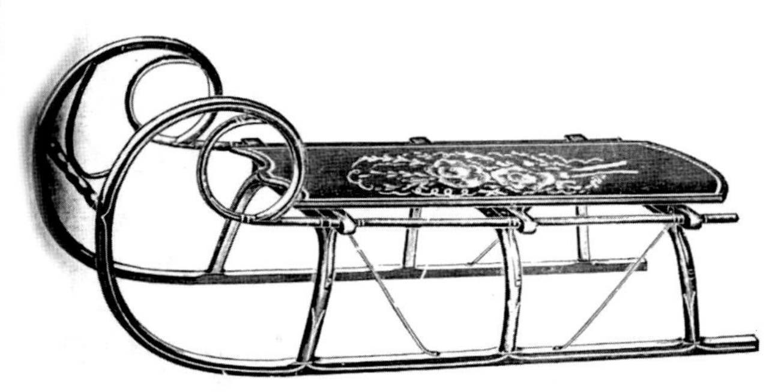

THE CHILDREN'S FAVORITE.

No. 58. Size 13x36 in. Three knees, brightly plated braces, half-oval shoes. This sleigh is made all of hardwood, very handsomely painted and decorated throughout.

Shipping weight 78 lbs. per dozen.

Packed one-sixth dozen in a bundle.

71

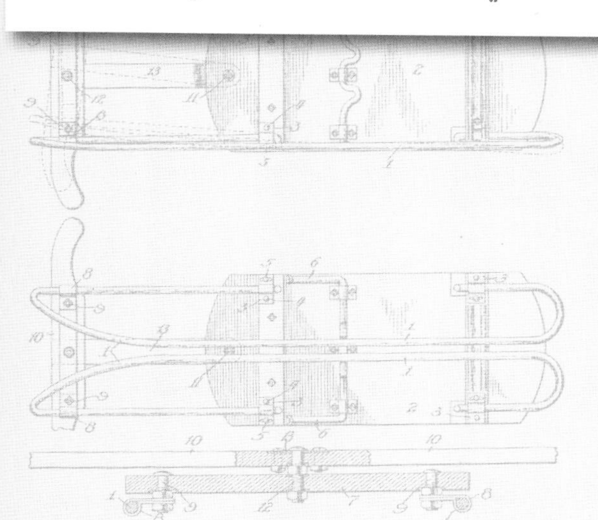

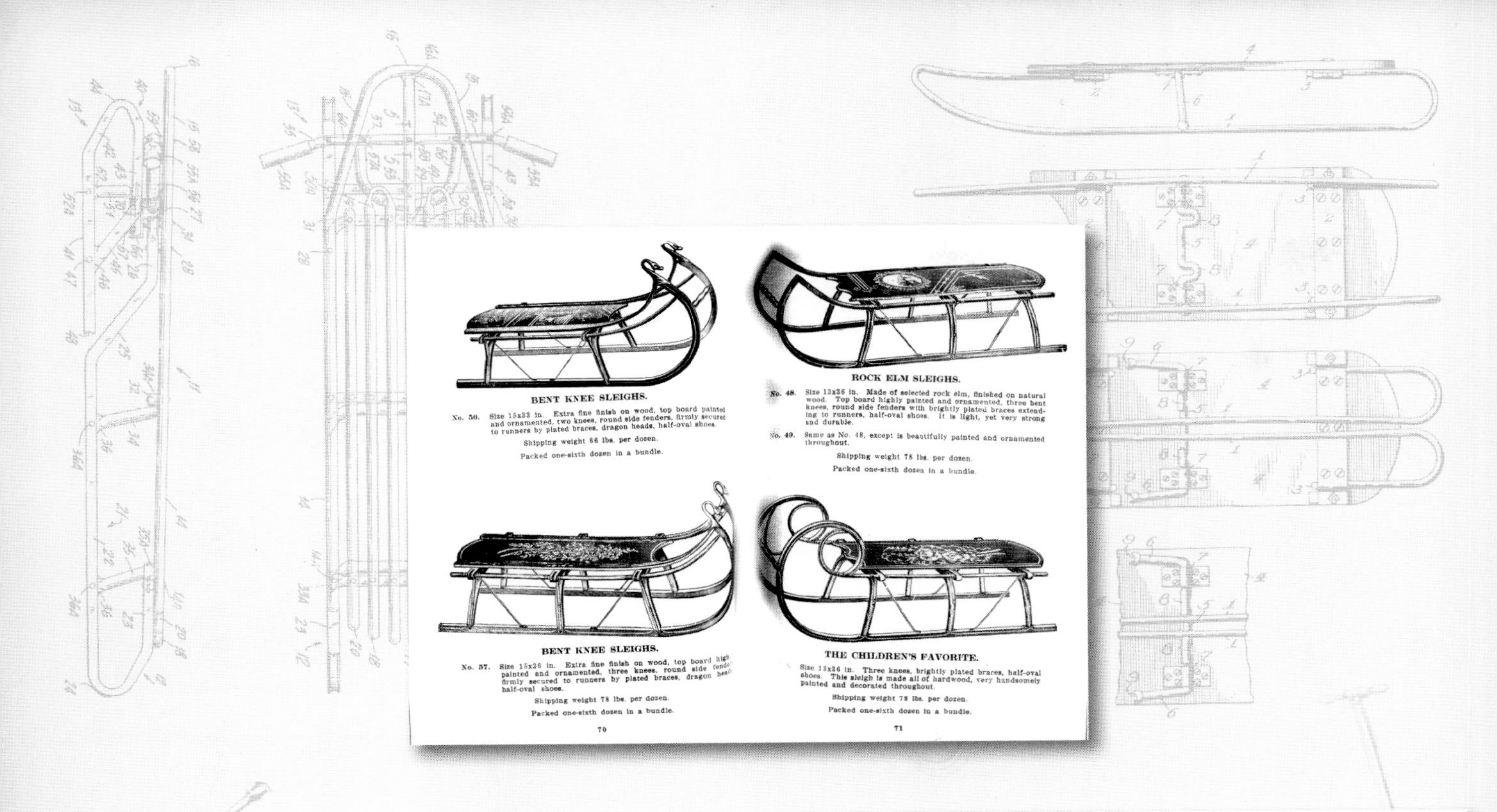

BENT KNEE SLEIGHS.

No. 56. Size 15x33 in. Extra fine finish on wood, top board painted and ornamented, two knees, round side fenders, firmly secured to runners by plated braces, dragon heads, half-oval shoes.

Shipping weight 66 lbs. per dozen.

Packed one-sixth dozen in a bundle.

ROCK ELM SLEIGHS.

No. 48. Size 13x36 in. Made of selected rock elm, finished on natural wood. Top board highly painted and ornamented, three bent knees, round side fenders with brightly plated braces extending to runners, half-oval shoes. It is light, yet very strong and durable.

No. 49. Same as No. 48, except is beautifully painted and ornamented throughout.

Shipping weight 78 lbs. per dozen.

Packed one-sixth dozen in a bundle.

BENT KNEE SLEIGHS.

No. 57. Size 15x36 in. Extra fine finish on wood, top board high painted and ornamented, three knees, round side fende firmly secured to runners by plated braces, dragon head half-oval shoes.

Shipping weight 78 lbs. per dozen.

Packed one-sixth dozen in a bundle.

THE CHILDREN'S FAVORITE.

Size 13x36 in. Three knees, brightly plated braces, half-oval shoes. This sleigh is made all of hardwood, very handsomely painted and decorated throughout.

Shipping weight 78 lbs. per dozen.

Packed one-sixth dozen in a bundle.

70

71

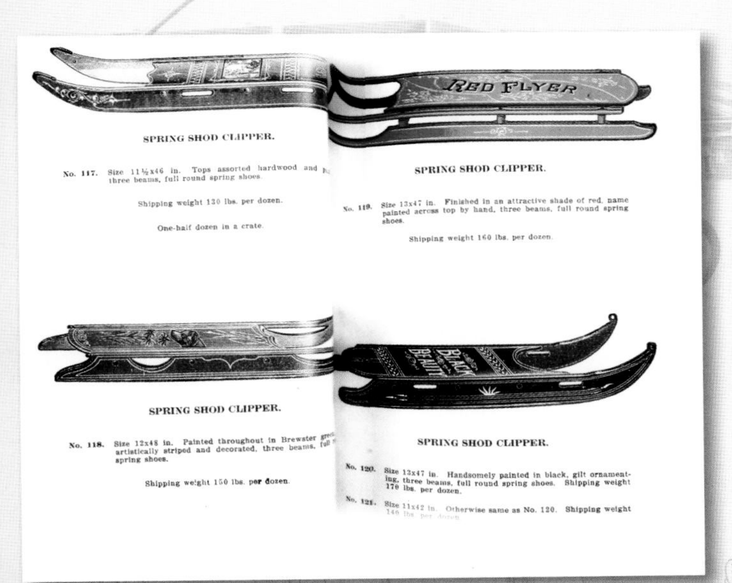

SPRING SHOD CLIPPER.

No. 117. Size 11½x46 in. Tops assorted hardwood and p three beams, full round spring shoes.

Shipping weight 130 lbs. per dozen.

One-half dozen in a crate.

SPRING SHOD CLIPPER.

No. 119. Size 13x47 in. Finished in an attractive shade of red, name painted across top by hand, three beams, full round spring shoes.

Shipping weight 160 lbs. per dozen.

SPRING SHOD CLIPPER.

No. 118. Size 12x48 in. Painted throughout in Brewster gre artistically striped and decorated, three beams, full spring shoes.

Shipping weight 150 lbs. per dozen.

SPRING SHOD CLIPPER.

No. 120. Size 13x47 in. Handsomely painted in black, gilt ornamenting, three beams, full round spring shoes. Shipping weight 170 lbs. per dozen.

No. 121. Size 11x42 in. Otherwise same as No. 120. Shipping weight 140 lbs. per dozen.

RIVERSIDE.CO.MIL.

POPULAR BEAUTY.

No. 59. Size 13x36 in. Plated braces, half-oval shoes. The cut is an exact representation of this sleigh, which is very attractively painted and ornamented throughout, making it one of the most popular fine sleighs manufactured.

Shipping weight 78 lbs. per dozen.

Packed one-sixth dozen in a bundle.

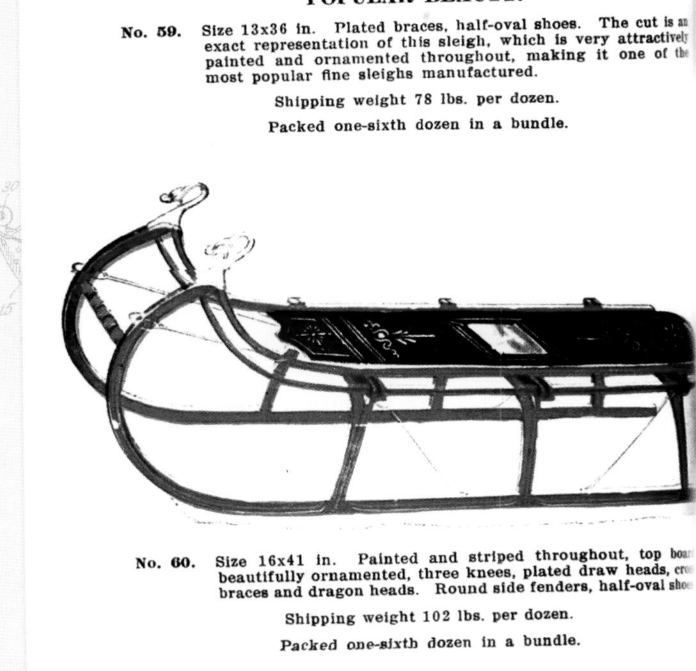

No. 60. Size 16x41 in. Painted and striped throughout, top boa beautifully ornamented, three knees, plated draw heads, cro braces and dragon heads. Round side fenders, half-oval shoe

Shipping weight 102 lbs. per dozen.

Packed one-sixth dozen in a bundle.

THE "ESKIMO" AND "DREADNAUGHT" STEERING SLEDS.

There positively is no better line made. Stamped Steel Knees. Best Crucible Spring Steel T Shaped Runners, so curved in front as to give a maximum length of steering surface.

Steering bar works perfectly, the sled responding instantly to slightest bend in steel runner without retarding speed of sled a particle.

Gear and Runners finished in Red Enamel. Tops of Rock Elm, beautifully painted and decorated.

We do not care what line you have been handling. The "Eskimo" and "Dreadnaught" are bound to give satisfaction, and the prices are right.

All Steering Sleds are Packed One-sixth Dozen in Bundle, Securely Fastened with Wire.

Garton Toy Badger Line Catalog, 1916-1917 sleds: Eskimo, Dreadnaught, Dreadnaught Special, Torpedo Racer. *Courtesy of Ann Huenink McIntyre.* (continued on opposite page)

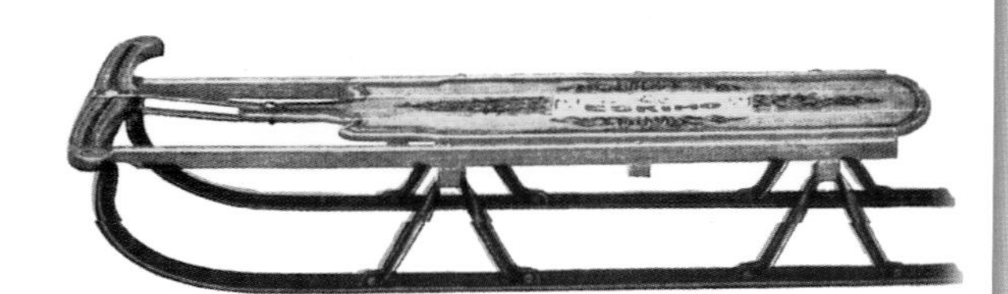

No. 130. 33 in. long, 11 in. wide, 6 in. high. Shipping weight 90 lbs. per dozen.

No. 131. 36 in. long, 12 in. wide, 6 in. high. Shipping weight 102 lbs. per dozen.

Carry one "Grown up."

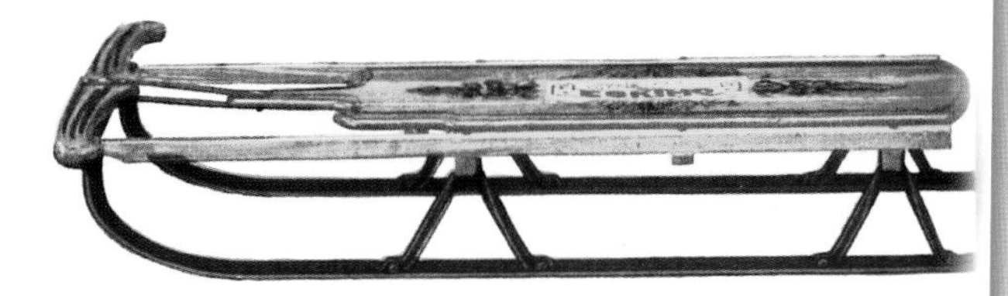

No. 132. 40 in. long, 13 in. wide, 6¼ in. high.

Shipping weight 114 lbs. per dozen.

Carries one "Grown up" or two "Kids."

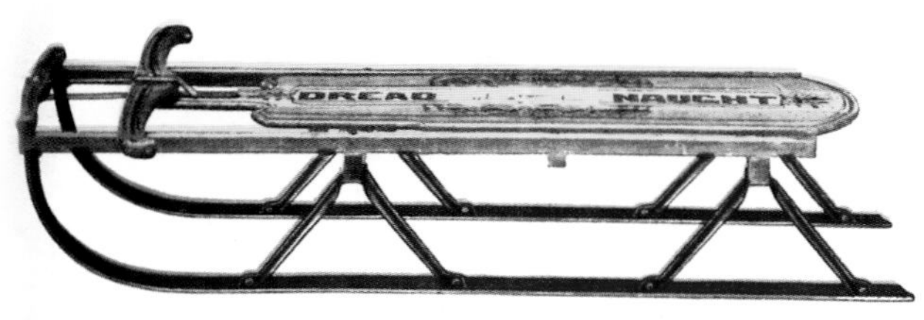

No. 133. 45 in. long, 14 in. wide, 8¼ in. high.

Shipping weight 168 lbs. per dozen.

Carries two "Grown ups" or three "Kids."

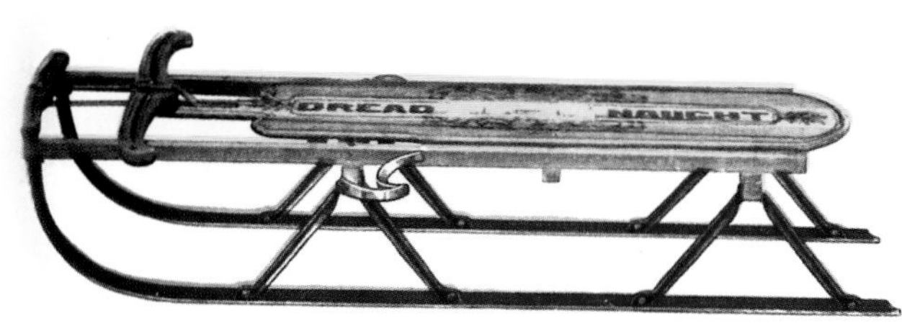

No. 134. 50 in. long, 14 in. wide, 8½ in. high.

Has two iron foot rests.

Shipping weight 192 lbs. per dozen.

Carries two "Grown ups" or three or four "Kids."

CHILD'S CUTTER.

No. 3. Size 14x27 in. Nicely painted and ornamented throughout. Large deep body and seat, push handle, two knees, round side fenders, brightly plated braces, half-oval shoes.

No. 3½. Same as No. 3, with white enamel finish.

No. 5. Same as No. 3, upholstered in velour.

No. 5½. Same as No. 5, with white enamel finish.

Packed one in crate.

Shipping weight 25 lbs. each.

65

SPECIAL "TORPEDO RACER."

No. 136. 56 in. long, 13 in. wide, 6¼ in. high.
Long, low, narrow, racy lines.
Built for speed.
Carries three good sized boys.
Shipping weight 168 lbs. per dozen.

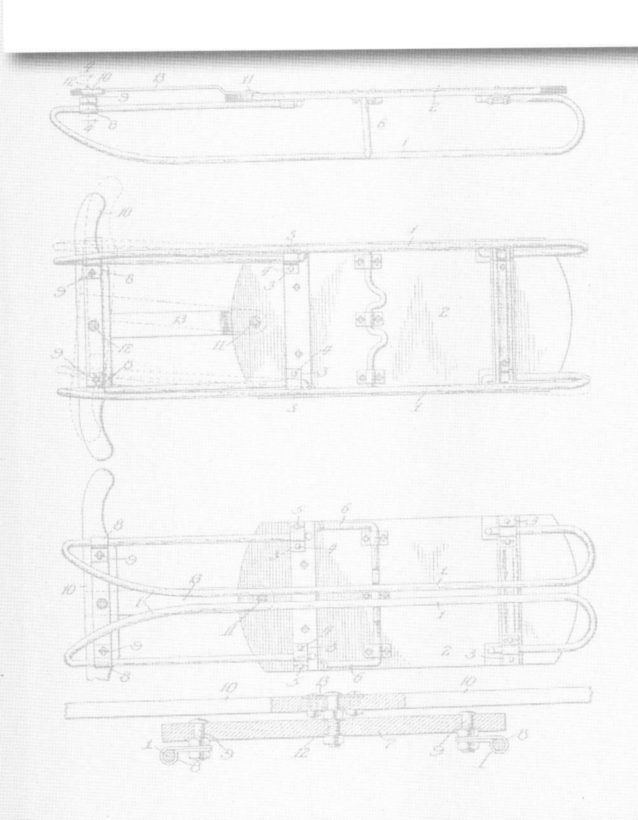

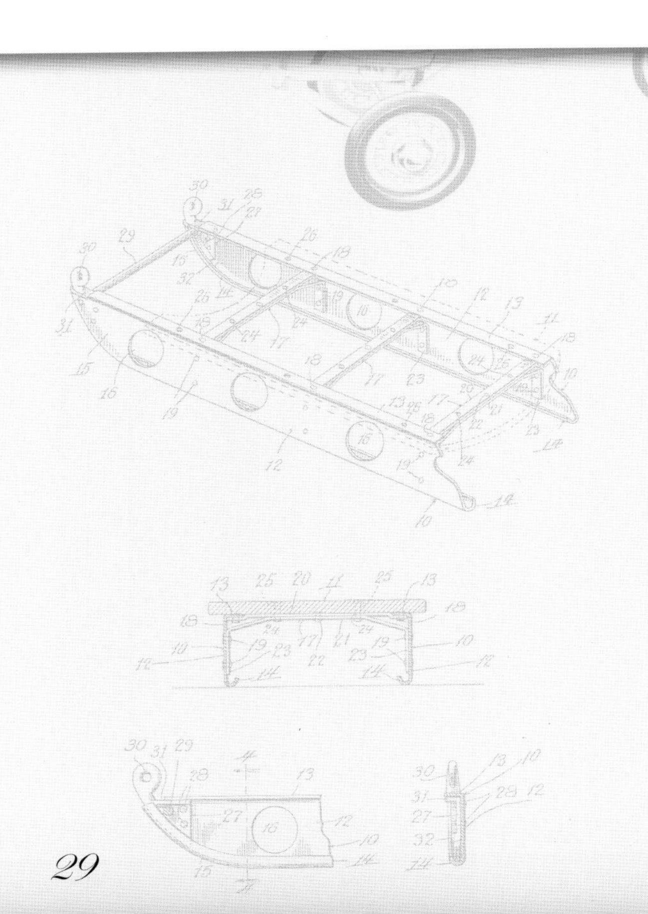

Garton Badger Line Juvenile Vehicles Catalog, 1924. Go-Din, Dreadnaught ball bearing coaster, Badger roller bearing coaster, Teenie Weenie, & Badger Coaster. *Courtesy of Ann Huenink McIntyre.* (continued on opposite page)

GARTON TOY COMPANY

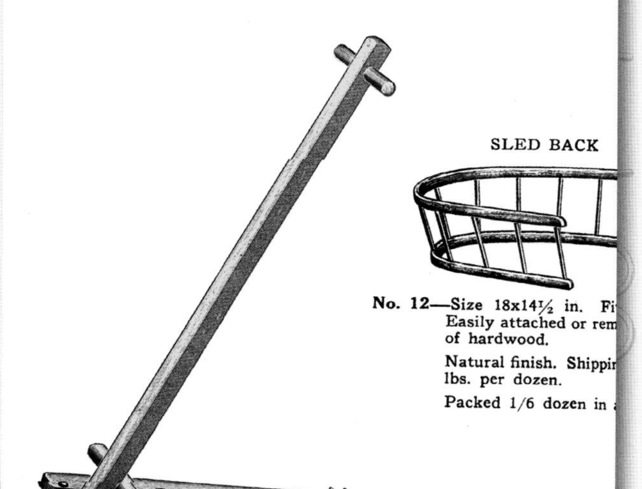

SLED BACK

No. 12—Size 18x14½ in. Fi
Easily attached or rem
of hardwood.

Natural finish. Shippi
lbs. per dozen.

Packed 1/6 dozen in

No. 1

GO-DINK

A SCOOTER ON RUNNERS

Varnished on natural wood.
Red enameled T-iron Runners.
Length Foot Board 25 in. Handle 28 in.
Packed four in a carton.
Weight per carton 30 lbs.

52

GARTON TOY COMPANY

"DREADNAUGHT" BALL BEARING DOUBLE DISC WHEEL COASTERS

Rubber Tires

Heavy Stamped Disc Wheels, finished in bright red and fitted with large rubber tires. Extra large heavy nickel plated hub caps. (This wheel guaranteed by the maker against defective workmanship. All wheels not giving satisfactory service replaced free of charge. Balls are in self-containers and cannot be lost.) Body and gear made of finest grade selected kiln dried hardwood, varnished on the natural wood, decorated in red and black.

No.	Body	Wheel	Tires
320⅝	13 x 28 in.	8 in.	⅝ in.
324⅝	14 x 32 in.	8 in.	⅝ in.
325⅝	14 x 32 in.	10 in.	⅝ in.
326⅝	16 x 36 in.	10 in.	⅝ in.
327⅝	18 x 40 in.	10 in.	⅝ in.

Packed one in a carton.

No. 320⅝ is of lighter construction throughout than other numbers, but strong serviceable.

6

THE BADGER LINE

BADGER COASTER WAGONS

Bodies—Constructed of well seasoned hardwood. Cleats under movable box.

Bolsters—Of No. 1 selected stock and well braced to body. Roun

Wheels—Strong and well built with heavy spokes and stampe

Finish—Varnished on the natural wood, with red and black Wheels painted bright red.

MADE IN FOUR SIZES:

No.	Body	Wheels
302	13 x 28 inches	8 inches
303	14 x 32 inches	10 inches
304	16 x 36 inches	10 inches
305	18 x 40 inches	10 inches

No. 305 packed one in a carton; Nos. 302, 303 and 304 packed two each in

No. 302 is of lighter construction throughout than other numbers, but stror able. Box not removable.

THE BADGER LINE

BADGER ROLLER BEARING COASTER WAGON

Heavy Channel Steel Gear, Strongly Braced

General body construction same as that of Dreadnaught Coaster, var-shed on the natural wood, decorated in red and black. Bolsters of 7/8 in. annel steel reinforced with one inch steel braces. Axles of cold rolled steel.

Wheels—Double disc with contained roller bearings, finished in bright and fitted with large rubber tires. Nickel plated hub caps.

Body	Wheels	Tires	Weight
16 x 36 in.	10 in.	5/8 in.	36 lbs.
16 x 36 in.	10 in.	1 in.	40 lbs.

Packed one in a carton

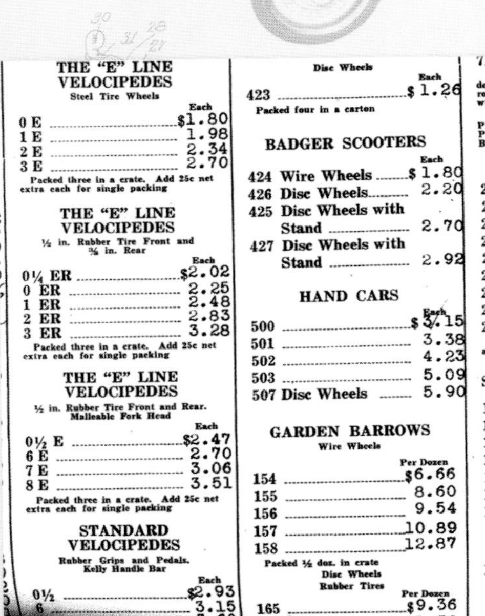

THE "E" LINE VELOCIPEDES
Steel Tire Wheels

	Each
0 E	$1.80
1 E	1.98
2 E	2.34
3 E	2.70

Packed three in a crate. Add 25c net extra each for single packing

THE "E" LINE VELOCIPEDES
½ in. Rubber Tire Front and ⅜ in. Rear

	Each
0¼ ER	$2.02
0 ER	2.25
1 ER	2.48
2 ER	2.83
3 ER	3.28

Packed three in a crate. Add 25c net extra each for single packing

THE "E" LINE VELOCIPEDES
½ in. Rubber Tire Front and Rear. Malleable Fork Head

	Each
0½ E	$2.47
6 E	2.70
7 E	3.06
8 E	3.51

Packed three in a crate. Add 25c net extra each for single packing

STANDARD VELOCIPEDES
Rubber Grips and Pedals. Kelly Handle Bar

	Each
0½	$2.93
6	3.15
[illegible]	3.60

Disc Wheels

	Each
423	$1.26

Packed four in a carton

BADGER SCOOTERS

	Each
424 Wire Wheels	$1.80
426 Disc Wheels	2.20
425 Disc Wheels with Stand	2.70
427 Disc Wheels with Stand	2.92

HAND CARS

	Each
500	$3.15
501	3.38
502	4.23
503	5.09
507 Disc Wheels	5.90

GARDEN BARROWS
Wire Wheels

	Per Dozen
154	$6.66
155	8.60
156	9.54
157	10.89
158	12.87

Packed ½ doz. in crate

Disc Wheels
Rubber Tires

	Per Dozen
165	$9.36

71 BBD	12.79

All Autos are priced K. D. If ordered set up, less wheels and steering rod, an extra charge of 50c net each will be made.

All Autos equipped with rubber pedals
PBW means Plain Bearing Wire Wheels
PBD means Plain Bearing Disc Wheels
BBD means Ball Bearing Disc Wheels

FOLDING TABLES

	Per Dozen
209	$9.59
210	11.25
211	13.50
209½	10.12
210½	12.10
211½	14.35
209 X	12.38
210 X	16.88
211 X	20.25

Nos. 209 and 210 packed ½ doz. in a carton.
No. 211 packed ⅓ doz. in a carton

STEEL STEERING SLEDS

	Per Dozen
130	$10.57
131	13.05
132	15.52
133	18.90
134	23.62
135	31.95
136	26.10

Packed 1/6 dozen in a bundle

SLED GUARDS

	Per Dozen
12	$7.43

Packed 1/6 dozen in a bundle

GO-DINK SCOOTER SLED

Garton 60th Anniversary Catalog, 1939. *Courtesy of Jane Dwyer Garton.* *(continued on opposite three pages)*

60th Anniversary 1879-1939

GARTON TOY COMPANY

SHEBOYGAN, WISCONSIN

GOOD TOYS Garton SINCE 1879

BRANCH OFFICES AND PERMANENT SHOWROOMS:

NEW YORK	SAN FRANCISCO	CHICAGO	DALLAS
1107 Broadway	Standard Toy Agencies	Merchandise Mart	D. D. Otstott
Space 407	718 Mission St.	Space 1417	Santa Fe Bldg.

CATALOG OF

WAGONS • VELOCIPEDES • PARK CYCLES • AUTOS

SCOOTERS • BABY WALKERS • BUDDIE BIKES • SLEDS

CROQUET AND DOLL STAND CIRCULARS ON REQUEST

60th Anniversary 1879-1939

GARTON TOY COMPANY

4.94

1312

BOX—Heavy kiln dried lumber, 16x36 inches, reinforced with cleats. Ends mortised to sides.

GEAR—Bolsters heavy channel steel. Fifth wheel construction.

WHEELS—10-inch roller bearing artillery.

TIRES—¾-inch rubber.

FINISH—Box natural varnish, racks green, wheels red.

PACKING—One in carton.

WEIGHT—41 pounds.

8.70

1349

BOX—Heavy kiln dried lumber, 16x40 inches, reinforced with cleats. Sides mortised to bottom. Ends mortised to sides. Ends drawn together with box rods.

GEAR—Bolsters heavy channel steel. Fifth wheel construction.

WHEELS—Rimrold. 10-inch ball bearing artillery.

TIRES—1-inch rubber.

FINISH—High lustre durable enamel and varnish. Box, ladders and hose reel natural varnish. Stakes, seat, footboards, handle and wheels red decorated ivory. Undergear black.

EQUIPMENT—Hose reel. Ladders. Foot gong. Emergency lanterns. Removable seat and steps front and rear.

PACKING—One in carton.

WEIGHT—57 pounds.

2

SHEBOYGAN, WISCONSIN

60th Anniversary 1879 · 1939

6.94

1.33

1336 SIX WHEEL TRUCK

BOX—Heavy kiln dried hardwood, 18x40 inches, 1-inch sides, ends reinforced with box rods.
GEAR—Massive stamped gear. Turntable construction completely assembled.
WHEELS—Rimrold. 10-inch roller bearing disc.
TIRES—1-inch rubber.
FINISH—Box varnished natural. Rear racks natural, front racks red decorated ivory. Wheels red, striped ivory.
PACKING—One in carton.
WEIGHT—58 pounds.

1399

SHAFTS—Can be used for dog or goat and will fit any truck or wagon. Made of selected hardwood. Finished red. Complete with necessary parts for regulation harness.

3

GARTON TOY COMPANY

60th Anniversary 1879 · 1939

3.53

4.23

1431

BOX—Heavy kiln dried lumber, 16x36 inches, reinforced with cleats.
GEAR—Stamped gear with fifth wheel device.
WHEELS—10-inch roller bearing disc.
TIRES—¾-inch rubber.
FINISH—Box varnished natural, decorated red, stencilled black. Wheels red striped ivory. Gear black.
PACKING—One in carton.
WEIGHT—32 pounds.

1440

BOX—Heavy kiln dried hardwood, 16x36 inches. 1-inch paneled sides, ends reinforced with box rods.
GEAR—Massive stamped gear. Turntable construction completely assembled.
WHEELS—Rimrold. 10-inch ball bearing disc.
TIRES—¾-inch rubber.
FINISH—Box varnished natural. Wheels green striped white.
PACKING—One in carton.
WEIGHT—39 pounds.

4

SHEBOYGAN, WISCONSIN

60th Anniversary 1879 · 1939

3.69

4.21

1624

BOX—16x36 inches seasoned lumber reinforced with cleats. Slatted bottom.
GEAR—Massive stamped gear with turntable fifth wheel construction.
WHEELS—10-inch roller bearing artillery. ¾-inch tires.
HANDLE—Tubular steel.
FINISH—Box varnished, sides red. Gear and handle black. Wheels red striped ivory.
PACKING—One in carton.
WEIGHT—30 pounds.

1651

TOP—Seasoned lumber reinforced with cleats. Length 42-inches. Width at front 14 inches tapering to 12½ inches at rear. Polished, plated metal shield on front.
GEAR—Massive stamped gear with turntable fifth wheel construction.
WHEELS—10-inch roller bearing artillery. 1-inch rubber tires.
HANDLE—Plated tubular steel.
FINISH—Center slat and outside rails red decorated ivory, two side slats natural varnish decorated red. Gear aluminum. Wheels red striped ivory.
PACKING—One in carton.
WEIGHT—31 pounds.

1652

Same as 1651 but with brake. 4.34

5

GARTON TOY COMPANY

60th Anniversary 1879 · 1939

4.05

1.56

1627

BOX—Seasoned hardwood, 16x36 inches. Racer style, with heavy side rails.
GEAR—Massive stamped gear. Turntable construction completely assembled.
WHEELS—Rimrold. 10-inch ball bearing disc.
TIRES—1-inch rubber.
FINISH—Varnished natural, with side rails and red wheels decorated ivory.
PACKING—One in carton.
WEIGHT—31 pounds.

2117 TRAILER

Can be attached to any standard wagon or velocipede.
SIZE—18x14 inches.
WHEELS—8-inch wire spoke.
TIRES—⅝-inch rubber.
FINISH—Baked enamel. Red. Wheels and stencil ivory.
PACKING—One in carton.
WEIGHT—12 pounds.

6

SHEBOYGAN, WISCONSIN

60th Anniversary 1879 · 1939

.82

1.26

1.77

2401

BODY—Size 10½x21x2¼ inches. Stamped from one piece auto body steel.
GEAR—Stamped from heavy steel.
WHEELS—5-inch double disc.
TIRES—½-inch rubber.
FINISH—Red decorated ivory.
PACKING—Six in carton.
WEIGHT—45 pounds per carton.

2409

BODY—Size 12 x 24½ x 3 inches. Stamped from one piece auto body steel.
GEAR—Stamped from heavy steel.
WHEELS—6-inch double disc.
TIRES—½-inch rubber.
FINISH—Two color box, black outside, marigold inside. Wheels marigold striped black.
PACKING—Four in carton.
WEIGHT—49 pounds per carton.

2408 1.17

Same as 2409 but finished red decorated ivory.

2426

BODY—Size 14½x31x3¼ inches. Streamlined rounded front. Stamped from one piece auto body steel.
GEAR—Stamped from heavy steel.
WHEELS—7-inch double disc.
TIRES—⅝-inch rubber.
FINISH—Red decorated ivory.
PACKING—One in carton.
WEIGHT—24 pounds.

7

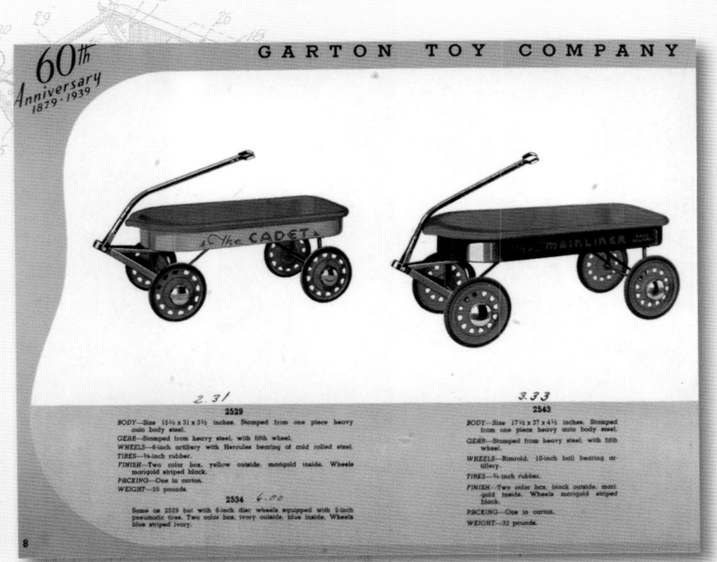

GARTON TOY COMPANY

60th Anniversary 1879 · 1939

2.31

3.33

2529

BODY—Size 15½ x 31 x 3½ inches. Stamped from one piece heavy auto body steel.
GEAR—Stamped from heavy steel, with fifth wheel.
WHEELS—8-inch artillery with Hercules bearing of cold rolled steel.
TIRES—⅝-inch rubber.
FINISH—Two color box, yellow outside, marigold inside. Wheels marigold striped black.
PACKING—One in carton.
WEIGHT—25 pounds.

2534 6.00

Same as 2529 but with 8-inch disc wheels equipped with 2-inch pneumatic tires. Two color box, ivory outside, blue inside. Wheels blue striped ivory.

2543

BODY—Size 17½ x 37 x 4½ inches. Stamped from one piece heavy auto body steel.
GEAR—Stamped from heavy steel, with fifth wheel.
WHEELS—Rimrold. 10-inch ball bearing artillery.
TIRES—¾-inch rubber.
FINISH—Two color box, black outside, marigold inside. Wheels marigold striped black.
PACKING—One in carton.
WEIGHT—32 pounds.

8

SHEBOYGAN, WISCONSIN

60th Anniversary 1879 · 1939

3.86

2550

BODY—Size 17½x37x4½ inches. Stamped from one piece heavy auto body steel.

GEAR—Massive stamped gear. Turntable construction completely assembled.

WHEELS—Rimrold. 10-inch ball bearing artillery.

TIRES—¾-inch rubber.

EQUIPMENT—Plated bumper.

FINISH—Baked enamel. Green decorated ivory.

PACKING—One in carton.

WEIGHT—35 pounds.

9

60th Anniversary 1879 · 1939

GARTON TOY COMPANY

4.60

2700

BODY—Stamped from one piece heavy auto body steel, correctly reinforced, not welded. Size overall 17½x42 inches. Inside dimensions—14¼x33¾x3¾ inches.

GEAR—Massive stamped gear. Turntable construction completely assembled.

EQUIPMENT—Rubber cushions between gear and body.

WHEELS—Rimrold. 8-inch roller bearing disc.

TIRES—¾-inch rubber.

FINISH—Baked enamel. Box and wheels red decorated ivory. Gear and handle black.

PACKING—One in carton.

WEIGHT—40 pounds.

10

SHEBOYGAN, WISCONSIN

60th Anniversary 1879 · 1939

ONE PIECE BODY—Drawn from a single piece of metal correctly reinforced; not welded. Practically indestructible.

QUICK DETACHABLE FEATURE—The one-piece body is attached to the running gear with eight bolts. The undergearing is die-formed and completely assembled at the factory.

HEADLIGHTS recessed in the body, as in aero-dynamic automobiles.

NEW STEERING PRINCIPLE—Patented construction provides for steering through slot in body. Gives positive, direct, accurate steering for the first time, with handle resting almost flat against the body—and a stronger, more rigid front end than ever before. Large fifth wheel is die-formed, turntable style: no front braces required.

7.15

2750

BODY—Stamped from one piece heavy auto body steel, correctly reinforced, not welded. Size overall 20x48 inches. Usable top surface 14x38 inches.

GEAR—Massive stamped gear. Turntable construction completely assembled.

EQUIPMENT—Plated rail and guard on front. Electric head lights less battery.

WHEELS—Rimrold. 8½-inch ball bearing disc.

TIRES—1-inch rubber.

FINISH—Baked enamel. Box and wheels red decorated ivory. Handle aluminum. Gear black.

PACKING—One in carton.

WEIGHT—48 pounds.

11.95

2752

Same as No. 2750 but with 8-inch ball bearing disc wheels and 2½-inch pneumatic tires.

11

60th Anniversary 1879 · 1939

GARTON TOY COMPANY

2.78

2828

BODY—Size 36x17¾ inches. Stamped one piece heavy auto body steel. Streamlined rounded front and rear. Width at front 17¾ inches tapering to 16 inches at rear. Oval ledge around top of box 1¾ inches wide at front tapering to 1¼ inches at rear. Embossed bottom.

GEAR—Massive stamped steel gear with turntable fifth wheel construction.

WHEELS—10-inch roller bearing artillery.

TIRES—¾-inch rubber.

FINISH—Red decorated ivory.

PACKING—One in carton.

WEIGHT—30 pounds.

3.22

2831

BODY—Stamped one piece heavy auto body steel, 36x18 inches.

GEAR—Stamped gear with fifth wheel device.

EQUIPMENT—Plated rail around front of box.

WHEELS—10-inch roller bearing artillery.

TIRES—¾-inch rubber.

FINISH—Two color box. White outside, blue inside. Wheels blue striped white.

PACKING—One in carton.

WEIGHT—32 pounds.

2832

Same as 2831 but with two color box, black outside, red inside. Wheels red striped ivory.

12

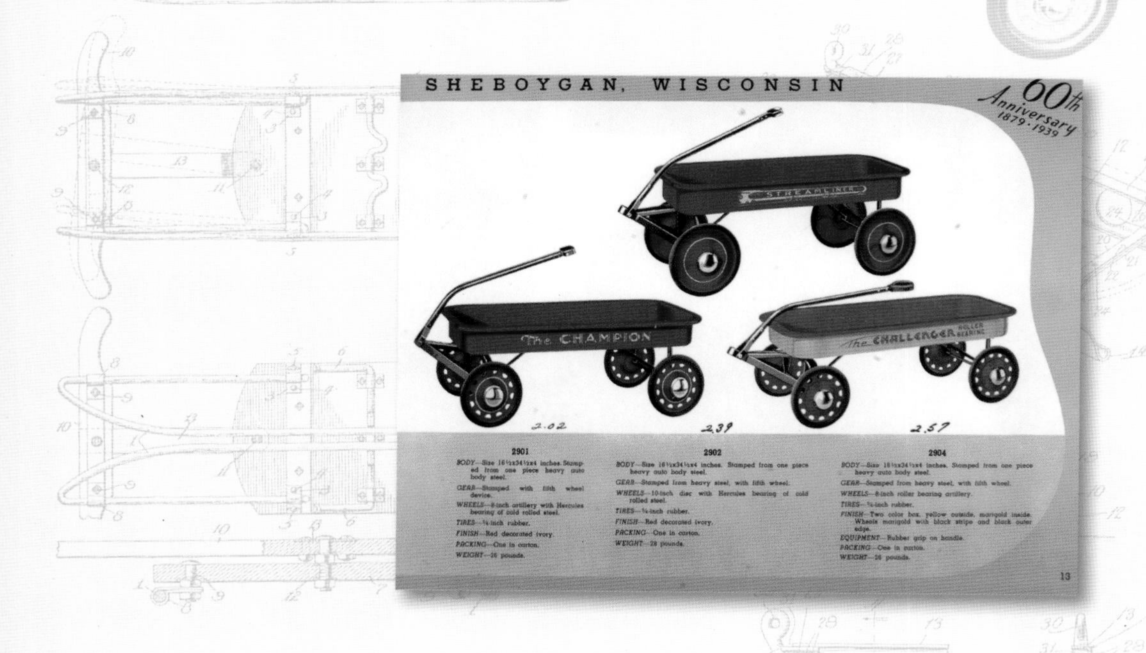

SHEBOYGAN, WISCONSIN

60th Anniversary 1879 · 1939

2.02

2901

BODY—Size 16½x34½x4 inches. Stamped from one piece heavy auto body steel.

GEAR—Stamped with fifth wheel device.

WHEELS—8-inch artillery with Hercules bearing of cold rolled steel.

TIRES—¾-inch rubber.

FINISH—Red decorated ivory.

PACKING—One in carton.

WEIGHT—26 pounds.

2.39

2902

BODY—Size 16½x34½x4 inches. Stamped from one piece heavy auto body steel.

GEAR—Stamped from heavy steel, with fifth wheel.

WHEELS—10-inch disc with Hercules bearing of cold rolled steel.

TIRES—¾-inch rubber.

FINISH—Red decorated ivory.

PACKING—One in carton.

WEIGHT—28 pounds.

2.57

2904

BODY—Size 16½x34½x4 inches. Stamped from one piece heavy auto body steel.

GEAR—Stamped from heavy steel, with fifth wheel.

WHEELS—8-inch roller bearing artillery.

TIRES—¾-inch rubber.

FINISH—Two color box, yellow outside, marigold inside. Wheels marigold with black stripe and black outer edge.

EQUIPMENT—Rubber grip on handle.

PACKING—One in carton.

WEIGHT—28 pounds.

13

2908

BODY—Size 16½x34½x4¾ inches. Stamped from one piece heavy auto body steel with bead around bottom.
GEAR—Stamped from heavy steel with fifth wheel.
WHEELS—Rimrold. 10-inch ball bearing artillery.
TIRES—¾-inch rubber.
FINISH—Two color box, maroon outside, red inside. Wheels red striped ivory.
PACKING—One in carton.
WEIGHT—33 pounds.

2911

BODY—Size 16½x34½x4¾ inches. Stamped from one piece heavy auto body steel with bead around bottom.
GEAR—Stamped from heavy steel, with fifth wheel.
WHEELS—Rimrold. 8½-inch ball bearing artillery.
TIRES—¾-inch rubber.
EQUIPMENT—Plated side rails.
FINISH—Two color box, black outside, red inside. Wheels red striped ivory.
PACKING—One in carton.
WEIGHT—34 pounds.

2912 3.92

Same as 2911 but with brake.

2932

BODY—Size 16½x34½x4¼ inches. Stamped from one piece heavy auto body steel with bead around bottom.
GEAR—Massive stamped gear. Turntable construction completely assembled.
WHEELS—Rimrold. 10-inch ball bearing artillery.
TIRES—1-inch rubber.
EQUIPMENT—Canvas knee pad. Plated side rails, bumper and handle. Rubber guard on handle.
FINISH—Baked enamel. Two-color box. White outside, green inside. Wheels green striped white.
PACKING—One in carton.
WEIGHT—38 pounds.

14

2933

BODY—Size 16½x34½x4 inches. Stamped from one piece heavy auto body steel.
GEAR—Stamped from heavy steel, with fifth wheel.
WHEELS—10-inch roller bearing artillery.
TIRES—2-inch pneumatic.
FINISH—Two color box, ivory outside, red inside. Wheels red striped ivory.
PACKING—One in carton.
WEIGHT—35 pounds.

2938

BODY—Size 18½x34½x4¾ inches. Stamped from one piece heavy auto body steel with bead around bottom.
GEAR—Massive stamped gear. Turntable construction completely assembled.
EQUIPMENT—Canvas knee pad. Plated side rails, handle and bumper. Rubber guard on handle.
WHEELS—10-inch ball bearing.
TIRES—2¾-inch pneumatic.
FINISH—Baked enamel. Two color box, black outside, green inside. Green wheels.
PACKING—One in carton.
WEIGHT—37 pounds.

15

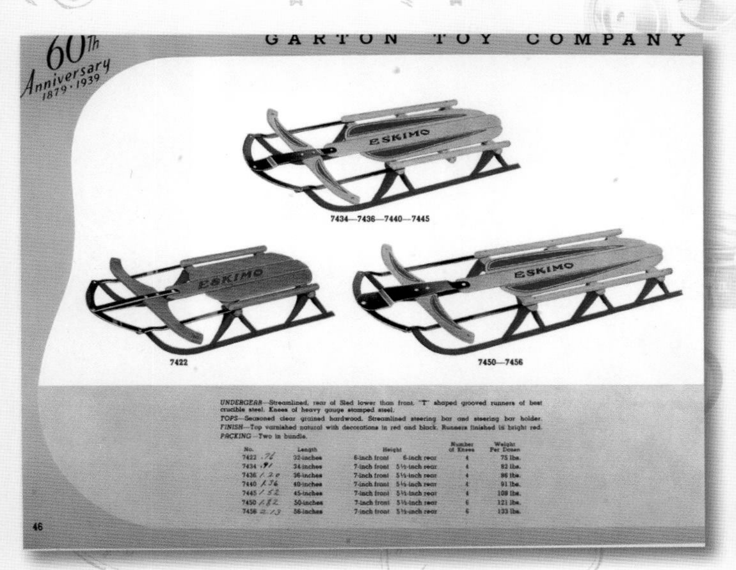

7434—7436—7440—7445

7422

7450—7456

UNDERGEAR—Streamlined, rear of Sled lower than front. "T" shaped grooved runners of best crucible steel. Knees of heavy gauge stamped steel.
TOPS—Seasoned clear grained hardwood. Streamlined steering bar and steering bar holder.
FINISH—Top varnished natural with decorations in red and black. Runners finished in bright red.
PACKING—Two in bundle.

No.	Length	Height		Number of Knees	Weight Per Dozen
7422 .76	32-inches	6-inch front	6-inch rear	4	75 lbs.
7434 .91	34-inches	7-inch front	5½-inch rear	4	82 lbs.
7436 1.20	36-inches	7-inch front	5½-inch rear	4	86 lbs.
7440 1.36	40-inches	7-inch front	5½-inch rear	4	91 lbs.
7445 1.52	45-inches	7-inch front	5½-inch rear	4	108 lbs.
7450 1.82	50-inches	7-inch front	5½-inch rear	6	121 lbs.
7456 2.13	56-inches	7-inch front	5½-inch rear	6	133 lbs.

46

7541—7547—7553

7200

SIZE—Length 19 inches, width 14 inches, height 17 inches.
CONSTRUCTION—Collapsible to fit any sled. Made of seasoned hardwood.
FINISH—Varnished natural.
PACKING—½-dozen in carton.
WEIGHT—21 pounds.

7541—7547—7553

UNDERGEAR—Streamlined, rear of Sled lower than front. "T" shaped grooved runners of best crucible steel. Knees of heavy gauge stamped steel.
HEIGHT—7-inches at front, 5-inches at rear.
TOP—Seasoned clear grained heavy hardwood. Streamlined stamped steel front. Streamlined side rails. Improved steering device responding to slightest pressure.
FINISH—Side rails and center board in high lustre black decorated in white. Natural varnished side slats decorated in black. Streamlined steel front, polished cadmium plated. Runners and knees finished in aluminum.
PACKING—Two in bundle.

No.	Length Inches	Number of Knees	Weight Per Doz.
7541	42	4	150 lbs.
7547	47	4	156 lbs.
7553	53	6	176 lbs.

7254

UNDERGEAR—Streamlined rear of Sled lower than front. "T" shaped grooved runners of best crucible steel. 6 knees of heavy gauge stamped steel.
HEIGHT—7-inch front 5½-inch rear.
LENGTH—47-inches.
TOPS—Seasoned clear grained hardwood. Streamlined steering bar and steering bar holder.
FINISH—Top varnished natural with decorations in red and black. Runners finished in bright red.
PACKING—Two in bundle, weight—per dozen 155 pounds.

47

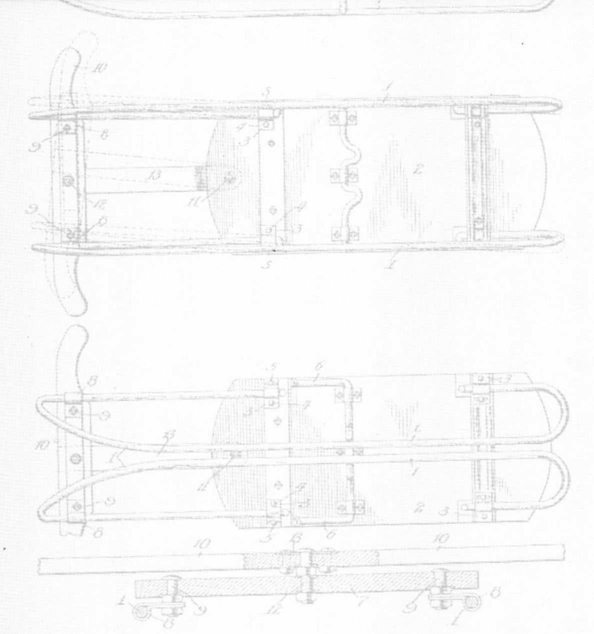

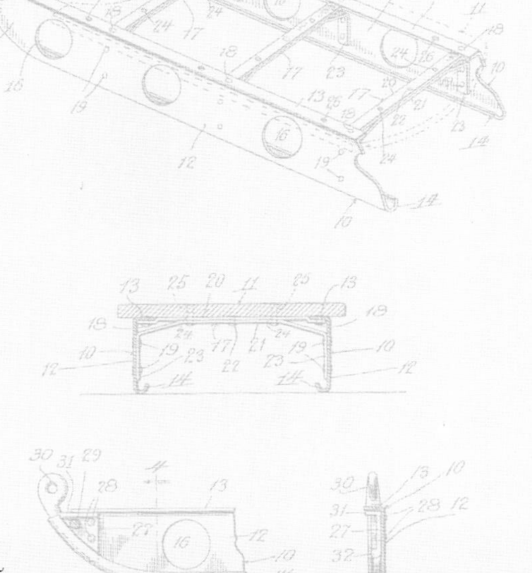

AGAIN Globe leads the wheel goods industry in further developing a new conception of **design** and **utility** — a salable combination of sound mechanics and psychological appeal . . . Again Globe adheres to its policy of giving retailers **something more** to sell—at prices unequalled for value received. The whole line has been rounded out with the addition of original, new numbers and accessories that reflect the spirit of youth. • Priced to sell in volume, Globe Biltwell Vehicles are designed and built with the belief that nothing is too good for Young America. Careful workmanship, honest materials, and real engineering go into all Globe wheel goods—a vital factor in selling parents, who naturally demand sturdy practical products.

THE GLOBE COMPANY

Since 1850

SHEBOYGAN, WISCONSIN

NEW YORK SHOWROOM • 1140 BROADWAY

Globe Biltwell Line, 1937 Catalog. *Courtesy of Jane Dwyer Garton.* *(continued on the following four pages)*

the Aeroflite

THE WAGON THAT REVOLUTIONIZED THE COASTER WAGON INDUSTRY

now heads a COMPLETE *Globe* FLEET OF *streamlined* WAGONS

Body is drawn from a single piece of metal, producing these marked advantages over welding or any other method of construction.

1. **Much greater strength. The body is practically indestructible.**
2. **Made absolutely square, without distortion. Undergearing runs absolutely true.**
3. **No seams.**
4. **No sharp metal edges—a safety feature that helps you sell.**
5. **Rubber cushions between body and gearing assure silent operation.**

NEW STEERING PRINCIPLE

Patented construction provides for steering through slot in body. Gives positive, direct, accurate steering for the first time, with handle resting almost flat against the body—and a stronger, more rigid front end than ever before. Large fifth wheel is die formed, turntable style; no front braces required.

No. 3000

No. 3000. Finish—Red enamel baked on. Green or blue optional. **Wheels**—Ball bearing. **Equipment**—Head lights, including bulbs, reflectors, wiring, switch and battery box but without batteries; plated front rail. Electric horn or plated handle optional at small extra cost. **Packing**—One in a corrugated shipping carton.

No. 3001 is same as **No. 3000** except with pneumatic tires.

No. 3002 is same as **No. 3000** except with spoke type wheels.

No.	Body Size	Usable Top Surface	Wheels	Tires	Axles	Weight
3000	20x48"	14x38"	8½"	1"	½"	49
3001	20x48"	14x38"	8"	2½"	½"	49
3002	20x48"	14x38"	8½"	1"	½"	49

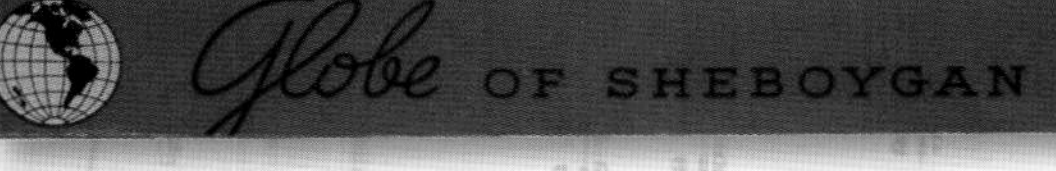

Globe OF SHEBOYGAN Page 2

1 THE AEROFLITE

SQUARE STEEL HANDLE

No. 3000

ONE PIECE BODY. Drawn from a single piece of metal correctly reinforced; not welded. Practically indestructible.

QUICK DETACHABLE FEATURE. The one-piece body is attached to the running gear with eight bolts. The under-gearing is die-formed and completely assembled at the factory.

Ball Bearing Equipped

HEADLIGHTS recessed in the body, as in aero-dynamic automobiles.

NEW STEERING PRINCIPLE. Patented construction provides for steering through slot in body. Gives positive, direct, accurate steering for the first time, with handle resting almost flat against the body—and a stronger, more rigid front end than ever before. Large fifth wheel is die-formed, turntable style; no front braces required.

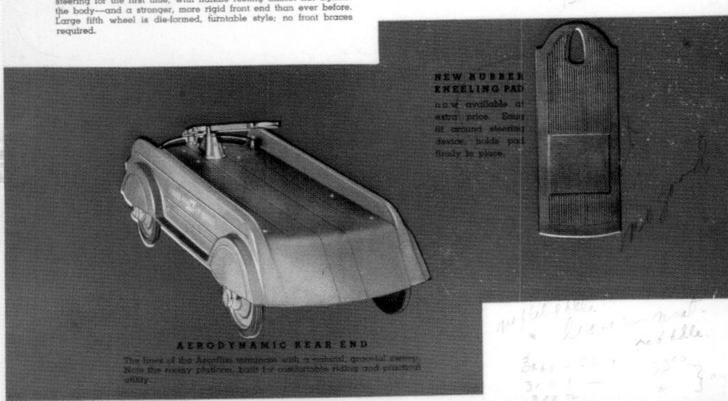

NEW RUBBER KNEELING PAD now available at extra price. Snug fit around steering device, holds pad firmly in place.

AERODYNAMIC REAR END

The lines of the Aeroflite terminate with a natural, graceful sweep. Note the roomy platform, built for comfortable riding and practical utility.

Page 3 *Globe* OF SHEBOYGAN

1 THE AEROFLITE

SQUARE STEEL HANDLE

No. 3000

ONE PIECE BODY. Drawn from a single piece of metal correctly reinforced; not welded. Practically indestructible.

QUICK DETACHABLE FEATURE. The one-piece body is attached to the running gear with eight bolts. The under-gearing is die-formed and completely assembled at the factory.

Ball Bearing Equipped

HEADLIGHTS recessed in the body, as in aero-dynamic automobiles.

NEW STEERING PRINCIPLE. Patented construction provides for steering through slot in body. Gives positive, direct, accurate steering for the first time, with handle resting almost flat against the body—and a stronger, more rigid front end than ever before. Large fifth wheel is die-formed, turntable style; no front braces required.

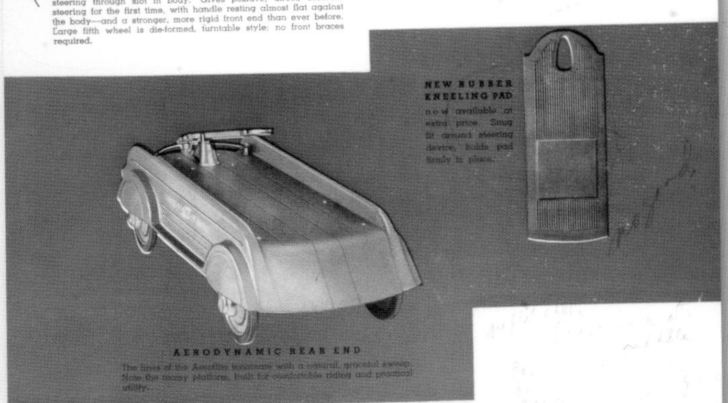

NEW RUBBER KNEELING PAD now available at extra price. Snug fit around steering device, holds pad firmly in place.

AERODYNAMIC REAR END

The lines of the Aeroflite terminate with a natural, graceful sweep. Note the roomy platform, built for comfortable riding and practical utility.

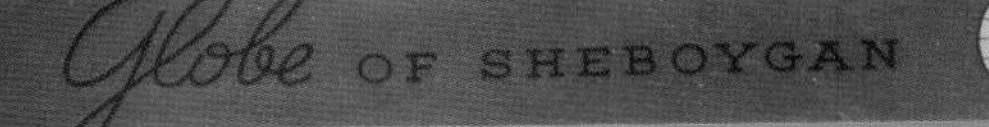

Globe OF SHEBOYGAN

2 THE AIR CRUISER

—a modified stream-lined coaster with full utility box . . .

No. 5000 Disc Wheels
No. 5001 Pneumatic Wheels
No. 5002 Spoke Wheels

This popular GLOBE coaster of modified streamlined design—companion to the Aeroflite—is built for those who won't buy a wagon for coasting alone. The body is drawn from a single piece of metal, giving it great strength, no distortion, no seams, no sharp edges, all rattles eliminated. • Like every Globe product, the Air Cruiser is sturdy and built to stand the gaff—a practical **utility wagon** you can be proud to sell. Very moderate price.

No. 5000. Body—One-piece steel, correctly reinforced, not welded. **Wheels**—Ball bearing. **Finish**—Red enamel baked on. Green or blue optional. **Equipment**—Tubular steel handle and rubber cushions between gearing and body. **Packing**—One in a corrugated shipping carton.

No. 5001 is same as **No. 5000** except wheels.

No. 5002 is same as **No. 5000** except wheels.

No.	Outside	Inside Dimensions	Wheels	Tires	Axles	Weight
5000	17½x42"	14⅛x33¾x3¾"	8"	¾"	½"	40
5001	17½x42"	14⅛x33¾x3¾"	8"	2"	½"	40
5002	17½x42"	14⅛x33¾x3¾"	8½"	¾"	½"	40

Globe OF SHEBOYGAN Page 4

3 THE MAINLINER

» » » a moderately-priced streamlined coaster to satisfy demand at competitive price levels » » »

No. 5050 With Bumper
No. 5055 Without Bumper
Ball Bearing Equipped

Here's a new wagon, with round front, styled along the lines of the Air Cruiser, but without aprons over the wheels. It has the "flash" to make it stand out on the retail floor . . . yet it has the quality materials, sound design, and rigid construction which the trade expects of every Globe product. Built to retail at a very moderate price, it is a wagon which you can sell in substantial volume.

No. 5050

Body—One-piece steel. **Finish**—Green enamel baked on. Blue or red optional. Can be finished in two-tone combinations at small extra cost. **Wheels**—Spoke type, ball bearing. **Equipment**—Nickel plated bumper. **Packing**—One in a corrugated shipping carton.

No. 5055 is same as **No. 5050** except without bumper.

No.	Body Size	Wheels	Tires	Axles	Weight
5050	17½x37x4½"	10"	¾"	½"	35
5055	17½x37x4½"	10"	¾"	½"	34

No. 1228

Body—One-piece steel. **Finish**—Red enamel baked on. Blue or green optional. **Packing**—One in a corrugated shipping carton.

No. 1230 is same as **No. 1228** except with spoke type wheels.

No.	Body Size	Wheels	Tires	Axles	Weight
1228	15½x31x3½"	8"	⅝"	½"	26
1230	15½x31x3½"	8"	⅝"	½"	26

4 The CHALLENGER

BUILT TO HANDLE LIKE A FULL SIZE COASTER

No. 1228
No. 1230

The junior model with full-size handle and undergearing. Easy for adults to pull in normal stride.

Globe OF SHEBOYGAN

THE GLOBE RACER

the standard of quality in the wheel goods industry

No. 2405
No. 2405-1
Ball Bearing Equipped

No. 2405

Body—One-piece steel. **Finish**—Red enamel baked on. Green or blue optional. Can be finished in red and white, or other two-tone combinations. **Wheels**—spoke type, ball bearing. **Packing**—One in a corrugated shipping carton.

No.	Body Size	Wheels	Tires	Axles	Weight
2405	16½x34½x4¾"	10"	¾"	½"	33
2405-1	16½x34½x4¾"	10"	1"	½"	34

Can be equipped with Brake as shown on **No. 2511.**

No. 2410—Ball Bearing Equipped

Body—One-piece steel. **Finish**—Red enamel baked on. Green, blue or orange optional. **Wheels**—Ball bearing. **Packing**—One in a corrugated shipping carton.

No.	Body Size	Wheels	Tires	Axles	Weight
2410	16½x34½x4¾"	10"	¾"	½"	33
2410-1	16½x34½x4¾"	10"	1"	½"	34

Can be equipped with Brake as shown on **No. 2511.**

Hound is of heavy steel, all one piece with fifth wheel formed in hound.

Page 6

Globe OF SHEBOYGAN

No. 2511—Ball Bearing Equipped

Body—One-piece steel. **Finish**—Blue enamel baked on. Green, red or orange optional. **Equipment**—Bumper and Brake. **Wheels**—Ball bearing. **Packing**—One in a corrugated shipping carton.

No.	Body Size	Wheels	Tires	Axles	Weight
2511	16½x34½x4¾"	10"	1⅛"	½"	36

No. 2517—Ball Bearing Equipped

Body—One-piece steel. **Finish**—Blue enamel baked on. Red or green optional. Can be finished in two-tone combinations at small extra cost. **Equipment**—Canvas knee pad, nickel plated side rails, bumper, and handle. **Wheels** — Ball bearing. **Packing**—One in a corrugated shipping carton.

No.	Body Size	Wheels	Tires	Axles	Weight
2517	16½x34½x4¾"	10"	1"	½"	39

Can be equipped with Brake as shown on **No. 2511.**

No. 2519—Ball Bearing Equipped

Body—One-piece steel. **Finish**—Red enamel baked on. Green or blue optional. Plated handle. **Wheels**—Ball bearing with pneumatic tires. **Packing**—One in a corrugated shipping carton.

No.	Body Size	Wheels	Tires	Axles	Weight
2519	16½x34½x4¾"	10"	2¾"	½"	34

Page 7

Globe OF SHEBOYGAN

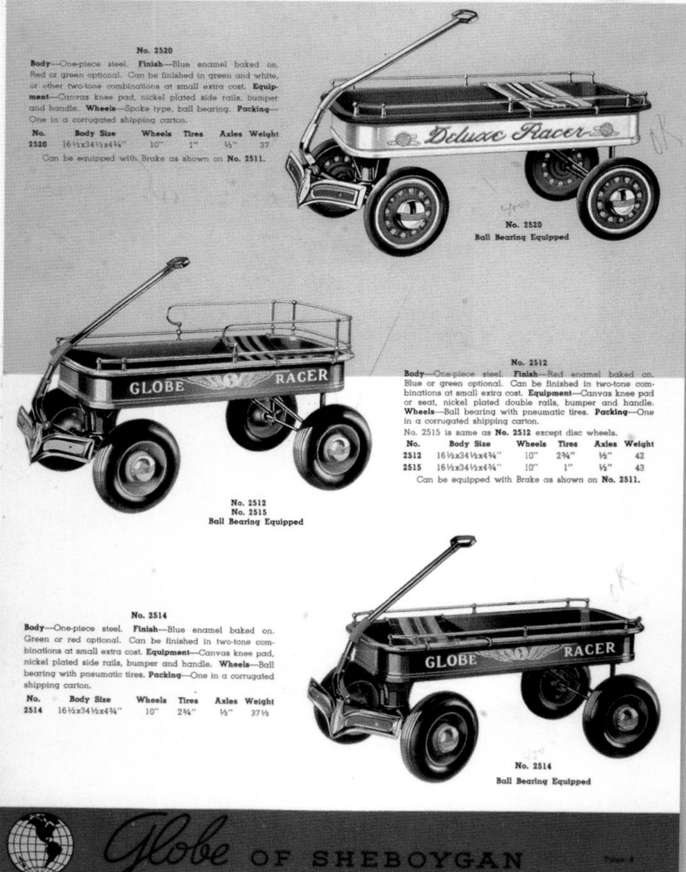

No. 2520

Body—One-piece steel. **Finish**—Blue enamel baked on. Red or green optional. Can be finished in green and white, or other two-tone combinations at small extra cost. **Equipment**—Canvas knee pad, nickel plated side rails, bumper and handle. **Wheels**—Spoke type, ball bearing. **Packing**—One in a corrugated shipping carton.

No.	Body Size	Wheels	Tires	Axles	Weight
2520	16½x34½x4¾"	10"	1"	½"	37

Can be equipped with Brake as shown on **No. 2511.**

No. 2520
Ball Bearing Equipped

No. 2512

Body—One-piece steel. **Finish**—Red enamel baked on. Blue or green optional. Can be finished in two-tone combinations at small extra cost. **Equipment**—Canvas knee pad or seat, nickel plated double rails, bumper and handle. **Wheels**—Ball bearing with pneumatic tires. **Packing**—One in a corrugated shipping carton.

No. 2515 is same as **No. 2512** except disc wheels.

No.	Body Size	Wheels	Tires	Axles	Weight
2512	16½x34½x4¾"	10"	2¾"	½"	42
2515	16½x34½x4¾"	10"	1"	½"	43

Can be equipped with Brake as shown on **No. 2511.**

No. 2512
No. 2515
Ball Bearing Equipped

No. 2514

Body—One-piece steel. **Finish**—Blue enamel baked on. Green or red optional. Can be finished in two-tone combinations at small extra cost. **Equipment**—Canvas knee pad, nickel plated side rails, bumper and handle. **Wheels**—Ball bearing with pneumatic tires. **Packing**—One in a corrugated shipping carton.

No.	Body Size	Wheels	Tires	Axles	Weight
2514	16½x34½x4¾"	10"	2¾"	½"	37½

No. 2514
Ball Bearing Equipped

Globe OF SHEBOYGAN

No. 7001

Body—One-piece steel. **Finish**—Red enamel baked on. Green optional. **Gearing**—Channel bolster with special fifth wheel steering device. **Wheels** — Roller bearing. Ball bearing optional. **Packing**—One in a corrugated shipping carton.

No.	Body Size	Wheels	Tires	Axles	Weight
7001	16½x34½x4¾"	8"	¾"	½"	30

No. 7001 GEARING
Channel steel bolster with special fifth wheel steering device.

No. 7001-S

Body—One-piece steel. **Finish**—Red enamel baked on. Green optional. Can be finished in two-tone combinations at small extra cost. **Gearing**—Channel bolster with special fifth wheel steering device. **Wheels**—Spoke type, ball bearing. **Packing**—One in a corrugated shipping carton.

No.	Body Size	Wheels	Tires	Axles	Weight
7001-S	16½x34½x4¾"	8½"	¾"	½"	30

Can be equipped with Brake as shown on **No. 2511.**

No. 7001-S

No. 7025

Body—One-piece steel. **Finish**—Red enamel baked on. Green or blue optional. Can be finished in two-tone combinations at small extra cost. **Equipment**—Nickel plated side rails. **Wheels**—Spoke type, ball bearing. **Packing**—One in a corrugated shipping carton.

No.	Body Size	Wheels	Tires	Axles	Weight
7025	16½x34½x4¾"	8½"	1"	½"	32

Can be equipped with Brake as shown on **No. 2511.**

No. 7025

Globe OF SHEBOYGAN

No. 7015
No. 7015-1
No. 7020

No. 7015 GEARING
Double channel steel bolster with special fifth wheel steering device.

Body—One-piece steel. **Finish**—Red enamel baked on. Green optional. Can be finished in two-tone combinations at small extra cost. **Gearing**—Double channel bolster with special fifth wheel steering device. **Wheels**—Spoke type, ball bearing. **Packing**—One in a corrugated shipping carton.

No. 7020 is same as **No. 7015** except with disc wheels.

No.	Body Size	Wheels	Tires	Axles	Weight
7015	16½x34½x4¾"	10"	¾"	½"	33
7015-1	16½x34½x4¾"	10"	1"	½"	34
7020	16½x34½x4¾"	10"	¾"	½"	33

Can be equipped with Brake as shown on **No. 2511.**

No. 7003

Body—One-piece steel. **Finish**—Red enamel baked on. Green optional. **Gearing**—Channel bolster with special fifth wheel steering device. **Wheels** — Roller bearing. Ball bearing optional. **Packing**—One in a corrugated shipping carton.

No.	Body Size	Wheels	Tires	Axles	Weight
7003	16½x34½x4¾"	10"	¾"	½"	33

No. 7003

Globe OF SHEBOYGAN Page 10

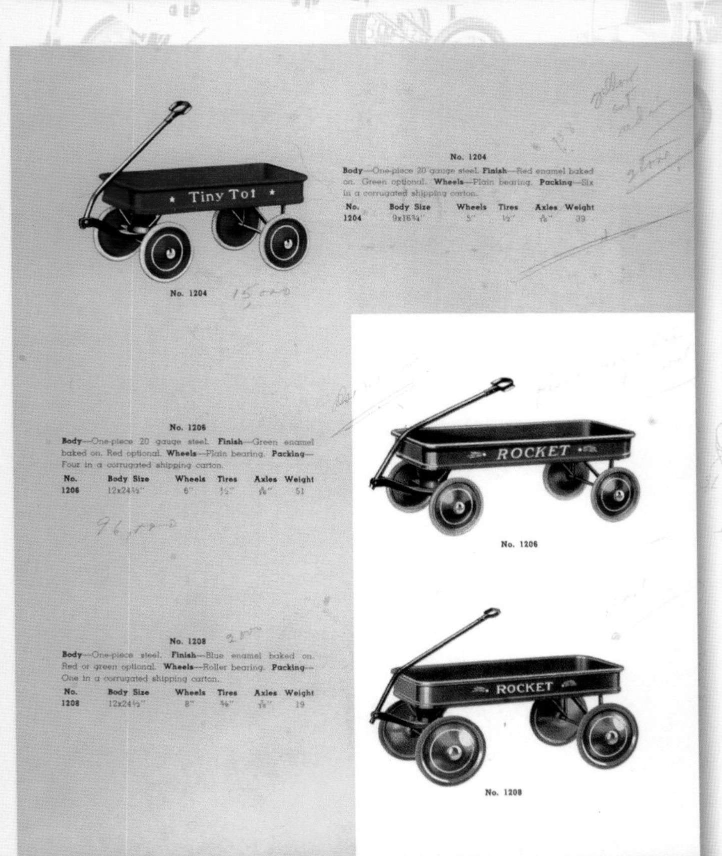

No. 1204

Body—One-piece 20 gauge steel. **Finish**—Red enamel baked on. Green optional. **Wheels**—Plain bearing. **Packing**—Six in a corrugated shipping carton.

No.	Body Size	Wheels	Tires	Axles	Weight
1204	9x16¾"	5"	½"	⅜"	39

No. 1204

No. 1206

Body—One-piece 20 gauge steel. **Finish**—Green enamel baked on. Red optional. **Wheels**—Plain bearing. **Packing**—Four in a corrugated shipping carton.

No.	Body Size	Wheels	Tires	Axles	Weight
1206	12x24½"	6"	½"	⅜"	51

No. 1206

No. 1208

Body—One-piece steel. **Finish**—Blue enamel baked on. Red or green optional. **Wheels**—Roller bearing. **Packing**—One in a corrugated shipping carton.

No.	Body Size	Wheels	Tires	Axles	Weight
1208	12x24½"	8"	⅝"	⅜"	19

No. 1208

Globe OF SHEBOYGAN

THE GLOBE ROCKET
in wood

Ball Bearing Equipped

No. 1005
No. 1005-1

Body—Seasoned hardwood, 1" paneled sides, ends reinforced with tie rods. **Finish**—Box varnished natural with blue panel (green or black optional) with white stencil and wheels to match. **Handle**—Tubular steel with loop grip. **Wheels**—Ball bearing. **Packing**—One in a corrugated shipping carton.

No.	Body Size	Wheels	Tires	Axles	Weight
1005	16x36"	10"	¾"	½"	36
1005-1	16x36"	10"	1"	½"	37

Can be equipped with Brake as shown on **No. 1000.**

★Like all Globe products this sturdy wagon is built "to take it". Has plenty of eye-appeal, and sells on sight to the great numbers of people who still prefer a wood wagon.

Globe OF SHEBOYGAN Page 12

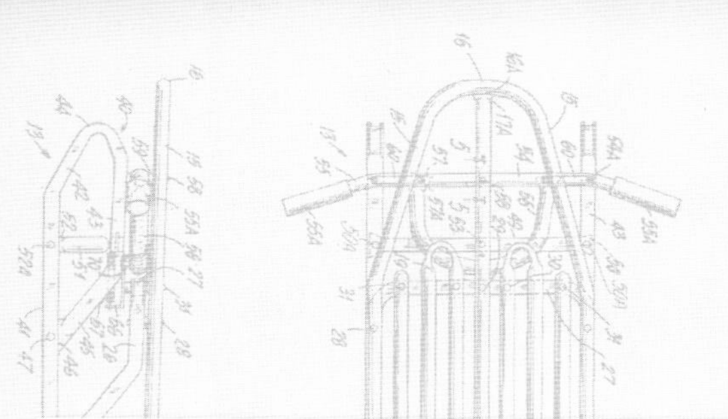

THE GLOBE FLASH
in wood

No. 1012
Ball Bearing Equipped

No. 1009—Ball Bearing Equipped

Body—Seasoned hardwood, racer style, with heavy side rails. **Finish**—Varnished natural, with red side rails and red wheels. **Handle**—Tubular steel with loop grip. **Packing**—One in a corrugated shipping carton.

No. 1009 is furnished with bumper, with ball bearing wheels with 2¾" pneumatic tires and nickel plated handle.

No. 1011 is furnished without bumper, with ball bearing wheels with 1" tires.

No. 1080 is same as **No. 1009** except length 80".

No.	Body Size	Wheels	Tires	Axles	Weight
1009	16x52"	10"	2¾"	½"	40
1011	16x52"	10"	1"	½"	40
1080	16x80"	10"	2¾"	½"	65

Can be equipped with Brake as shown on **No. 1000.**

No. 1012

Body—Seasoned hardwood, racer style, with heavy side rails. **Finish**—Varnished natural, with red side rails and red wheels. **Handle**—Tubular steel with loop grip. **Packing**—One in a corrugated shipping carton.

No. 1012 is furnished without bumper, with ball bearing wheels with 1" tires.

No. 1010 is furnished with bumper, ball bearing wheels with 2¾" pneumatic tires.

No.	Body Size	Wheels	Tires	Axles	Weight
1012	16x36"	10"	1"	½"	34½
1010	16x36"	10"	2¾"	½"	36

Can be equipped with Brake as shown on **No. 1000.**

★This racer-style coaster always has a substantial following. Made of selected hardwood with standard Globe undergearing, it stands up in the toughest service. Natural wood finish trimmed in colors.

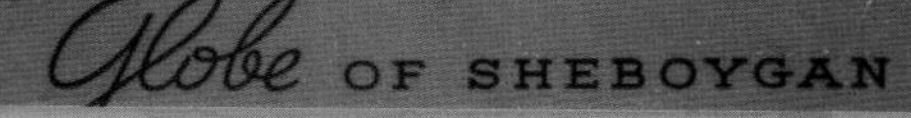

THE GLOBE EXPRESS

No. 1004-6
Ball Bearing Equipped

No. 1004-6 Dual Wheels in Rear
No. 1004-4 Single Wheels in Rear
No. 1014 4 Wheels, **Pneumatic Tires**

Body—Seasoned hardwood, 1" sides, ends reinforced with tie rods. **Finish**—Wheels red, box varnished natural, red trimming. **Handle**—Tubular steel with loop grip. **Wheels**—Ball bearing. **Packing**—One in a corrugated shipping carton.

No.	Body Size	Wheels	Tires	Axles	Weight
1004-6	16x40"	10"	1"	½"	50
1004-4	16x40"	10"	1"	½"	45
1014	16x40"	10"	2¾"	½"	42

No. 1014 is same as **No. 1004-4** but with ball bearing wheels and 2¾" pneumatic tires.

No. 1006—Ball Bearing Equipped

Body—Seasoned hardwood, 1" sides, ends reinforced with tie rods. **Finish**—Wheels red, box varnished natural, red trimming. **Handle**—Tubular steel with loop grip. **Wheels**—Ball bearing. **Packing**—One in a corrugated shipping carton.

No.	Body Size	Wheels	Tires	Axles	Weight
1006	16x36"	10"	1"	½"	47½

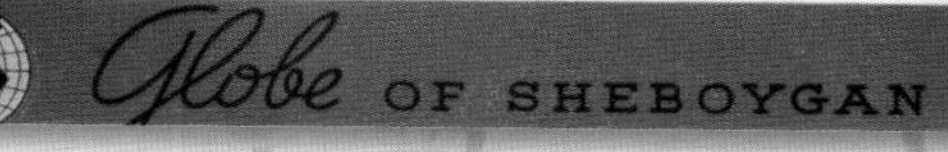

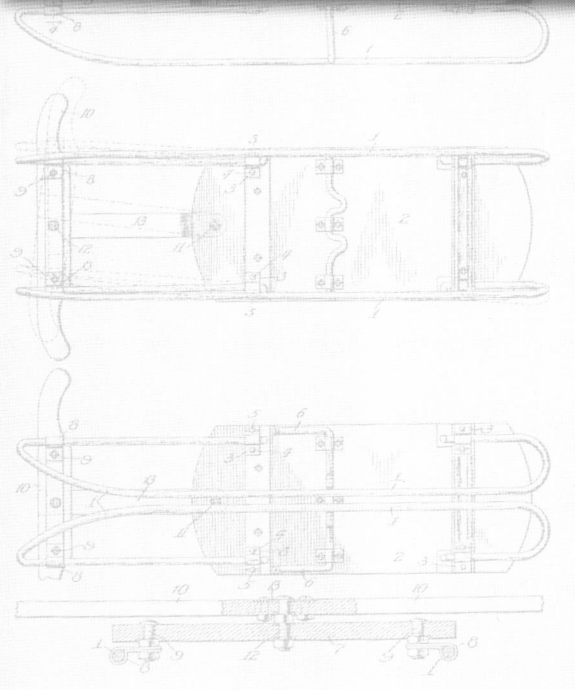

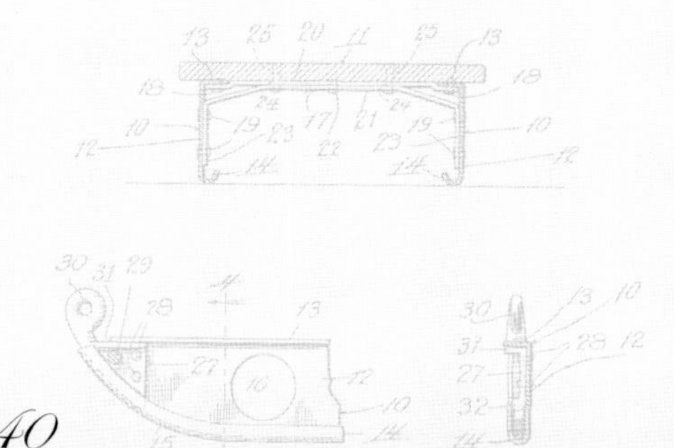

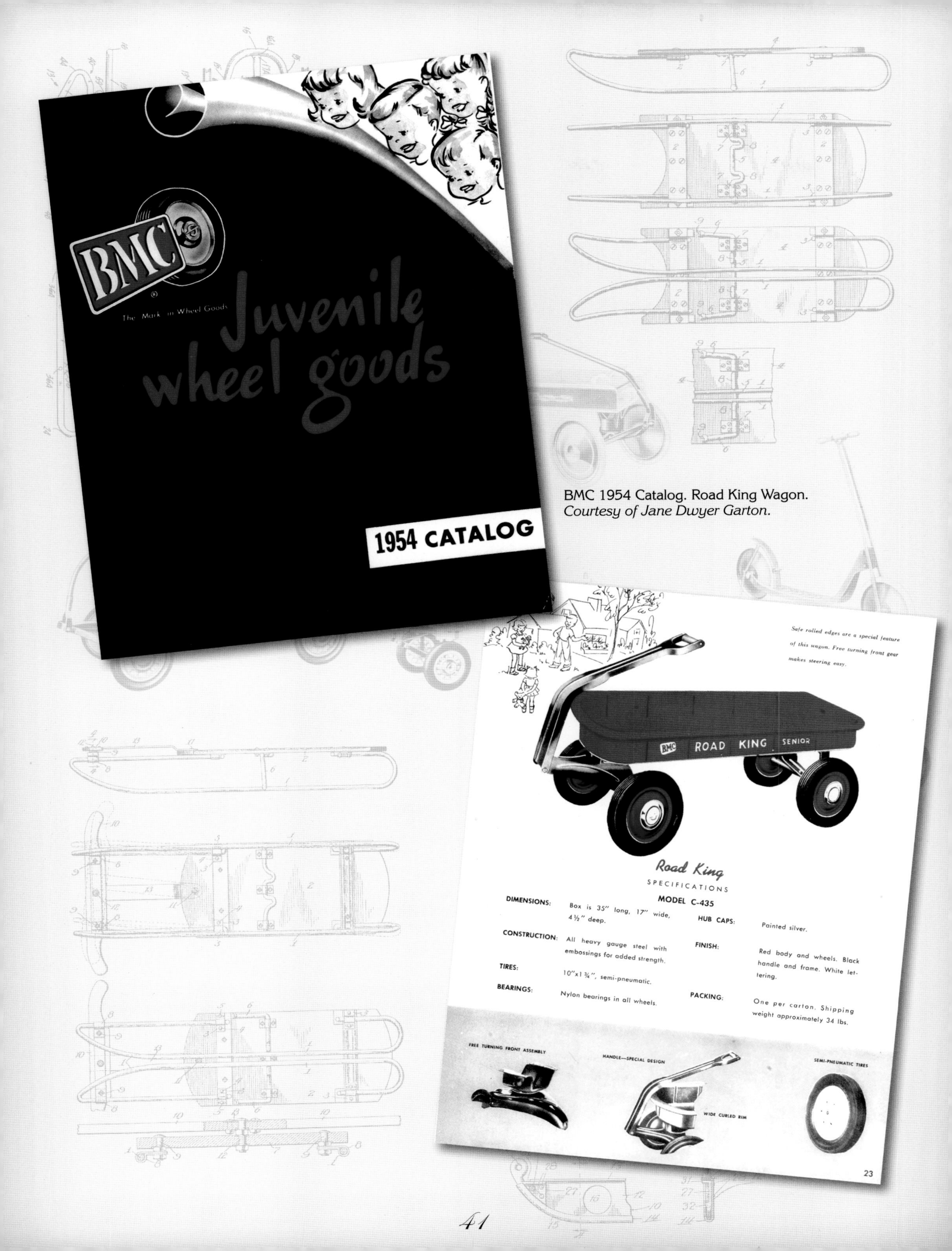

BMC 1954 Catalog. Road King Wagon. *Courtesy of Jane Dwyer Garton.*

The Line that
"SETS THE PACE"
in Wheelgoods Innovations

FEATURES THE LEADER IN SLEDS

THE 

SILVER STREAK

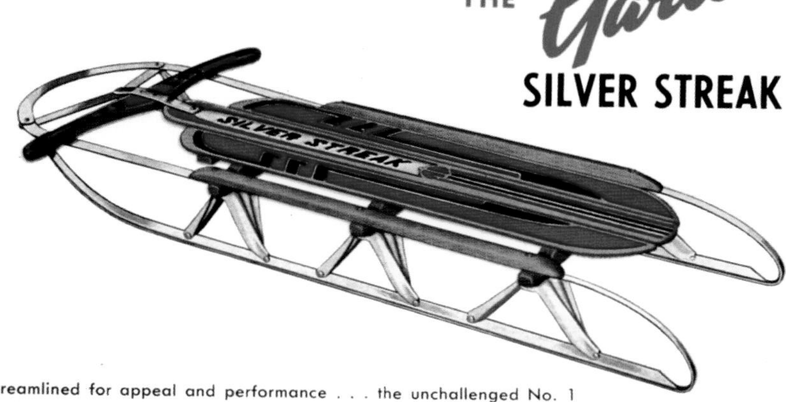

Streamlined for appeal and performance . . . the unchallenged No. 1 leader in sled sales from coast-to-coast . . . the "SILVER STREAK" features the famous patented "Y" steering that brings high flexibility to sled steering for the first time. For safety and streamlined appearance, runner ends are closed . . . rear of sled is lower than front. "T" shaped grooved rnnners of high carbon crucible steel and patented knees of heavy gauge stamped steel. Made of selected hardwoods, finished in red, blue and silver color-fast enamels, it's coated with weather resistant varnish. All metal parts are finished in bright baked-on aluminum. Offered in 40", 46", 52" and 58" lengths. Here is your feature seller for the 1954-55 Winter Season . . . priced to sell—*fast!*

Everything on Wheels **FOR YOUNG AMERICA!**

For everything that is "new and different" in wheelgoods—for the one line that features everything on wheels that every child has dreamed about—write for the new GARTON Catalog today. The pace-setting GARTON Line is designed to spell BIG SALES and priced to spell BIG PROFITS. Complete Catalog sent on request.

SHOW THE LINE THAT SELLS ITSELF

Garton TOY COMPANY
SHEBOYGAN, WISCONSIN

SINCE *1879* LEADERS IN MANUFACTURE OF WHEEL GOODS, CROQUET SETS AND SLEDS

AUTOMOBILES
SIDEWALK BIKES
VELOCIPEDES
PEDAL BIKES
SCOOTERS
WAGONS
BABY WALKER
SLEDS

Garton Silver Streak advertisemnt from Toys & Novelties, October 1954. $5-7.

Everyone Has the Most Fun Using Garton Toys

Using Garton Toys Everyone Has the Most Fun

Everyone Has the Most Fun Using Garton Toys

Using Garton Toys Everyone Has the Most Fun

Garton Toy Catalog, 1959. $10-20.

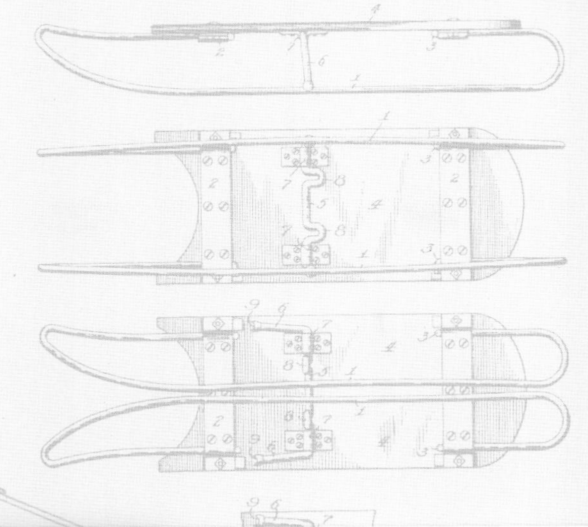

Garton SLEDS

7000 — SNO-FLAKE

Stamped out of 20 gauge auto body steel. 27 inch diameter by 3½ inch depth at center. Edges rolled for extra strength and added safety. Handles of plastic webbing. Two holes for tow rope.

Packed six in a carton, two finished red, two in blue and two in yellow. Weight 50 lbs. per carton. Also available finished red, packed one in a carton at extra charge, weight 10 lbs. each.

"ROYAL RACER" SLEDS

Streamlined, rear of sled lower than front. "T" shaped grooved runners. Big curl on runner ends for safety. Decks of selected hardwoods, coated with weather-resistant varnish.
Two in a bundle.

No.	Length	No. of Knees	Wt. per bundle
7136	36-inch	4	14 lbs.
7140	40-inch	4	16 lbs.
7145	45-inch	4	18 lbs.
7154	54-inch	6	22 lbs.

"SILVER STREAK" SLEDS

Streamlined, rear of sled lower than front. "T" shaped grooved runners. Closed runner ends for safety. Highest flexibility for steering with patented "Y" steering.

Decks of selected hardwoods, coated with weather-resistant varnish. Two in a bundle.

No.	Length	No. of Knees	Wt. per bundle
8940	40-inch	4	16 lbs.
8946	46-inch	4	20 lbs.
8952	52-inch	6	22 lbs.
8958	58-inch	6	24 lbs.
8964	64-inch	8	27 lbs.

"COMET" SLEDS

Streamlined, rear of sled lower than front. "T" shaped grooved runners. Closed runner ends for safety. Highest flexibility for steering with patented "Y" steering. Tubular steel steering bar with grips, designed and shaped to protect hands.

Decks of one-piece five ply veneer, waterproof and weatherproof. Coated with weather-resistant varnish. Two in a bundle, well wrapped for shipping.

No.	Length	No. of Knees	Wt. per bundle
9048	48-inch	4	28 lbs.
9054	54-inch	6	30 lbs.
9060	60-inch	6	32 lbs.
9066	66-inch	8	34 lbs.

Everyone Has the Most Fun Using Garton

14

Garton WAGONS

1325—WOOD WAGON

Box — 16 x 36 x 3 inches, reinforced with cleats. Ends drawn together with box rods. Stamped steel gear with fifth wheel. Racks removable.

10 inch artillery nylon bearing wheels with 1.75 inch semi-pneumatic tires.

One in carton, 39 lbs. each.

2709 — STEEL WAGON

Body — 17 x 35 inches, stamped from one piece heavy steel. Stamped steel gear with fifth wheel.

10 inch artillery nylon bearing wheels with 1.75 inch semi-pneumatic tires.

One in carton, 30 lbs. each.

7300 — PUSHER TYPE SLED GUARD

Fits easily onto any size sled, making sled into pusher-type for taking small child out for airing.

Height 30 inches. Made of tubular and strip steel with wood back. Wood parts finished in weather resistant varnish.

Six per carton, weight 28 lbs. per carton. Also available packed one in carton at extra charge, weight 6 lbs. each.

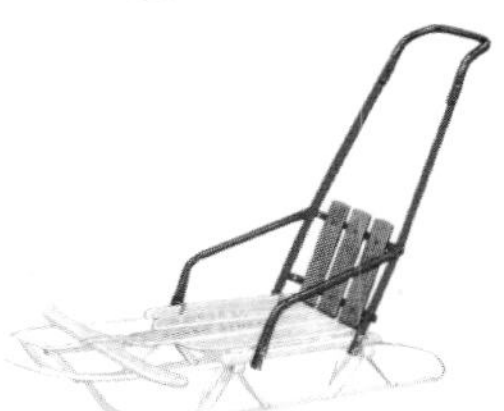

Toys—80 Years of Play for Girls and Boys

13

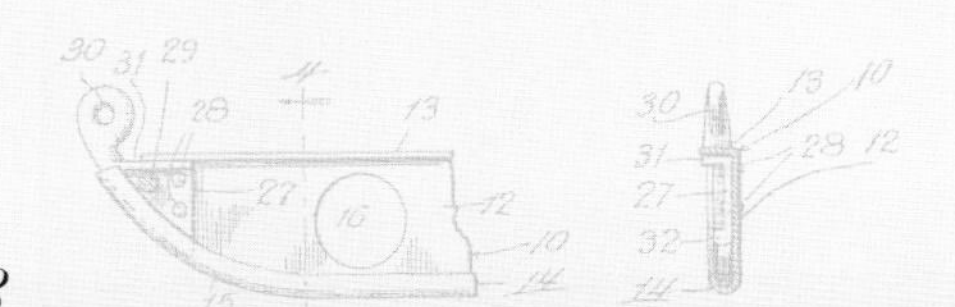

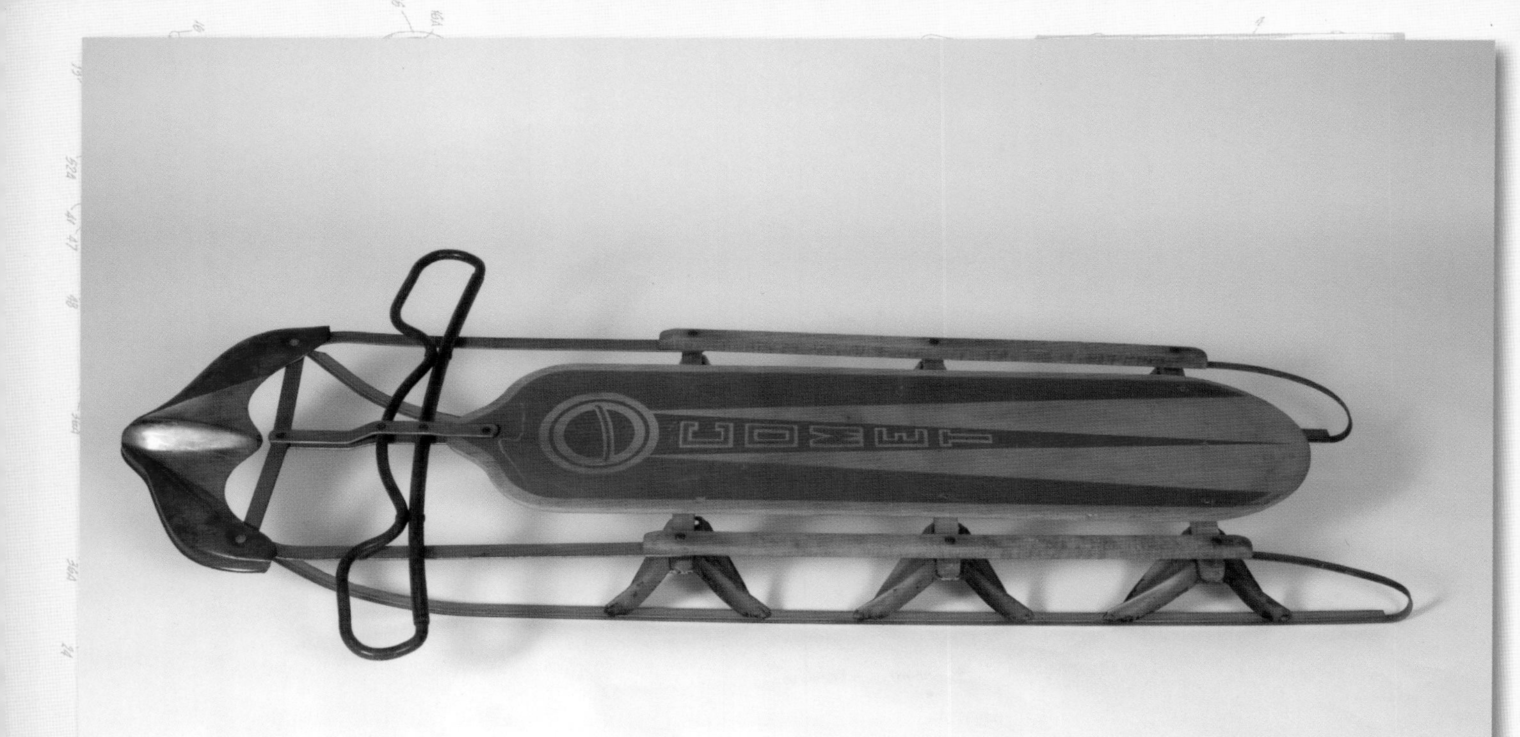

Garton Comet #9060. Seen in the 1959 catalog. The ornamental design for the sled bumper was granted in 1937. $50-100. *Courtesy of Paul Cote*.

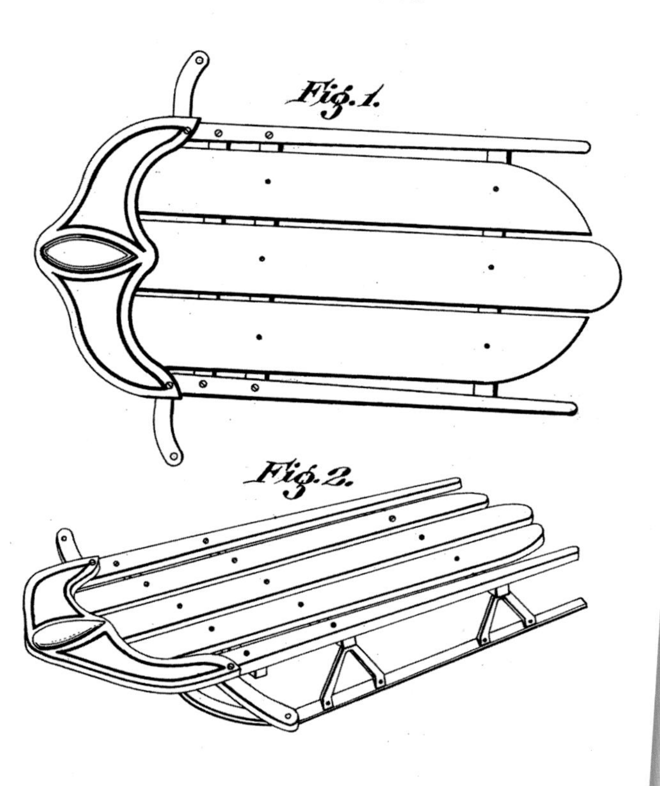

Patent drawing, 1937, with the ornamental bumper design.

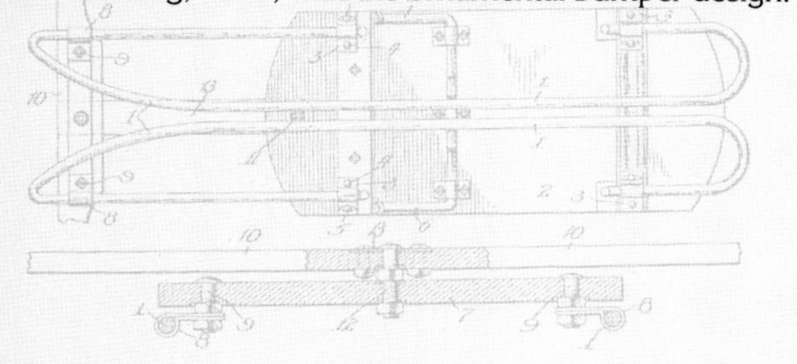

Hunt, Helm, Ferris, & Company

Sometimes the thrill of discovery is overshadowed by the reality of what is revealed. Stop for a moment and think how different our childhood memories would be if S.L. Allen had not invented the Flexible Flyer. This new invention crossed the gender lines. It was marketed to every boy and girl and guaranteed parents no more colds or doctor bills from wet feet, money saved on shoe leather, and most of all good health. It was the ingenious marketing techniques that made Flexible Flyer a household name, and the eagle, shield, and ribbon the most recognizable trademark in the world, even today. Imagine my surprise when I stumbled upon a copy of the *Scientific American* dated August 20, 1864 and staring me right in the face was "Hunt's Coasting Sled." Could this be the same H.C. Hunt of Hunt, Helm, Ferris & Company? Further investigation confirmed my suspicions. Mr. Hunt was granted a patent in October 1864, for a new Coasting Sled that allowed the rider to steer and also to stop the sled by leaning against a spring brake. Remember this was twenty-five years prior to S.L. Allen's patent and nineteen years before Hunt & Helm formed a partnership with Ferris.

We can only speculate as to why this sled was never marketed, and if it had been, would the Cannonball equipped with the wings of Mercury, enhanced with the words, "Cannonball Beats em' All" have the same impact as Flexible Flyer "Sled of the Nation?" Only you can be the judge of that.

In recollections entitled "Starline as I Remember" by H. J. Ferris, written at the age of ninety-two, Mr. Ferris states "With the formation of the Steel Angle Sled Company in Milwaukee, Wisconsin, the market for Steel Coaster Sleds would end for Hunt, Helm, & Ferris." Up until this time 150,000 sleds per year were being sold. With sales dropping, patents for Coaster Wagons were granted (late 1800s) and proved to be extremely successful. For the next fifteen years 150 wagons were sold per day. As the patent rights expired, the newly formed Auto Wheel Wagon Company of Milwaukee began ruining the market by going too far with price reduction forcing them to close their doors. Hunt, Helm, Ferris & Company continued to flourish, producing Star, Overland, & Cannonball wagons. A 1925 Advertisement boasts "Life-size cut-out of Cannonball Boy, fits large size wagon, and sent with order of $50.00 or more." I would give anything to know if any of these cut-outs exist today. It is today's collector that will keep the legacy of Hunt, Helm, Ferris & Company alive and the "Star" in Star Steel Coasters and Wagons glowing for generations to come.

Scientific American.

A WEEKLY JOURNAL OF PRACTICAL INFORMATION IN ART, SCIENCE, MECHANICS, CHEMISTRY AND MANUFACTURES.

Vol. XI.—No. 8. (NEW SERIES.) | NEW YORK, AUGUST 20, 1864. | $3 PER ANNUM (IN ADVANCE)

Improved "Coasting" Sled.

There is no more exhilarating winter pastime than sliding down hill, or "coasting," as Boston and New Brunswick people call it. The phrase is doubtless derived from the Acadians, who use the old French language in which *coté* or *coste* signifies hill-side. Old and young pursue it with the utmost zest, and a heavy fall of snow is anxiously looked for where the natural facilities are good for this sport. Not unfrequently, however, terrible accidents occur from carelessness or mismanagement, and broken heads or fractured limbs, caused by sleds colliding with each other or with obstructions, are by no means uncommon. It is superfluous to say that the possibility of danger very much detracts from the pleasure, and the prospect of being jumped upon by an over-loaded, mismanaged sled, or turned over by one's own awkwardness, deters very many from seeking the hill-side.

If people would only use the sled here shown there would be fewer accidents from mismanagement, for the track can be kept and the sled steered to avoid obstructions as easily as a horse is guided, and in much the same way. The reader will see two ropes, A and B, which the boy is holding, these ropes run in opposite directions, after they pass through the holes, C, so that by drawing on A, the forward runners are turned to the left, thus diverting the whole body of the sled in the direction desired. The arrangement of the runners is shown in Fig. 2; the forward runners being pivoted at D so that they turn freely; the ends are connected by the cross-bar, E, through which the steering lines pass.

At the back of the rider is a spring brake, E, which, when leaned against, throws the points, F, into the snow and checks the progress immediately. A convenient brace, G, forward is provided for the feet to rest against. This seems to be an excellent sled, and we doubt not that boys and children of a larger growth will be exceedingly pleased with it. It was invented by H. C. Hunt, of Amboy, Lee county, Ill., and a patent is now pending through the Scientific American Patent Agency. For further information address the inventor as above.

HUNT'S "COASTING" SLED.

Checking Perspiration.

If while perspiring, or while something warmer than usual, from exercise or a heated room, there is a sudden exposure in stillness to a still, cold air, or to a raw, damp atmosphere, or to a draught, whether at an open window or door or street-corner, an inevitable result is a violent and instantaneous closing of the pores of the skin, by which waste and impure matters, which were making their way out of the system, are compelled to seek an exit through some other channel, and break through some weaker part, not the natural one, and harm to that part is the result. The idea is presented by saying that the cold has settled in that part. To illustrate: A lady was about getting into a small boat to cross the Delaware; but wishing first to get an orange at a fruit-stand, she ran up the bank of the river, and on her return to the boat found herself much heated, for it was summer, but there was a little wind on the water, and the clothing soon felt cold to her; the next morning she had a severe cold, which settled on her lungs, and within the year she died of consumption. A stout, strong man was working in a garden in May; feeling a little tired about noon, he sat down in the shade of the house and fell asleep; he waked up chilly; inflammation of the lungs followed, ending, after two years of great suffering, in consumption. On opening his chest, there was such an extensive decay that the matter was scooped out by the cupful.

A Boston ship-owner, while on the deck of one of his vessels, thought he would "lend a hand" in some emergency; and pulling off his coat, worked with a will, until he perspired freely, when he sat down to rest awhile, enjoying the delicious breeze from the sea. On attempting to rise, he found himself unable, and was so stiff in his joints, that he had to be carried home and put to bed, which he did not leave until the end of two years, when he was barely able to hobble down to the wharf on crutches. A lady, after being unusually busy all day, found herself heated and tired toward sundown of a summer's day. She concluded she would rest herself by taking a drive to town in an open vehicle. The ride made her uncomfortably cool, but she warmed herself up by an hour's shopping, when she turned homeward; it being late in the evening, she found herself more decidedly chilly than before. At midnight she had pneumonia (inflammation of the lungs), and in three months had the ordinary symptoms of confirmed consumption. A lady of great energy of character lost her cook, and had to take her place for four days; the kitchen was warm, and there was a draft of air through it. When the work was done, warm and weary, she went to her chamber, and laid down on the bed to rest herself. This operation was repeated several times a day. On the fifth day she had an attack of lung fever; at the end of six months she was barely able to leave her chamber, only to find herself suffering with all the more prominent symptoms of confirmed consumption; such as quick pulse, night and morning cough, night-sweats, debility, short breath, and falling away.

A young lady rose from her bed on a November night, and leaned her arm on the cold window-sill to listen to a serenade. Next morning she had pneumonia, and suffered the horrors of asthma for the remainder of a long life. Multitudes of women lose health and life every year, in one of two ways; by busying themselves in a warm kitchen until weary, and then throwing themselves on a bed or sofa, without covering, and perhaps in a room without fire; or by removing the outer clothing, and perhaps changing the dress for a more common one, as soon as they enter the house after a walk or a shopping. The rule should be invariable to go at once to a warm room and keep on all the clothing at least for five or ten minutes, until the forehead is perfectly dry. In all weathers, if you have to walk and ride on any occasion, do the riding first.—*Hall's Journal of Health.*

To Prevent Prints from Fading.

The dress should be washed in lather, and not by applying the soap in the usual way direct upon the muslin. Make a lather by boiling soap and water together; let it stand until it is sufficiently cool for use, and previously to putting the dress into it, throw in a handful of salt; rinse the dress without wringing it, in clear, cold water, into which a little salt has been thrown; remove it and rinse it in a fresh supply of clear water and salt. Then wring the dress in a cloth and hang it to dry immediately, spreading as open as possible, so as to prevent one part lying over another. Should there be any white in the pattern, mix a little blue in the water.

The great drought prevailing in this country extends over England also.

The Scientific American, August 20, 1864.
Hunt's Coasting Sled, $25-50

Hunt, Helm, Ferris & Co. catalog, circa 1890. $100-150. (continued on next two pages)

Star Steel Sled. Style No. 00

PATENTED

ON the opposite page we present an accurate photographic reproduction in actual colors of our No. 00 sled. This is the best steel sled on the market for the money. It is all steel except the top. The runners are made of the celebrated round back angle steel used on all Star Steel Sleds. It has wood top, well painted, varnished, and nicely decorated.

		LENGTH	WIDTH	HEIGHT	KNEES	PRICE PER DOZ.
No. 00	Sled	27 Inches	9 Inches	6 Inches	2	

Star Steel Sled. Style No. 30

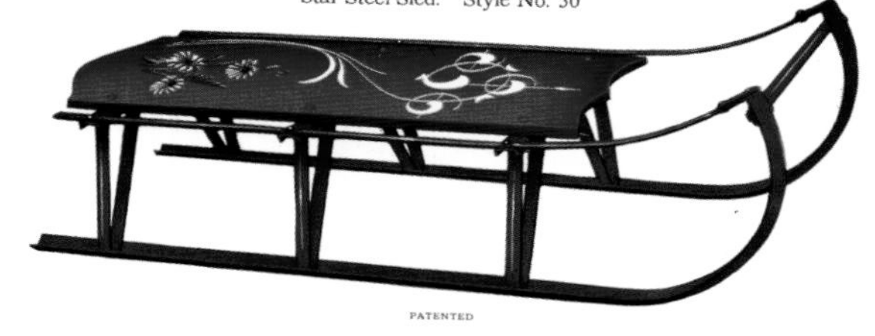

PATENTED

FOR several seasons these have been our most popular selling sleds. They are constructed with raves, the same as the most expensive sleds which we make. All tops are handsomely hand decorated. The running gears are finished in assorted colors, artistically striped. We are now making the beams on these sleds of channel steel instead of wood, as heretofore.

		LENGTH	WIDTH	HEIGHT	KNEES	PRICE PER DOZ.
No. 10	Sled	30 Inches	12 Inches	7½ Inches	2	
No. 20	Sled	33 "	12 "	7½ "	2	
No. 30	Sled	36 "	12 "	7½ "	3	

Star Steel Sled. Style No. 30

PATENTED

FOR several seasons these have been our most popular selling sleds. They are constructed with raves, the same as the most expensive sleds which we make. All tops are handsomely hand decorated. The running gears are finished in assorted colors, artistically striped. We are now making the beams on these sleds of channel steel instead of wood, as heretofore.

		LENGTH	WIDTH	HEIGHT	KNEES	PRICE PER DOZ.
No. 10	Sled	30 Inches	12 Inches	7½ Inches	2	
No. 20	Sled	33 "	12 "	7½ "	2	
No. 30	Sled	36 "	12 "	7½ "	3	

Star Steel Sled. Style No. 130

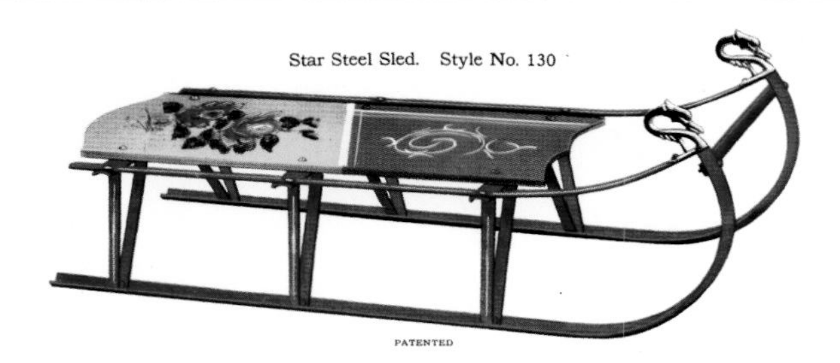

PATENTED

GOOSE-NECK sleds have always been favorites with the girls, and as this is the handsomest sled of this style on the market, it is bound to be a favorite. The Goose-necks are made of malleable iron, and tinned, and make a most attractive appearance in connection with the handsome body colors. The running gears are painted in assorted colors and artistically striped. We are now making the beams on these sleds of channel steel instead of wood, as heretofore.

		LENGTH	WIDTH	HEIGHT	KNEES	PRICE PER DOZ.
No. 110	Sled	30 Inches	12 Inches	7½ Inches	2	
No. 120	Sled	33 "	12 "	7½ "	2	
No. 130	Sled	36 "	12 "	7½ "	3	

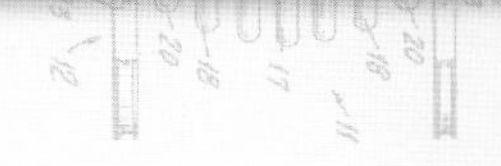

Star Steel Sled. Style No. 35

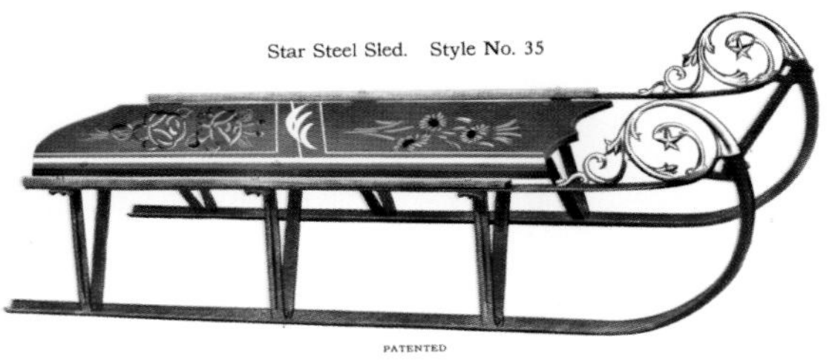

PATENTED

FOR beauty, style, correctness in proportions, for lightness and for strength, our Scroll Tip Sled cannot be equaled. The Scroll Tip is made of malleable iron and tinned, and makes a handsome finish and stylish effect. We believe this to be the most attractive sled ever designed. It is a beauty, and you will say so when you see it. Raves made of half-round hard wood are furnished on these sleds in addition to the steel raves, adding much to the beauty of these sleds. The running gear is finished in assorted colors, artistically striped. The beams on this sled are made of channel steel instead of wood, as heretofore.

		LENGTH	WIDTH	HEIGHT	KNEES	PRICE PER DOZ.
No. 25	Sled	33 Inches	12 Inches	7½ Inches	2	
No. 35	Sled	37 "	12 "	7½ "	3	

Star Steel Coaster. Style No. 11

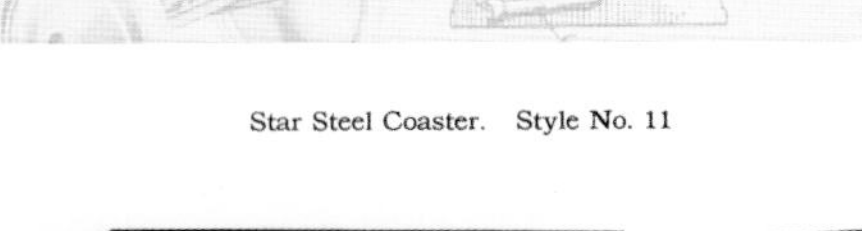

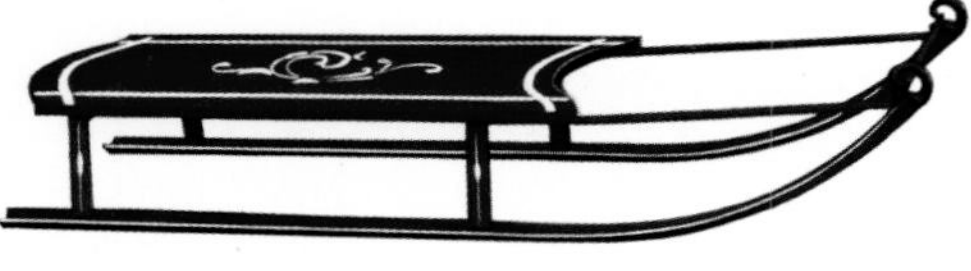

PATENTED

THE cut shown on the opposite page represents in actual colors our Coaster No. 11. This is the best steel coaster ever put on the market for the money. All steel except the top. The runners are made of the celebrated round back angle steel used on all Star Steel Sleds. It has wood top, well painted, varnished and nicely decorated.

		LENGTH	WIDTH	HEIGHT	KNEES	PRICE PER DOZ.
No. 11	Coaster	32 Inches	9 Inches	4½ Inches	2	

Star Steel Coaster. Style No. 108

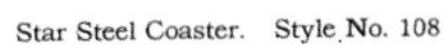

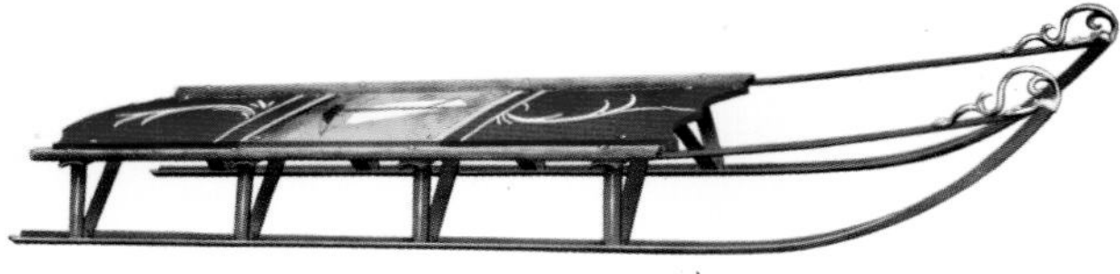

PATENTED

IT is with pride that we present to you our line of fancy coasters. They are the climax of boys' coasters and far excel anything of the kind ever before attempted. The scroll tips that add so much to their attractiveness are made of malleable iron and tinned, and form a contrast with the body colors that can only be appreciated when seen. The beams on these coasters are made of channel steel instead of wood, as heretofore.

		LENGTH	WIDTH	HEIGHT	KNEES	PRICE PER DOZ.
No. 106	Coaster	46 Inches	12 Inches	4½ Inches	3	
No. 108	Coaster	48 "	12 "	4½ "	4	

The American Boy Bobs. Style No. 202

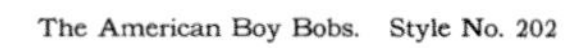

PATENTED

OUR "American Boy" Bobs have been so generally introduced to the trade that we do not consider any extended description necessary. They are without question the best bob-sleds ever offered for the money. They ride easier than other bob-sleds, because both forward and rear runners are made to oscillate, thereby going over uneven ground more smoothly. The runners are made of our special round back angle steel, which gives all Star Steel Sleds such speed. The knees are constructed according to our patented bridge truss principle, which makes our sleds so strong. The runners are finished in assorted colors, artistically striped. The tops are handsomely hand-decorated.

		LENGTH	WIDTH	HEIGHT	PRICE PER DOZ.
No. 201	Bob	40 Inches	10½ Inches	5½ Inches	
No. 202	Bob	44 "	11 "	5½ "	
No. 203	Bob	48 "	11½ "	5½ "	

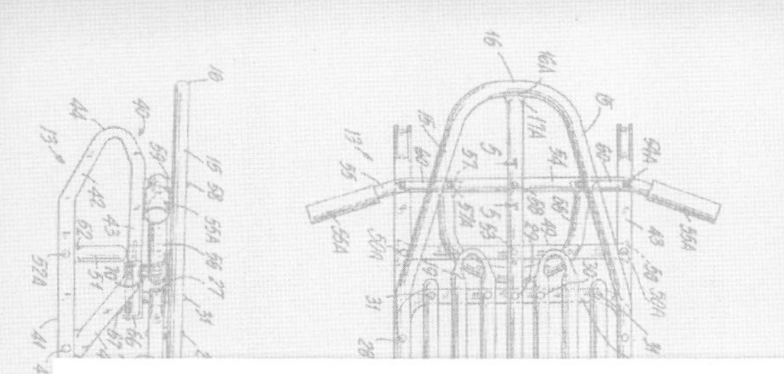

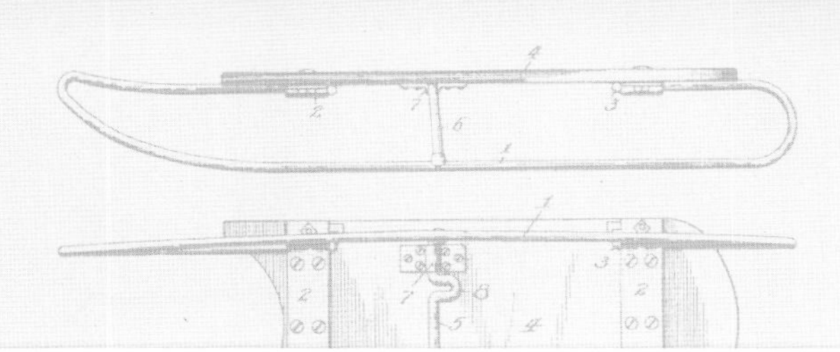

The American Boy Bobs. Style No. 206

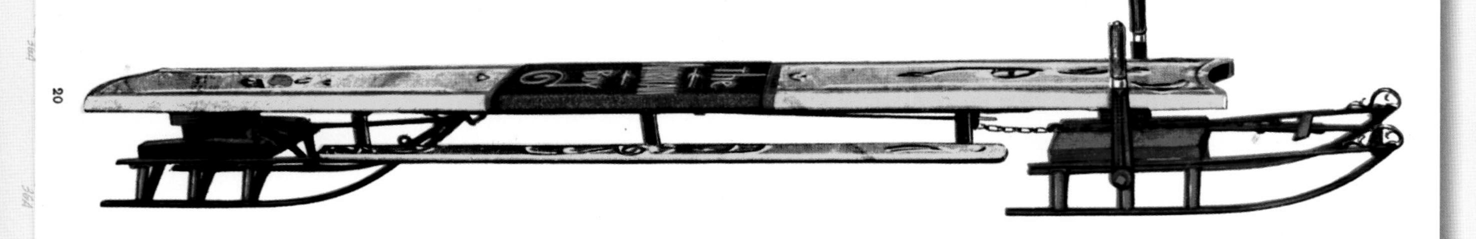

PATENTED

THIS photograph represents the finest bob ever built. In construction it is as strong as steel can make it; in finish it is unexcelled. Our patent steering and brake attachments give the rider complete control of the bob and preclude the possibility of an accident. The malleable handles, with which this bob is guided, are located on each side of the operator and are readily dropped down out of the way when desired. This steering device is much more safe and convenient than those that are made rigid and stand directly in front of the operator. The brake attachment is simple in construction, yet very effective, and is placed under the sled board entirely out of the way of the passengers. The pressure is applied to the brake with the feet of the operator. Tops are handsomely finished in colors and beautifully hand decorated. Running gears are finished in colors, artistically striped. The seating capacity given below refers to adults.

		LENGTH	WIDTH	HEIGHT	KNEES	PRICE EACH
No. 204	Four passengers, without brake	84 Inches	14 Inches	9½ Inches	2	
No. 205	Four passengers, with brake	84 "	14 "	9½ "	2	
No. 206	Six passengers, with brake	108 "	14 "	9½ "	3	

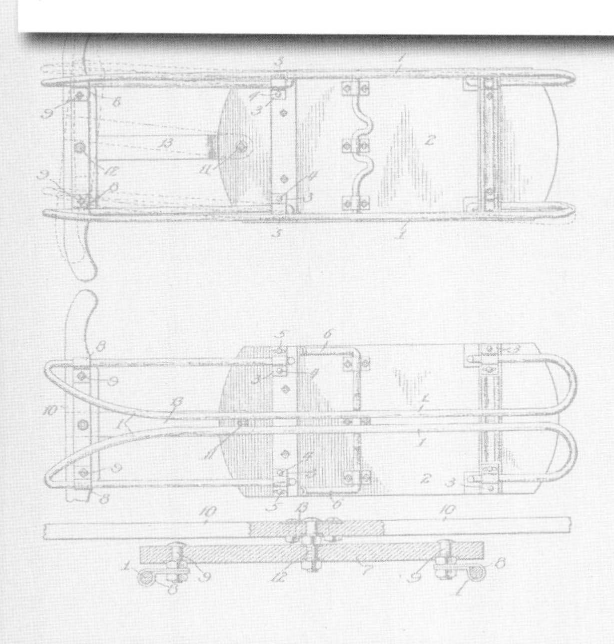

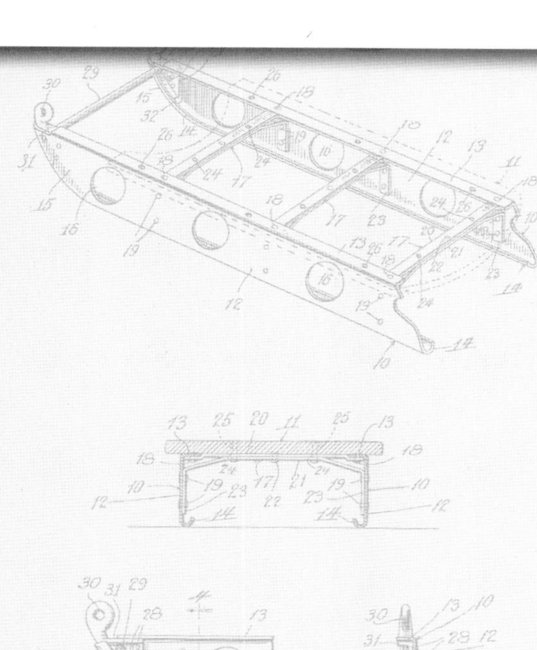

★ STAR LINE EQUIPMENT ★

Overland Coaster Wagon

Fig. 275

SPECIFICATIONS

BODY—Clear hardwood, natural finish, trimmed in red and stenciled in black and green.

GEARS—Channel arch truss steel construction, enameled black.

FIFTH WHEEL—Extra large, made of steel.

AXLES—One-half inch round steel, firmly braced front and rear.

WHEELS—Heavy iron hub into which straight, smooth, kiln-dried hardwood spokes are driven. Felloes and tires of heavy steel, electrically welded, with edges curled in to hold the ends of the spokes.

CONTAINED BEARINGS—Each wheel fitted with eleven cold rolled steel bearings, held in place by a special washer that holds bearings in place and prevents them from dropping out when wheel is removed.

HUB CAPS—Fastened firmly over outer ends of hubs—keep the grease in and the dirt out.

TONGUE—Hard, straight maple which bends back and allows wagon to be steered from box.

BRAKES—Malleable iron.

Page Three Hundred Twenty-nine

Star Line Equipment advertisement, c. 1900. "Overland Coaster Wagon." *Courtesy of the Greater Harvard Area Historical Society.*

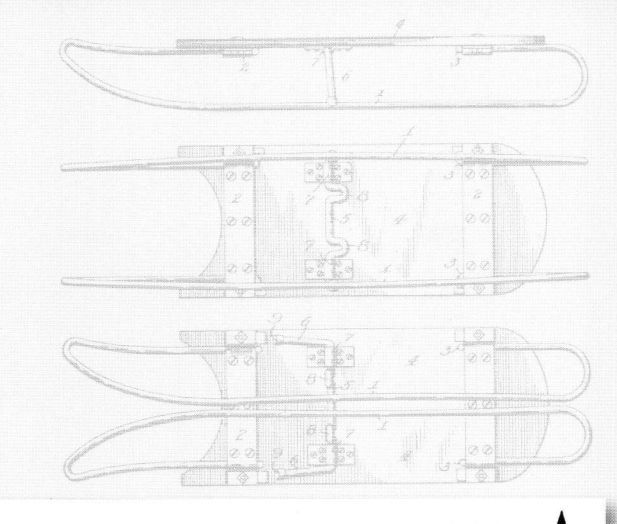

★ STAR LINE EQUIPMENT ★

Automatic Wagon Salesman

Goods Attractively Displayed Are More Than Half Sold

And these Wagon Display Fixtures make the wagons look extra good to the folks who are going to buy them.

Put a rack of our wagons out in front of your store—let our goods do their own talking—sales will come almost of themselves. Results from the use of Display Fixtures will show you why dealers who have used it, call it the "AUTOMATIC WAGON SALESMAN."

Star Line Equipment advertisement, c. 1900. "Automatic Wagon Salesman." *Courtesy of the Greater Harvard Area Historical Society.*

★ HUNT · HELM · FERRIS & CO · ★

A Word About Wagons

THE two BIG things about a wagon are—STRENGTH—LOOKS. The appearance of Overland Wagons sells them on sight. Clear, clean, hardwood boxes finished with two fine coats of implement coach varnish, covered on the bottom as well as on the sides—an honest job of finishing, artistically striped, scrolled and stenciled. Every single part is finished carefully, no daub or "slab" work on our goods.

This, because we realize that "looks" make the first sale and our quality shows at a glance or on minute inspection.

But while "looks" may make the first sale, "durability" is what makes the repeat orders. And from a "wear" standpoint, our wagons are in a class by themselves. The rim of the wheels is pressed around the spokes, making it impossible for them to loosen. An all-steel construction below the box and a brace from the bottom of the bed to the front axle so when a boy runs into the curb or telegraph pole, his wagon is still in service. This is an exclusive feature with us.

The full roller bearing axles make our wagons an easy pull for a three-year-old. In our construction are embodied all modern improvements and many exclusive features which make Hunt, Helm, Ferris & Co. Wagons the most saleable and satisfactory on the market today.

Overland Coaster

Page Three Hundred Twenty-eight

Hunt, Helm, Ferris & Co. advertisement, c. 1900. *Courtesy of the Greater Harvard Area Historical Society.*

HUNT, HELM, FERRIS & CO., HARVARD, ILLINOIS 81

THE SHOOTING STAR SLED

Fig. 389

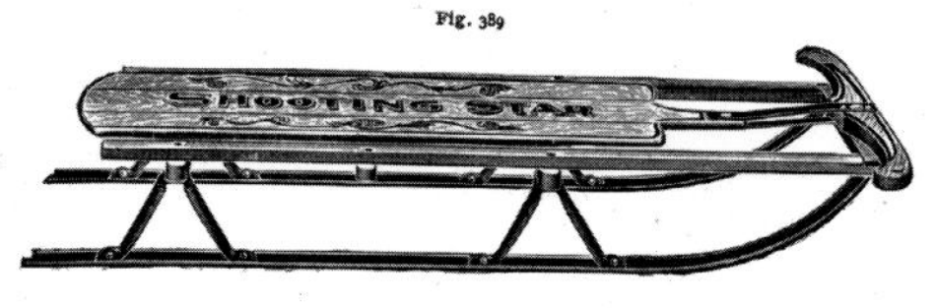

	LENGTH	WIDTH	HEIGHT	WEIGHT PER DOZ.
No. 301	36 in.	12 in.	6 in.	100 lbs.
No. 302	40 in.	13 in.	6¼ in.	120 lbs.

Fig. 394

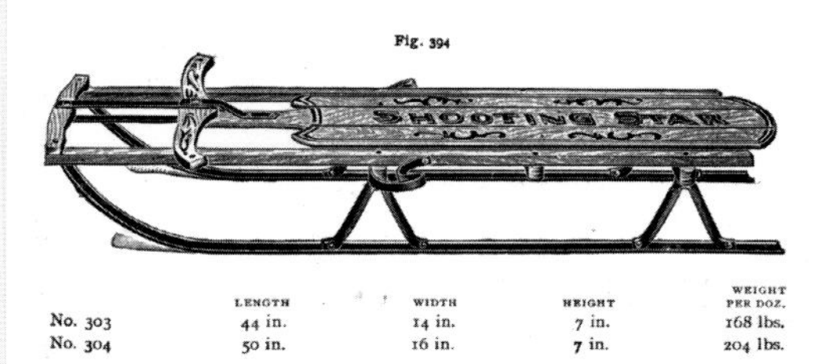

	LENGTH	WIDTH	HEIGHT	WEIGHT PER DOZ.
No. 303	44 in.	14 in.	7 in.	168 lbs.
No. 304	50 in.	16 in.	7 in.	204 lbs.

Packed two in a crate

To meet the demand of the trade we have added to our celebrated line of Star Steel Sleds, Coasters and Bobs the line of flexible steering sleds illustrated in the above cuts. These sleds combine features well known to the trade, such as the steering bar, which saves the necessity of retarding the sled by dragging the feet. The materials used are the best obtainable. The runners are made of spring steel, the knees of pressed steel and the frame and top of straight grained hardwood. The sleds are light in weight, yet practically indestructible. Handsomely finished.

82 HUNT, HELM, FERRIS & CO., HARVARD, ILLINOIS

STAR COASTER WAGONS

Fig. 272

White ash buckboard bottom and skeleton type removable express box. 6-inch solid hardwood wheels, with extension hubs and ¾-inch strap tire. 1⅛x2-inch rock maple axles, steel hounds, reliable steering mechanism, hard maple pole, equipped with hand brake, attractively finished.

No.	Size of Bottom, Inches	Wheel Diameter, Inches	Packed K. D. No. in Crate	Weight Pounds Dozen
5	11x30	6	2	120

Fig. 266

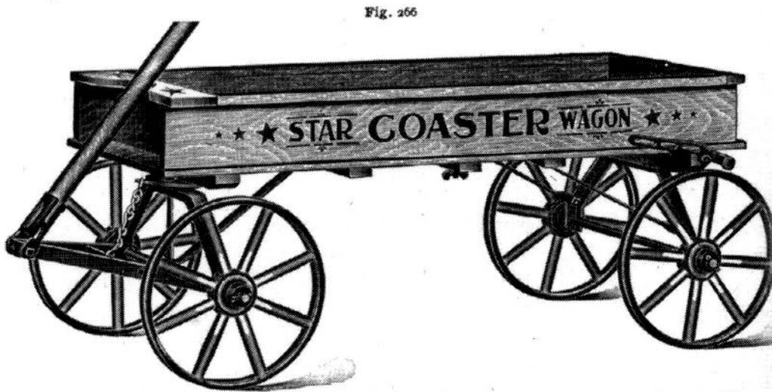

Steel gear, angle steel construction, front and rear axles strongly braced, axles ½-inch round steel, steel hounds, malleable fifth wheel, reliable steering mechanism, wheels extra strong, heavy iron hub, steel felloes and tire, hard steel roller bearings, ⅝-inch hardwood spokes, white ash buckboard bottom, with removable skeleton-type wagon box, attractively finished, equipped with hand brake.

No.	Size of Bottom, Inches	Wheel Diameter, Inches	Packed K. D. Dozen Crate	Weight Pounds Dozen
10	14¼x36	8	⅙	336
20	14¼x36	11	⅙	408
25	14¼x36	11	⅙	420
30	16¼x40	11	⅙	468
130	16¼x40	11	⅙	516
40	18¼x44	11	⅙	492
140	18¼x44	11	⅙	540

For further description see our Special Coaster Wagon Catalogue.

HUNT, HELM, FERRIS & CO., HARVARD, ILLINOIS 83

STAR COASTER WAGON, SHOWING BOX REMOVED

Fig. 273

The box can be removed from all sizes of Star Coaster Wagons, as shown in the above cut, which leaves a nice, smooth bottom, suitable for coaster purposes. The top can be instantly removed or replaced by merely turning the wooden strip pivoted to the under side of the bottom, which securely locks the box in place when it is desired to use it as a wagon instead of a coaster. Note the different ways in which this wagon can be used.

STAR COASTER WAGON ON RUNNERS

Fig. 272

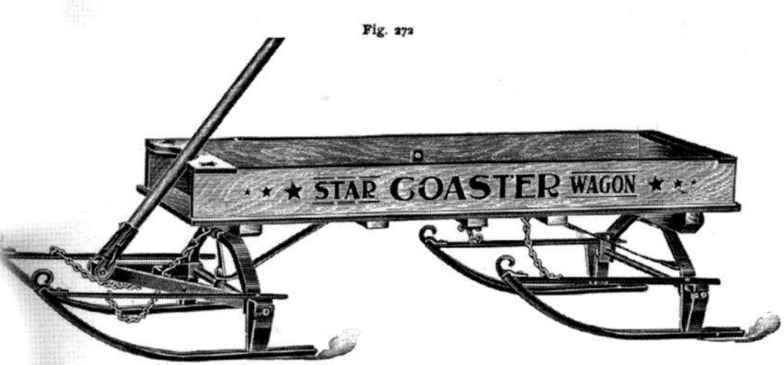

The above cut illustrates our Special Sled Runners, constructed to fit all sizes of Star Coaster Wagons. Any boy can remove the wheels and put these runners in place of same in a very few minutes. The Runner is made of round back angle steel. The knee is so formed as to be rigidly braced. The runners are nicely finished in red, artistically striped. The top can also be removed as shown above.

For further information, see our special Coaster Wagon Catalogue.

84 HUNT, HELM, FERRIS & CO., HARVARD, ILLINOIS

OVERLAND COASTER WAGONS

Fig. 33

White ash buckboard bottom and skeleton type removable express box; 6-inch solid hardwood wheels, with extension hubs and ¾-inch strap tire; 1⅛x2-inch rock maple axles; steel hounds; reliable steering mechanism; hard maple pole; equipped with hand brake; attractively finished.

Number	Size of Bottom, Inches	Wheel Diameter, Inches	Packed K. D. No. in Crate	Weight, Pounds, Dozen
0A	11x30	6	2	120

Fig. 275

Has perfect steering mechanism, which operates equally well with express box on or off. Steel gear; channel steel arch construction. Front and rear axles strongly braced. Axles, ½-inch round steel. Large and strong malleable fifth wheel. Extra heavy 11-inch wheels with heavy iron hubs. Steel felloes and tires. Hard steel roller bearings. 9-16 inch hard maple spokes. White ash bottom with removable skeleton type express box. Hard maple pole. Equipped with hand brake. Attractively finished.

Number	Size of Bottom, Inches	Wheel Diameter, Inches	Packed K. D. No. in Crate	Weight, Pounds, Dozen
1A	14¼x36	8	2	336
2A	14¼x36	11	2	408
2½A	14½x36	11	2	444
3A	16¼x40	11	2	468
4A	16¼x40	11	2	516
5A	18¼x44	11	2	492
6A	18¼x44	11	2	540

For further description see our Special Coaster Wagon Catalogue.

Hunt, Helm, Ferris & Co. Catalog #35, circa 1910. $75-125.

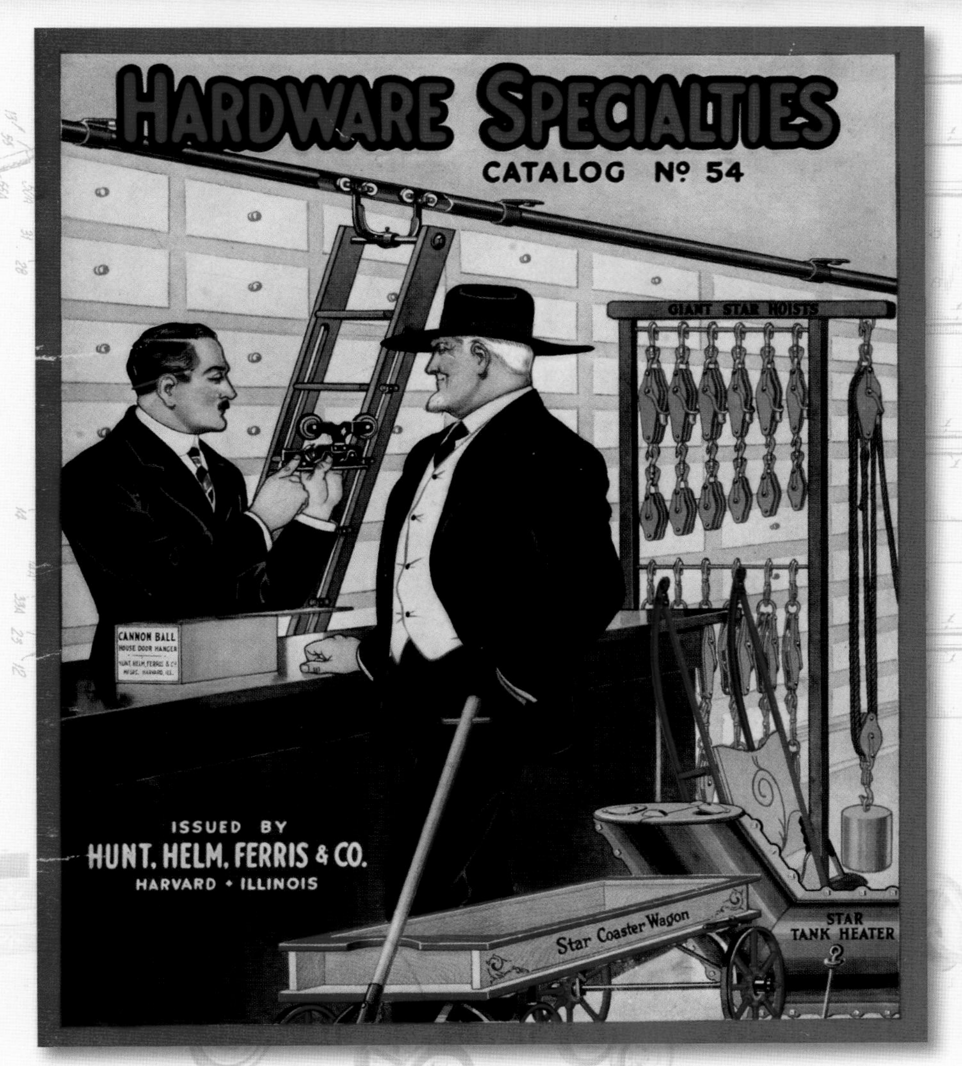

Hardware Specialties, Hunt, Helm, Ferris & Co. Catalog #54, 1916. $50-100.

Star Coaster Wagon

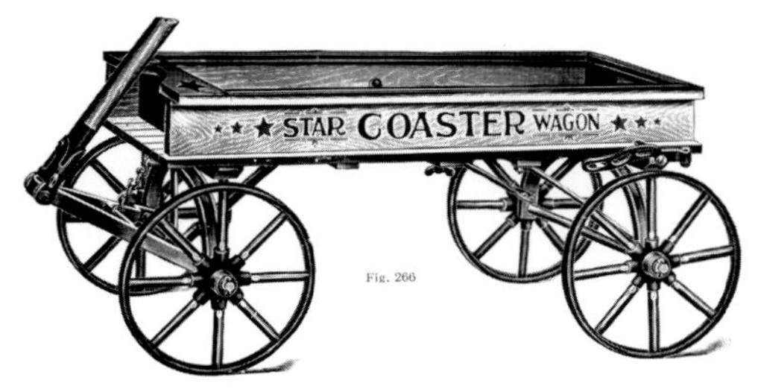

BODY—Clear white ash, natural finish, trimmed in red and stenciled in black and green.

GEARS—Channel arch truss steel construction, enameled black.

FIFTH WHEEL—Extra large, made of malleable iron.

AXLES—One-half inch round steel, firmly braced front and rear.

WHEELS—Heavy iron hub into which straight, smooth, kiln-dried hardwood spokes are driven. Felloes and tires of heavy steel, electrically welded, with edges curled in to hold the ends of the spokes.

BEARINGS—Each wheel fitted with eleven cold rolled steel bearings, held in place by a special washer that does not wear cotter pin.

TONGUE—Hard, straight maple which bends back and allows wagon to be steered from box.

BRAKES—Malleable iron. Larger size wagons provided with foot and hand brakes.

(For prices see next page.)

Star Coaster Wagon, Showing Box Removed

THE BOX on all Star Coaster Wagons is instantly removable and wagon can be changed from Express to Coaster in a minute. A simple connection, easily operated, securely locks the box to the bed, but at the same time provides for the instant removal of the box when desired.

No.	Size Bottom Inches	Wheel Diameter Inches	Style Brake	Weight, Lbs. Per Doz.	List Per Doz.
10	14¼ x 36	8	Hand	336	$60.00
20	14¼ x 36	11	Hand	408	72.00
25	14½ x 36	11	Hand	420	76.00
30	16¼ x 40	11	Hand	468	80.00
40	18¼ x 44	11	Hand	492	86.00
130	16¼ x 40	11	Foot and hand	516	86.00
140	18¼ x 44	11	Foot and hand	540	94.00

Star Coaster Wagon No. 5

Fig. 272

WHITE ash buckboard bottom and skeleton type removable express box. 6-inch iron wheels, with extension hubs and ¾-inch tire. 1⅛ by 2-inch rock maple axles, steel hounds, reliable steering mechanism, hard maple pole. equipped with hand brake, attractively finished.

No.	Size Bottom Inches	Wheel Diameter Inches	Style Brake	Weight Lbs. Per Doz.	List Per Doz.
5	11 x 30	6	Hand	120	$30.00

Overland Coaster Wagon

BODY—Clear white ash, natural finish, trimmed in red and stenciled in black and green.

GEARS—Channel arch truss steel construction, enameled black.

FIFTH WHEEL—Extra large, made of malleable iron.

AXLES—One-half inch round steel, firmly braced front and rear.

WHEELS—Heavy iron hub into which straight, smooth, kiln-dried hardwood spokes are driven. Felloes and tires of heavy steel, electrically welded, with edges curled in to hold the ends of the spokes.

BEARINGS—Each wheel fitted with eleven cold rolled steel bearings, held in place by a special washer that does not wear cotter pin.

TONGUE—Hard, straight maple which bends back and allows wagon to be steered from box.

BRAKES—Malleable iron. Larger size wagons provided with foot and hand brakes.

(For prices see next page.)

Overland Coaster Wagon

With Box Removed

THE express box on all Overland Coaster Wagons can be instantly removed or replaced. The wagon may be changed from express to coaster in a minute.

A simple connection, easily operated, locks box firmly to the bed.

Overland Wagons are furnished in the following sizes:

No.	Box	Wheel	Weight Each	List Price Per Doz.
1A	14¼ x 36 in.	8 in.	28 lbs.	$60.00
2A	14¼ x 36 in.	11 in.	34 lbs.	72.00
2½A	14½ x 36 in.	11 in.	37 lbs.	76.00
3A	16¼ x 40 in.	11 in.	39 lbs.	80.00
4A	16¼ x 40 in.	11 in.	43 lbs.	86.00
5A	18¼ x 44 in.	11 in.	41 lbs.	86.00
6A	18¼ x 44 in.	11 in.	45 lbs.	94.00

Overland Coaster Wagon No. 0A

For Small Children

THIS wagon is designed particularly for the use of small children, but is firmly and strongly built.

Bottom of buckboard style is of white ash. Skeleton type box is removable same as on the larger wagons.

Axles are of 1½ x 2-inch rock maple. Wheels 6 inches in diameter, with ¾-inch tread, are of iron, light but strong.

Wagon has reliable steering mechanism, hard maple pole and hand brake. It is attractively finished, with box natural color, trimmed in red and stenciled in black. Wheels red.

No.	Box	Wheel	Weight Each	List Price Per Doz.
OA	11x30	6	10	$30.00

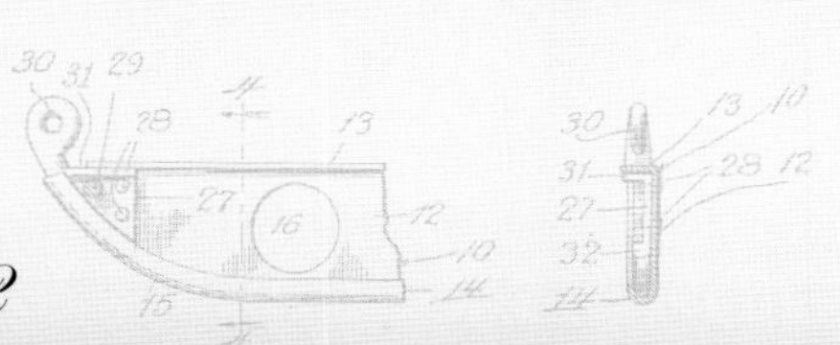

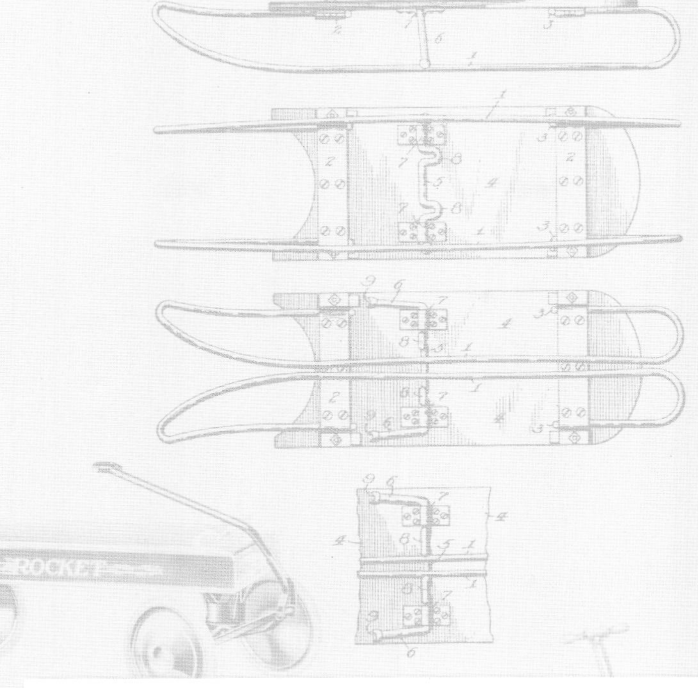

August 3, 1922 HARDWARE AGE 85

The Overland Coaster Wagon

With Disc Wheels and Rubber Tires

A Fast Seller—A Big Trade Builder

The minute a boy sees an "Overland" with its shiny, varnished body and nifty disc wheels with rubber tires, he ***wants*** one. And when he gets it every other boy in the block wants one, too.

Keep a rack of Overland Coasters prominently on display and you will find a steady sale for them the year around.

Has Looks—Strength—Speed

Clear, clean, hardwood boxes finished with two coats of implement varnish—artistically striped, scrolled and stenciled.

Built for long wear. All steel construction below the box and a brace from bottom of bed to front axle—an exclusive Star feature.

The full roller bearings make it easy for a three-year-old to pull. Bigger boys can get all the thrills of real speed because it runs so easily.

Heavy pressed, 10-in. Steel Disc Wheels with double reinforcing plates riveted to body sheet. Automobile hub caps prevent exposure of working parts. ¾-in. rubber tire for 10-in. wheels with 14x34 in. and 16x38 in. beds; ⅝-in. rubber tires for 8-in. wheels with 12x28 in. and 14x32 in. beds.

Write Today for Circular and Prices.

Hunt, Helm, Ferris & Co.

HARVARD, ILLINOIS ALBANY, N. Y.

Los Angeles—San Francisco—Portland, Oregon

Star Line Overland Coaster Wagon advertisement, August 3, 1922. Features new disc wheels and rubber tires. $10-15.

December 3, 1925 HARDWARE AGE 81

CANNON BALL BEATS 'EM ALL

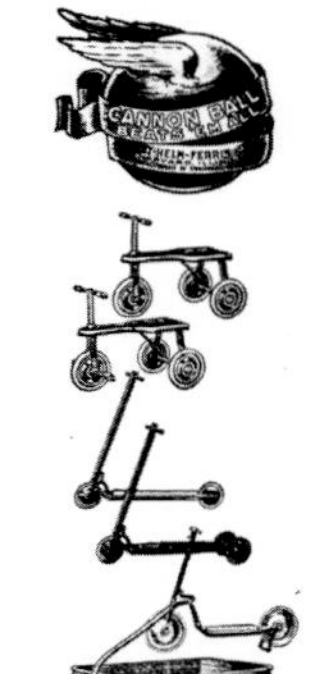

Saturday Evening Post Advertising

Attractive Posters

Window Cut-Outs

Movie Slides

Natural Attractiveness

Distinctive orange finish with green trim

Sturdy construction

— all are working together to help dealers sell Cannon Ball Wheel Goods on a profitable basis.

Cannon Ball Coaster Wagons in 3 sizes. Junior Wagons in 3 sizes. All steel wagons, Pedal Cars, Play Boys, Scooters and Skee-Skooters, Skeeboggans and Skee-Bobs are now on hand for prompt delivery.

Single or double column newspaper cuts of any illustration on this page, mailed on request. Life size cut-out of Cannon Ball Boy, fits large size wagon, sent with order for $50.00 or more.

Winter Thrillers and Heavy Sellers

A long winter is predicted with plenty of cold weather and snow; just the kind of weather to invite the use of Cannon Ball Skeeboggans and Skee-Bobs and help their sales. Cannon Ball Skeeboggan is made of heavy spring steel with runners formed in the sides. Finished in brilliant orange, green trim with rubber covered platform. Hinged handle folds flat — easily carried in car or bus.

CANNON BALL SKEE-BOBS

Cannon Ball Skee-Bob, No. 1146, has channel steel frame with hardwood rubber-covered running board. Easy to steer by turning handle. Can be quickly stopped by a full quarter turn. Operator sits on rubber topped wood platform.

No. 1145 Cannon Ball Skee-Bob. Same as No. 1146 except for a single instead of double front runner.

Phone, wire, or mail your fill-in orders to nearest branch for quick service and prompt delivery.

HUNT-HELM-FERRIS & CO.

Harvard, Illinois

Industrial Building Albany, New York	383 Brannan Street San Francisco, Cal.
22nd and Arch Streets Philadelphia, Pa.	422 Stinson Boulevard Minneapolis, Minn.
1811 E. 7th Street Los Angeles, Cal.	1322-1330 W. 13th St. Kansas City, Mo.

"Cannon Ball Beats 'Em All" 1925 advertisement for life-size cut-out of Cannon Ball Boy.

The Kalamazoo Sled Company

In 1968 the Gladding Corporation purchased Kalamzoo Sled Company and their famous Champion Sled was in for a face lift. The Gladding-Kalamazoo Company introduced the new Champion "Blue Racers" GTO series, along with the Champion Fast Back, a racing-style sled, complete with Blue Racer Runners, an air speed indicator, tachometer, fuel gauge and racing flags on the deck and steering bar. The newest economy model was the "Rocket Racer" series, fully equipped with grooved runners, safety heel construction, and red enameled steel parts. The center deck sported speed striping. The early models of the newly formed Gladding-Kalamazoo Company did not have Gladding on their decks. Company advertising indicates that by 1971 all models had Gladding on their decks, with the exception of the Champion Fast Back. In 1972 Gladding was added to its deck. It is not clear when the American Clipper came on board, although 1971 company advertising describes it "Like its nautical forerunner, offering smooth, fast sale-ing" fully decorated and offering a wave motif on the steering bar and the words Classic American Clipper on the deck. The marketing slogan reads, "Sales-Ahoy"

With the Gladding Corporation purchasing the Kalamazoo Sled Company, the familiar logo or trademark "K" and the Sno-ette Baby Sleigh virtually vanished from the market. The Sno-ette made its debut in the early 1950s, advertised as the "smartest thing on runners" It was Sno-Brite Orange and had a formed seat, back, and arms, constructed of moisture resistant plywood with a tubular handle designed for easy and effortless pushing. The standard Champion Kalamazoo Baby Sled continued to be manufactured through 1971 with a minor change. The handle could be pivoted from front to back allowing it to be pushed or pulled. It was adorned with the word Champion on a red and navy blue deck. The back support or sled guard was constructed of hardwood, with two navy blue snowflakes, coated with a clear weather resistant varnish, model # 507. Do not confuse the Baby Sled with the Convertible Stroller Sled, model #503A. The Convertible Stroller Sled had wheels, the Baby Sled did not. When sidewalks were snowy, the wheels could be retracted easily. The tubular steel push handle was stationary and not able to pivot as the handle on the Baby Sled. The year 1972 ushered in the new line of Gladding baby sleds, and said good-bye to the Champion models. It saw the "Sleighette" Convertible Baby Sled replacing the Champion Convertible Stroller Sled and the Sno- Flake replacing the Champion push/pull baby sled. Except for graphics, and the push/pull handle feature on both models they were identical. The deck now had Gladding Sno-Flake in red and navy blue lettering along with six pointed star-like graphics. The sled guard had an elephant balancing on a large ball, replacing the snow flakes on the Champion Models. Gladding, the oldest continuous leisure products manufacturing company, was fast becoming the largest, with Gladding- Kalamazoo purchasing Hedlund, makers of top quality toboggans, in 1969 to form Gladding-Hedlund, and finally purchasing the Paris Sled company in March 1970 to form the Gladding-Paris Corporation.

Updated Chronology of Kalamazoo Sled Company:

1968 • Gladding Corporation purchases Kalamazoo forming Gladding Kalamazoo
1969 • Gladding Corporation purchases Hedlund successor to the American Sled Company
1970 • Gladding Kalamazoo introduced "Blue Racers" GTO series the Champion Fast Back & Rocket Racers Gladding purchases Paris Mfg. Company
1971 • American Clipper Sleds introduced
1972 • Kalamazoo trademark "K" and Sno-ett Baby Sleigh discontinued. Kalamazoo Sled Co. slowly disappears from the market, leaving the Gladding Corporation

Kalamazoo Sled Company catalog, 1916, with price list. *Courtesy of Jim Pauzé.* *(continued on next two pages)*

ON THE OPPOSITE PAGE we present an accurate photographic reproduction in actual colors of our No. 87 sled. Forming the background are tops of some of our other numbers.

No. 87. Length, 36 inches; width, 12 inches. Three frame knees mortised into runners, with *square tenons* and set at **two** angles. No other sleds possess these features. Top painted various colors, with decorations and striping as shown in the color plate. Flat shoes...Per doz.

No. 88. Same as 87, except has brightly tinned malleable iron goose necks. Flat shoes, Per doz.. Six sleds in a crate.

8

No. 87. STRAIGHT KNEE SLED.

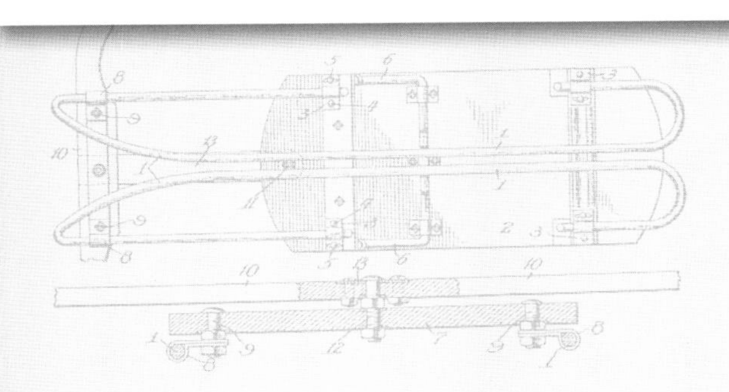

ON the opposite page in photographic colors we illustrate an example of our line of sleds having upholstered tops. These numbers embrace the most popular designs throughout our catalogue and where such style of sled is finished in this manner, the proper number will be found on the page giving the general description of each model as it would be impracticable to illustrate each one of these sleds. This method of finishing the top is a change from the usual decoration and seems to be very popular with our customers. The upholstering material consists of heavy duck in imitation of leather, thoroughly water proofed. Back of the sled we have placed a few tops of other numbers.

No. 126. As illustrated. Length, 36 inches; width, 12 inches. 3 bent knees. Gear finished on the wood. Top upholstered. Six full length tinned knee braces and tinned malleable goose necks, oval shoes...Per doz......

12

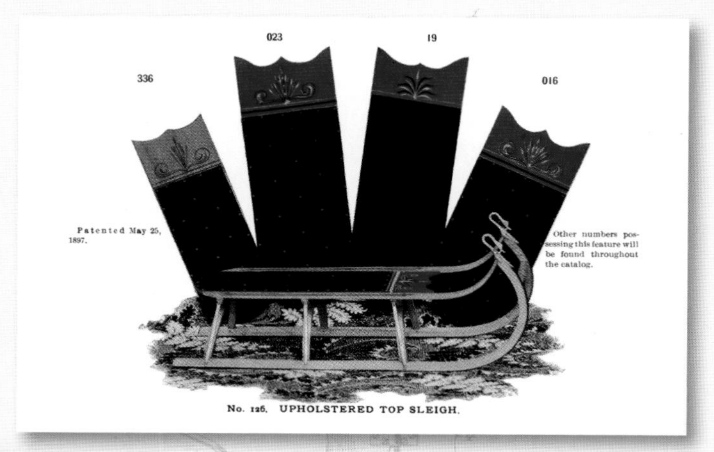

No. 126. UPHOLSTERED TOP SLEIGH.

NO. 125

PERFECTION BENT KNEE

GOOSE NECK SLEIGH.

All our patents will be protected.

No. 125. Length, 36 inches; width, 12 inches. Has three bent knees; gear finished on the wood; top handsomely hand decorated as illustrated on opposite page. Has six full length tinned braces and tinned malleable goose necks. Oval shoes..........Per doz..

No. 120. Same as No. 125, except has flat shoes. For character of decorations on top of this sled see corresponding number of top on page 17.................................Per doz..

Above sleds packed one-half dozen in crate, with two or more styles of decorations.

16

Patent May 25, 1897.

No. 125. PERFECTION BENT KNEE SLEIGH.

SCROLL TIP SLEIGHS.

WE CAN OFFER this design of Sleighs to the consideration of the most exacting buyer as an ideal large, but light and attractive model. These numbers, 350 and 360, are light enough for a small child, and at the same time in size and strength are suitable for those of ten or twelve years. Our design is attractive in appearance, correct in outline, and the finish and decorations are equally pleasing to the eye. Our illustration no longer does justice to our work, as we are making the present decorations much more elaborate than our plate shows.

No. 360. Length, 36 inches; width, 12 inches. Has three bent knees. Gear is oak, heavily braced with six heavy tinned knee braces and six round tinned runner braces. This sled is much better than the illustration.....................................Per doz......

No. 361. Same as No. 360, except has upholstered top as described on page 13..............Per doz......
Packed one-third dozen in crate.

28

No. 360. SCROLL TIP SLEIGH.

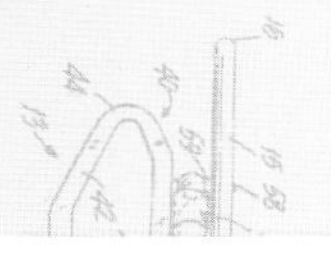
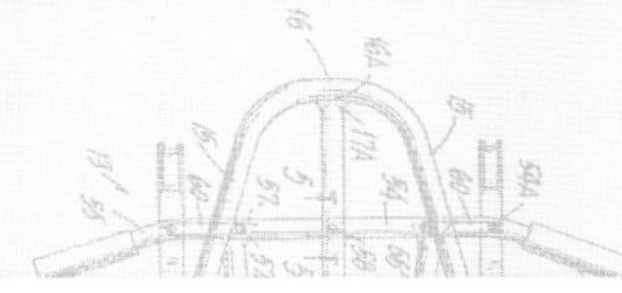

Fine coasters.

THE GENERAL OUTLINES of our medium priced Coasters are admitted by all to be the best in the market, and the new features in the way of better decorations, extra fine striping and increased curve of the runners, will be greatly appreciated. This popular priced line of Coasters has not its equal in the market. Patents were allowed Aug. 8, 1899, covering certain parts of the construction of these numbers, and infringements will not be tolerated.

No. 15. Length, 36 inches; width 11 inches. Gear nicely finished on the wood. Top painted bright colors and ornamented by hand with landscapes, flowers, etc. Four hand holes. Round spring shoes............Per doz.......

No. 16. Our present pattern has $2\frac{1}{2}$ inches more of upward curve on points of runners, and tops are very much better than the illustration shown on page 37. Length, 42 inches; width, 12 inches. Otherwise same as No. 15. Round spring shoes.......Per doz.......

No. 016. Same as No. 16, except has upholstered top as described on page 13...............Per doz.......

No. 17. Length, 48 inches; width, 12 inches. Shod with round spring shoes..............Per doz.......

No. 017. Same as No. 17, except has upholstered top as described on page 13...............Per doz.......

Tops of above sleds painted two colors and styles of decorations.
Packed one-half dozen in crate.

36

No. 16. CURVED RUNNER COASTER.
Patent August 8, 1899.

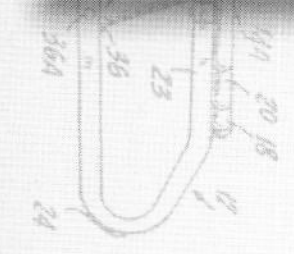
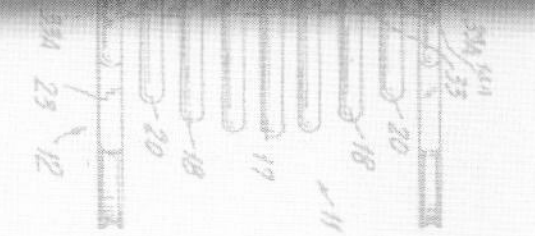

Coasting "Bobs."

No. 645. Size of sleds, 12 x 32, made extra heavy, each braced by four heavy angle irons; length of board, 6 feet with collapsing foot rest and brake.........Price of each complete......

No. 646. Same as No. 645, except top board is 8 feet long and trussed........Price of each complete......
When specially ordered, we can furnish extra lengths at an extra cost of $1.00 per foot net.

No. 650. Same as No. 645, except top board is upholstered in heavy duck in imitation of tan leather. Length, 6 feet..............................Price of each complete......

No. 651. Same as No. 650, except length, 8 feet..............................Price of each complete......
For each additional foot in length, add $1.50 net.
For additional description of above numbers, see page 45.

No. 635. Size of sleds, 12x32 each, braced by four angle irons. Length of top board 6 feet.
No brake or foot-rest..............................Price of each complete......
For each additional foot in length, add $1.00 to list. Packed two in crate unless otherwise ordered.

No. 655. For small children. The sleds are 24 inches long, shod with round steel, both oscillating and the top board is nicely finished and striped.
Size of sleds, 11x24. Length of top board, 4 feet, not illustrated....................Each......

44

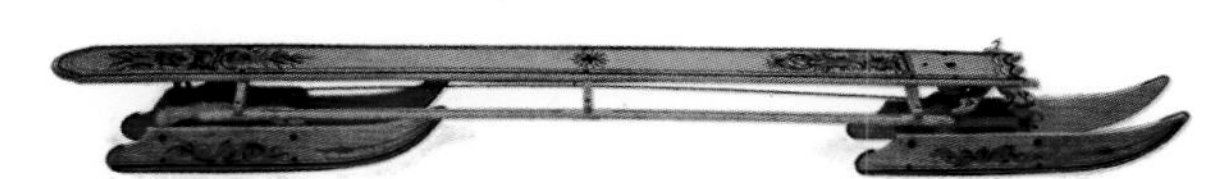

NO. 645. **COASTING "BOBS" OR "DOUBLE RUNNERS."**
Patented Aug. 8, 1899; Aug. 16, 1904.

The greatest care, combined with practical experience on this subject, has enabled us to produce unquestionably the most perfect design for coasting purposes, as illustrated above. It is impossible to bring out all the details or fine qualities of these bobs in our illustration, but it is described briefly as follows: The material throughout is of carefully selected hardwood stock, considerably heavier than our regular sleds. The gear is finished entirely in the natural wood and top boards in assorted colors, unless otherwise ordered. Entire "Bobs" are put together with rivets and bolts in a thoroughly substantial manner, both sleds being braced with riveted extra heavy corner irons and will stand the roughest class of coasting without injury. The back sled has our own design of oscillating movement, which, while giving it the necessary play will prevent its getting out of line with the front sled, and is provided with a powerful brake operated with a double compound lever, through the medium of the steersman's foot. This brake is so powerful that it will absolutely stop these "Bobs" at any time within a very few feet. The front or steering sled is provided with a suitable and perfectly safe device for guiding purposes. A collapsible foot-rest extends out and down from the board about 4 inches, which, in case of bad spill, folds underneath the board, allowing the occupants to slide off without danger of broken limbs. The arrangement of this foot-rest makes our "Bobs" about 22 inches wide when in use but the foot-rest may be folded underneath the board by a simple backward movement, when desired. You should order at least two of these "Bobs" for display. Do not let the price worry you. We have demonstrated that these numbers will sell. They are worth the money and just what the boy wants. One pair sold means many more orders.

For further description we refer you to opposite page.

45

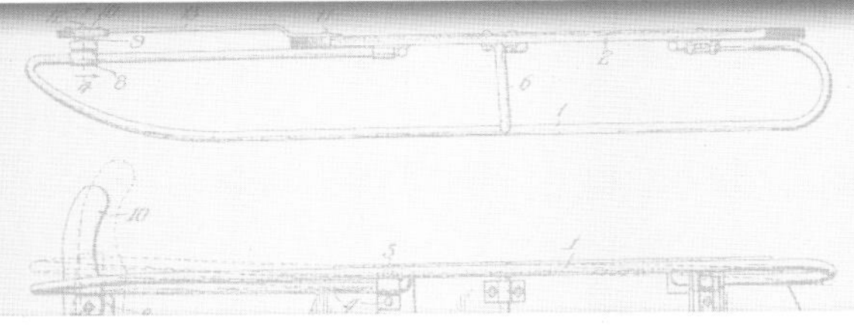

Sled novelties.

DOLL CUTTERS.

PLEASE NOTE that these two cutters are for doll's use only. While very nearly large enough for a baby, they are intended to be used as toys only. They are exact reproductions of our larger cutters in all but size, and are perfectly illustrated by color plate on opposite page.

No. 525. Length, 31 inches; width, $11\frac{1}{2}$ inches. No upholstery. Body painted in pleasing colors. Push handles and flat shoes..............................Per doz......

No. 530. Length, 31 inches; width, $11\frac{1}{2}$ inches. Bent dash. Body painted in bright colors and ornamented. Neatly upholstered. Push handles and flat shoes..............Per doz......
Both the above cutters packed in one-third dozen crates.
We can not furnish these cutters in **white,** except at an advance of $6.00 per doz. list.

48

DOLL CUTTERS
Can not be used for babies.

No. 530. DOLL CUTTERS No. 525.

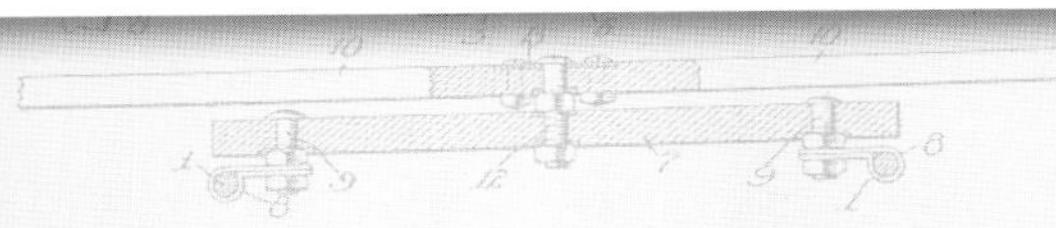

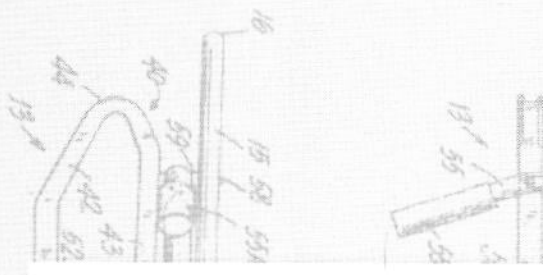

A MODEL DESIGN IN CHILDREN'S CUTTERS.

THERE IS NOTHING within the price better than our Nos. 550 and 555 Cutters. These Cutters have a bent dash high and wide enough to form a protection from the wind. The bodies are nicely painted in pleasing colors, and decorated and striped in a tasteful manner, as our illustration shows. If furnished in white, add 50 cents to list price.

No. 550. Length, 38 inches; width, 15 inches. No upholstery. Oval shoes.................... Each....

No. 555. Same as No. 550, except seat and back upholstered in damask....... ....Each.....

Packed one in a crate.

52

No. 555. CUTTER.

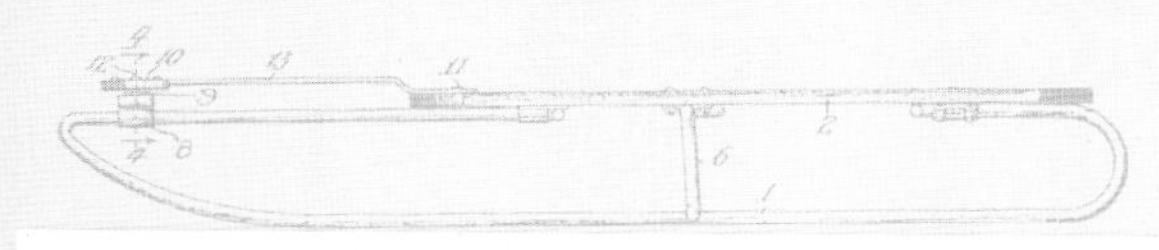

NOS. 600, 605, 610.

CHILDREN'S CUTTERS.

THE FAVORITE.

No. 605. Size over all, 17x46. Bent dash with fenders. Gear framed securely and made of best second growth timber. Knees shaped and ironed like a Portland cutter. Body has round corners on back, and entire cutter finished in both light and dark colors, artistically striped and decorated. Upholstered in plush, with plumes and dash rail but no arm rail.........Each.....

No. 610. Same as No. 605, except has plated rails, as shown in illustration opposite....Each......

No. 615. Same as No. 610 except it is fitted with the automatic wheel attachment illustrated and described on pages 56 and 57. This number is by far the most practical and desirable children's cutter on the market to-day.

Packed one in a crate.

We can furnish any of the above cutters in natural wood with light or dark Golden Oak finish at a cost of $1.00 extra list. This style of finish is made to order only and requires one week to produce.

60

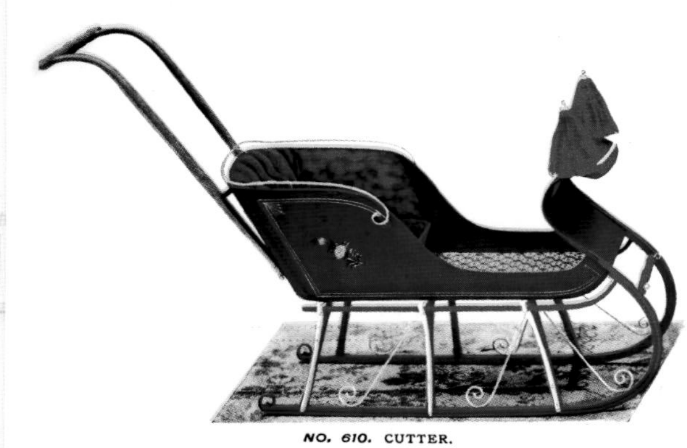

NO. 610. CUTTER.

Our present design is considerably larger than our cut indicates.

Kindly destroy all former Lists and order from this list only.

Use telegraph words for numbers when ordering by wire.

STEERING SLEDS

No. of Sled	Telegraph Word	List Price	Page in Catalogue
00	Rush	16 00 per doz.	4
01	Roam	20 00 per doz.	4
02	Robin	24 00 per doz.	4
03	Rocket	30 00 per doz.	4
04	Rapid	36 00 per doz.	4
05	Revel	40 00 per doz.	4
06	Racer	42 00 per doz.	4
03x	Roar not il'st'd size 13¼x 50x 6¼ inches high Six Knees,	36 00 per doz.	

COASTERS

No. of Sled	Telegraph Word	List Price	Page in Catalogue
1	Early	$6 25 per doz.	34
2	Ebony	6 75 per doz.	34
5	Easy	6 50 per doz.	34
7	Edit	8 00 per doz.	34
9 x	Exit, not illustrated	8 50 per doz.	
10	Eat	9 00 per doz.	35
12	Earth	13 50 per doz.	35
14	Echo	18 00 per doz.	35
15	Eden	15 00 per doz.	36
16	Edge	18 00 per doz.	36
016	Eagle	22 00 per doz.	39
17	Egg	22 50 per doz.	36
017	Elder	28 00 per doz.	36
18	Elbow	24 00 per doz.	42
19	Entry	30 00 per doz.	42
20	Elf	30 00 per doz.	42
21	Elope	36 00 per doz.	42
23	Era	36 00 per doz.	43
023	Evil	42 00 per doz.	43
24	Emit	24 00 per doz.	38
25	East	27 00 per doz.	38
26	Eckle	30 00 per doz.	38
27	Excel	48 00 per doz.	40
28	End	54 00 per doz.	40
30x	Eager	28 00 per doz.	39
31x	Exeter	36 00 per doz.	39
32x	Earl	42 00 per doz.	39

FRAME SLEDS

No. of Sled	Telegraph Word	List Price	Page in Catalogue
85	Fade	$6 25 per doz.	6
86	Fern	6 50 per doz.	6
87	Faun	8 00 per doz.	8
88	Fire	10 50 per doz.	8
90	—Discontinued		
95	—Discontinued		
100	Fish	6 75 per doz.	7
101	Flea	8 00 per doz.	7
102	Friend	10 00 per doz.	7
103	Folio	10 00 per doz.	10
104	Fry	12 50 per doz.	10
105	Field	11 50 per doz.	11
105x	Fashion, not il'st'd	9 50 per doz.	
106	Fund	13 50 per doz.	11
107	Fresh	14 00 per doz.	14
108	Farm	16 00 per doz.	14
110	Foul	14 00 per doz.	15
110x	Fork, not il'st'd	12 00 per doz.	
115	Frost	16 00 per doz.	15
120	Fruit	16 00 per doz.	16
125	Frank	18 00 per doz.	16
126	Flinch	22 00 per doz.	12
200	Sack	15 00 per doz.	26
210	Saul	17 50 per doz.	27
211	Slave	22 50 per doz.	27
250	Sago	17 50 per doz.	22
260	Soap	20 00 per doz.	23
261	Satan	24 00 per doz.	23
265	Sauce	24 00 per doz.	18
270	Sense	24 00 per doz.	24
271	Small	30 00 per doz.	24
325	Saber	15 00 per doz.	19
330	Sheep	18 00 per doz.	20
331	Slime	22 00 per doz.	20
335	Shoot	21 00 per doz.	20
336	Smile	25 00 per doz.	20
350	Sign	24 00 per doz.	30
351	Steep	28 00 per doz.	30
360	Siren	30 00 per doz.	28
361	Smart	34 00 per doz.	28
400	Snake	36 00 per doz.	31
401	Soul	42 00 per doz.	31
410	Solid	42 00 per doz.	32
411	Silk	48 00 per doz.	32

Kindly destroy all former Lists and order from this list only.

Use telegraph words for numbers when ordering by wire.

STEERING SLEDS

No. of Sled	Telegraph Word	List Price	Page in Catalogue
00	Rush	16 00 per doz.	4
01	Roam	20 00 per doz.	4
02	Robin	24 00 per doz.	4
03	Rocket	30 00 per doz.	4
04	Rapid	36 00 per doz.	4
05	Revel	40 00 per doz.	4
06	Racer	42 00 per doz.	4
03x	Roar not il'st'd size 13¼x 50x 6¼ inches high Six Knees,	36 00 per doz.	

COASTERS

No. of Sled	Telegraph Word	List Price	Page in Catalogue
1	Early	$6 25 per doz.	34
2	Ebony	6 75 per doz.	34
5	Easy	6 50 per doz.	34
7	Edit	8 00 per doz.	34
9 x	Exit, not illustrated	8 50 per doz.	
10	Eat	9 00 per doz.	35
12	Earth	13 50 per doz.	35
14	Echo	18 00 per doz.	35
15	Eden	15 00 per doz.	36
16	Edge	18 00 per doz.	36
016	Eagle	22 00 per doz.	39
17	Egg	22 50 per doz.	36
017	Elder	28 00 per doz.	36
18	Elbow	24 00 per doz.	42
19	Entry	30 00 per doz.	42
20	Elf	30 00 per doz.	42
21	Elope	36 00 per doz.	42
23	Era	36 00 per doz.	43
023	Evil	42 00 per doz.	43
24	Emit	24 00 per doz.	38
25	East	27 00 per doz.	38
26	Eckle	30 00 per doz.	38
27	Excel	48 00 per doz.	40
28	End	54 00 per doz.	40
30x	Eager	28 00 per doz.	39
31x	Exeter	36 00 per doz.	39
32x	Earl	42 00 per doz.	39

FRAME SLEDS

No. of Sled	Telegraph Word	List Price	Page in Catalogue
85	Fade	$6 25 per doz.	6
86	Fern	6 50 per doz.	6
87	Faun	8 00 per doz.	8
88	Fire	10 50 per doz.	8
90	—Discontinued		
95	—Discontinued		
100	Fish	6 75 per doz.	7
101	Flea	8 00 per doz.	7
102	Friend	10 00 per doz.	7
103	Folio	10 00 per doz.	10
104	Fry	12 50 per doz.	10
105	Field	11 50 per doz.	11
105x	Fashion, not il'st'd	9 50 per doz.	
106	Fund	13 50 per doz.	11
107	Fresh	14 00 per doz.	14
108	Farm	16 00 per doz.	14
110	Foul	14 00 per doz.	15
110x	Fork, not il'st'd	12 00 per doz.	
115	Frost	16 00 per doz.	15
120	Fruit	16 00 per doz.	16
125	Frank	18 00 per doz.	16
126	Flinch	22 00 per doz.	12
200	Sack	15 00 per doz.	26
210	Saul	17 50 per doz.	27
211	Slave	22 50 per doz.	27
250	Sago	17 50 per doz.	22
260	Soap	20 00 per doz.	23
261	Satan	24 00 per doz.	23
265	Sauce	24 00 per doz.	18
270	Sense	24 00 per doz.	24
271	Small	30 00 per doz.	24
325	Saber	15 00 per doz.	19
330	Sheep	18 00 per doz.	20
331	Slime	22 00 per doz.	20
335	Shoot	21 00 per doz.	20
336	Smile	25 00 per doz.	20
350	Sign	24 00 per doz.	30
351	Steep	28 00 per doz.	30
360	Siren	30 00 per doz.	28
361	Smart	34 00 per doz.	28
400	Snake	36 00 per doz.	31
401	Soul	42 00 per doz.	31
410	Solid	42 00 per doz.	32
411	Silk	48 00 per doz.	32

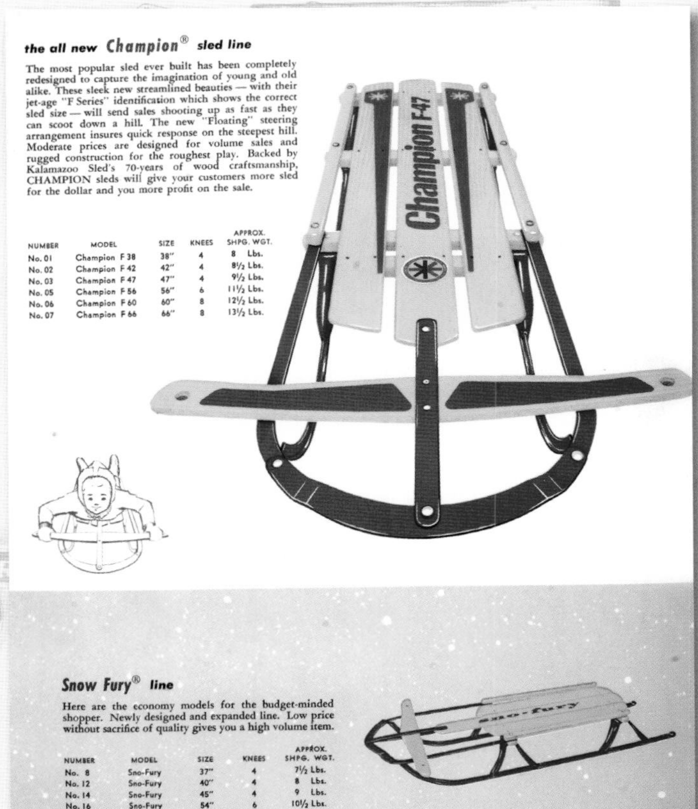

the all new Champion® sled line

The most popular sled ever built has been completely redesigned to capture the imagination of young and old alike. These sleek new streamlined beauties — with their jet-age "F Series" identification which shows the correct sled size — will send sales shooting up as fast as they can scoot down a hill. The new "Floating" steering arrangement insures quick response on the steepest hill. Moderate prices are designed for volume sales and rugged construction for the roughest play. Backed by Kalamazoo Sled's 70-years of wood craftsmanship, CHAMPION sleds will give your customers more sled for the dollar and you more profit on the sale.

NUMBER	MODEL	SIZE	KNEES	APPROX. SHPG. WGT.
No. 01	Champion F 38	38"	4	8 Lbs.
No. 02	Champion F 42	42"	4	8½ Lbs.
No. 03	Champion F 47	47"	4	9½ Lbs.
No. 05	Champion F 56	56"	6	11½ Lbs.
No. 06	Champion F 60	60"	8	12½ Lbs.
No. 07	Champion F 66	66"	8	13½ Lbs.

Snow Fury® line

Here are the economy models for the budget-minded shopper. Newly designed and expanded line. Low price without sacrifice of quality gives you a high volume item.

NUMBER	MODEL	SIZE	KNEES	APPROX. SHPG. WGT.
No. 8	Sno-Fury	37"	4	7½ Lbs.
No. 12	Sno-Fury	40"	4	8 Lbs.
No. 14	Sno-Fury	45"	4	9 Lbs.
No. 16	Sno-Fury	54"	6	10½ Lbs.

Kalamazoo Sled Company, 1964 catalog. $25-50.

Champion Mark® series

Here is the quality line in sleds — they'll make their sales mark wherever snow falls. Newly designed with easy steering, the MARK series are sleekly streamlined in the finest hardwood with bright chrome trim. New graphic designs provide colorful eye-catching appeal. There's a size to fit every youngster and the new MARK series model numbers give the size right on the sled.
Stock the Champion MARK series and watch profits rise.

NUMBER	MODEL	SIZE	KNEES	APPROX. SHPG. WGT.
No. 110	Champion Mark 48	48"	4	10 Lbs.
No. 114	Champion Mark 56	56"	6	12 Lbs.
No. 116	Champion Mark 60	60"	8	13½ Lbs.
No. 118	Champion Mark 66	66"	8	14½ Lbs.

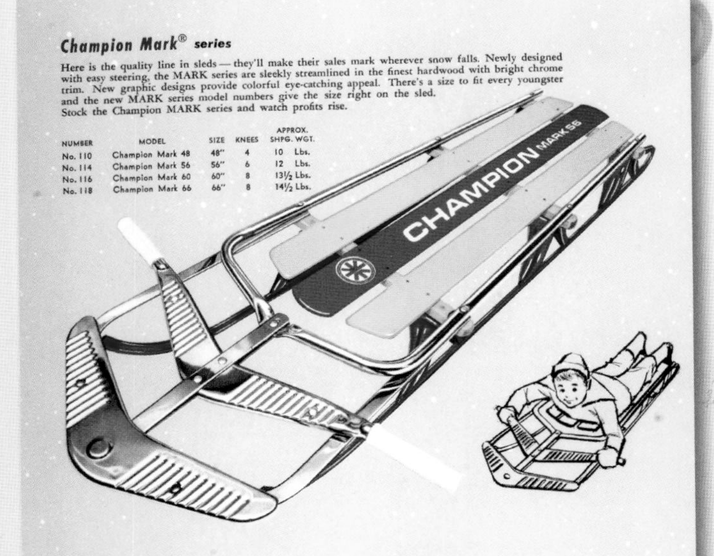

Champion® Flying-Disc

Filled with fun for even the smallest child. Molded of tough, durable Fiberglass — dents pop right out. Bright orange finish won't rub or wear off. Heavy webbed handle straps.

Model	Size	Approx. Shpg. Wt.	Pack
No. 100	26" Dia.	24 Lbs./Ctn.	6/Ctn.

extra large Snow Disc®

★ Molded ruby red color — won't rub off ★ Large 30" dia. — big enough for two children ★ Eye-catching point of sale label ★ Lightweight, tough, polyethylene ★ Withstands coldest weather... ★ Promotionally priced.

Model	Size	Approx. Shpg. Wt.	Pack
No. 105	30" Dia.	27 Lbs./Ctn.	12/Ctn.

Snow Disc®

This Sno-Disc of molded Fiberglass is a hot item for cold weather. Ruby-red finish won't rub or wear off. No sharp edges to cut or catch clothing on.

Model	Size	Approx. Shpg. Wt.	Pack
No. 95	25" Dia.	38 Lbs./Ctn.	12/Ctn.

Champion Baby sled

Tops for tots, this newly redesigned 28" sled is a steady mover. Removable guard and bright plastic pull handle. Red enameled runners and clear varnished wood parts.

MODEL	SIZE	APPROX. SHPG. WT.	PACK
No. 500	28"	9 Lbs.	1/Carton

Parcel Postable

Champion Sled Guard

Clamps easily onto most sleds for small tots' protection. Smooth, glossy finish of durable spar varnish withstands winter weather. Red enameled back support.

MODEL	APPROX. SHPG. WT.	PACK	
No. 501	14 Lbs./Carton	6/Carton	Parcel Postable

Baby Sled with wheels

★ Sleek Champion 2-color design
★ Fully retractable rubber tired wheels
★ Tubular steel push-type handle
★ Easy up and down wheel positioning
★ All wood guard, varnish finished
★ Red enameled runners

MODEL	APPROX. SHPG. WT.	PACK
No. 503A	13 Lbs./Carton	1/Carton

Length 30" Width 13"
Parcel Postable

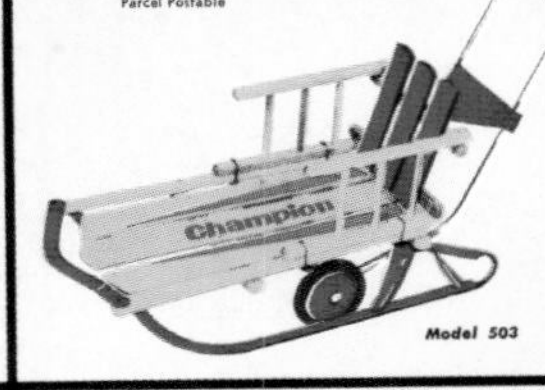

Model 503

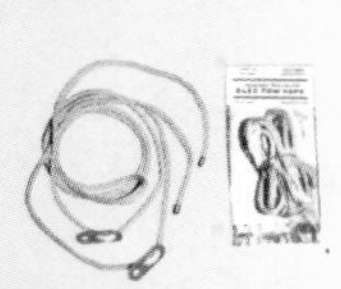

Tow Rope Model 505

Produced from high quality polyethylene. Sell one with each sled. Adjusts to fit any size sled. Highly resistant to weather. Individually packaged in poly bags; 24 bags per counter display carton.

MODEL	SIZE	APPROX. SHPG. WGT.	PACK
No. 505	9'	4 Lbs./Carton	24/Carton

Parcel Postable

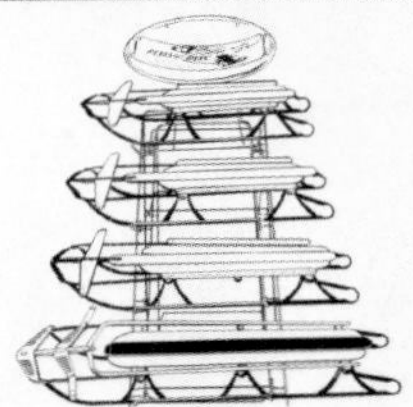

Eye Level Floor Display Model 600

Durably built, all-metal display stand holds up to 8 sleds, or other merchandise. Folds flat for storage. Width, 21". Height, 52". Weight, 24 lbs. Packed 1 per carton.

Sno-Boggan for snow fun Model 515

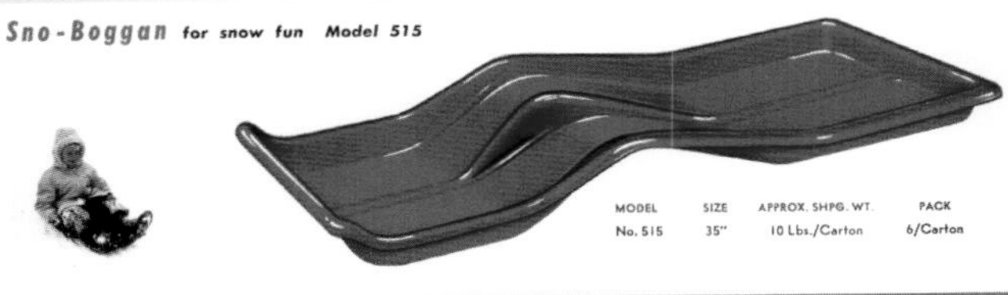

MODEL	SIZE	APPROX. SHPG. WT.	PACK
No. 515	35"	10 Lbs./Carton	6/Carton

the **CHAMPION** KZO

SWIFT AND SLEEK

A new, exciting addition to the Champion line!! Sleek — streamlined — with "floating steering" and a new modern look. The newest in children's sleds!! Sparkling chrome plated fronts and swept wing steering bar. Rugged hardwood decks — hi-carbon, red enameled steel runners — grooved for "tracking". Add more profits to your sled sales — order these best selling sizes.

NUMBER	MODEL	SIZE	KNEES	APPROX. SHPG. WGT.
NO. 103	CHAMPION KZO	47"	4	10 Lbs.
NO. 105	CHAMPION KZO	56"	6	12 Lbs.
NO. 106	CHAMPION KZO	60"	8	13 Lbs.

KALAMAZOO SLED & TOYS, INC. 844 CROSSTOWN PARKWAY, EAST / KALAMAZOO, MICHIGAN PHONE 344-6114 AC 616 49001

Champion KZ and Sno-Tryke, 1964 Kalamazoo Sled Company advertisement. $10-15.

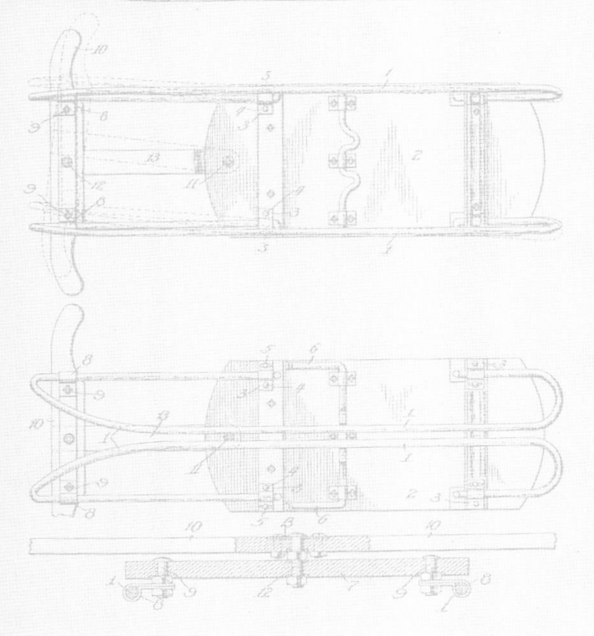

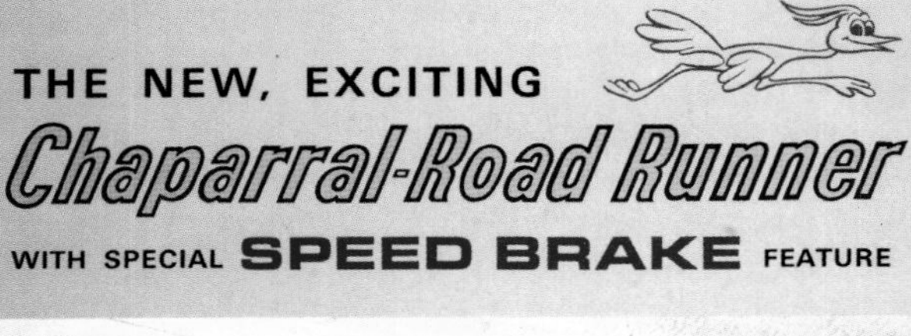

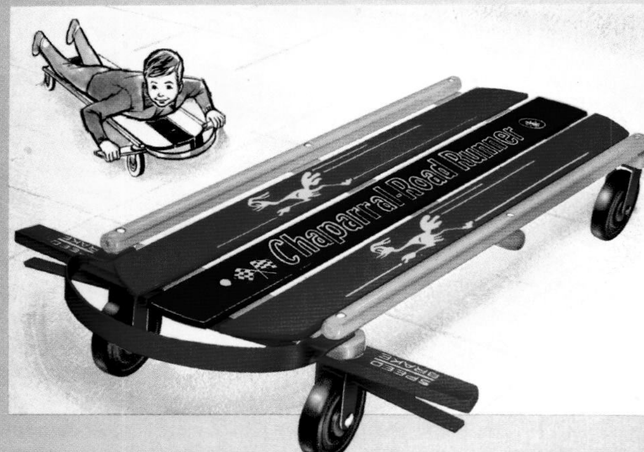

INTRODUCING *"Chaparral Road Runner"*

FROM THE KALAMAZOO SLED COMPANY

■ Built with traditional Kalamazoo Sled care, top quality materials, and featuring a new specially designed speed brake. This sled promises to be the most exciting toy of the year. Get the jump on sales and competition. Order today.

KALAMAZOO SLED & TOYS, INC.
844 CROSSTOWN PARKWAY, EAST KALAMAZOO, MICHIGAN
PHONE 344-6114 AC 616 49001

ONLY

Chaparral Road Runner

OFFERS THESE IMPORTANT FEATURES

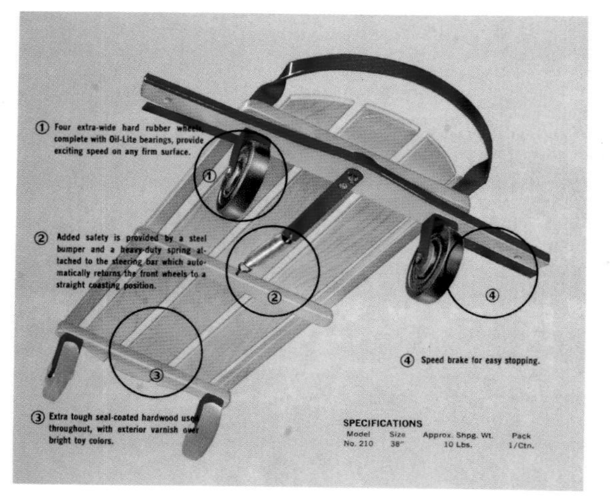

STOCK - DISPLAY And - SELL

The "Chaparral Road Runner" ■ A real profit maker

Chaparral-Road Runner, 1964 Kalamazoo Sled Company advertisement. $10-25.

GLADDING-KALAMAZOO

Champion® Quality Sleds

Our 76th Anniversary

GLADDING-KALAMAZOO

Champion® "Blue Racers"

Newest in Sled Profit-Makers from the World's Largest Quality Manufacturer

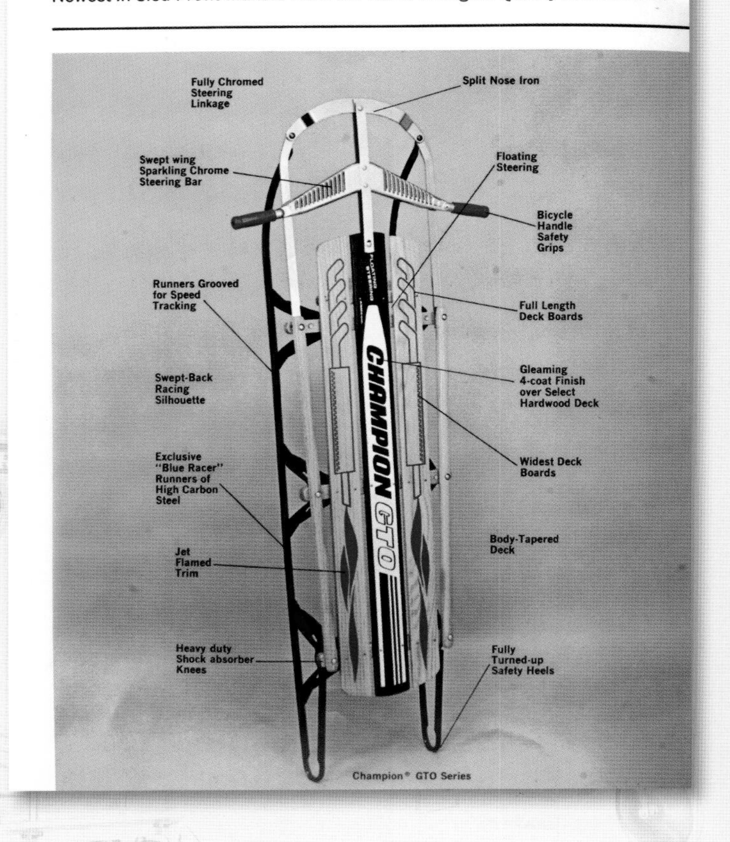

Champion® GTO Series

Gladding-Kalamazoo 1970 catalog. Note "Blue Racers'" Champion GTO deck. Gladding does not appear on the deck. $15-20.

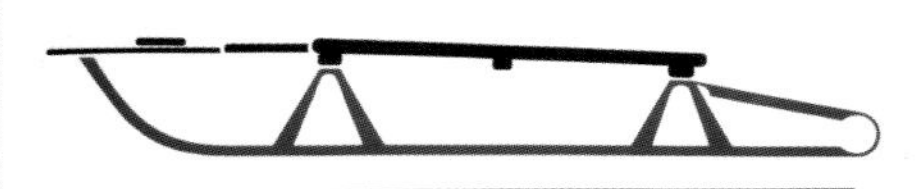

Gladding-Kalamazoo offers the world's most wanted sleds, PLUS geared-to-the-weather service—a combination that can't be topped for sure-fire sales. Because you can't depend on the weather, depend on Gladding-Kalamazoo to deliver your selection of famous Champion Sleds *when you need them.* And, our great new 75th anniversary line offers numerous features not found in other sleds.

Champion® GTO Series

Introducing the **exclusive "Blue Racers"**—a new standard of performance and prestige, with the features that generate selling excitement and profitable sales volume. The Champion GTO's modern design and styling are blended with 75 years of sled-making know-how to produce America's finest line of sleds. Seven popular sizes.

Style Number	Size	Shipping Weight (Each)
102	41"	9.0 lbs.
103	45"	10.0 lbs.
104	50"	11.5 lbs.
105	55"	13.0 lbs.
106	59"	14.5 lbs.
107	63"	15.5 lbs.

Champion® Fastback Series

America's best-selling sled line, now with **exclusive "Blue Racer"** runners and racing styling. "AIR-SPEED INDICATOR," "TACHOMETER" and "FUEL GAUGE" with "RACING FLAG" select hardwood steering bar. All the special features of the GTO, except chrome. Guaranteed to build record sales with more features than any other sled. Every Champion is backed by the famous Gladding-Kalamazoo reputation for quality construction and fastest service—geared to the weather.

Style Number	Size	Shipping Weight (Each)
01	37"	7.5 lbs.
02	41"	8.5 lbs.
03	45"	9.5 lbs.
04	50"	11.0 lbs.
05	55"	13.0 lbs.
06	59"	14.0 lbs.
07	63"	15.0 lbs.

Rocket Racer Series

Newest model in economy sledding. "Speed Striped" centerboard, grooved runners, full safety heel construction and red enameled, steel parts. A value-packed, volume item designed to offer every competitive advantage.

Style Number	Size	Shipping Weight (Each)
636	36"	7.0 lbs.
640	40"	7.75 lbs.
644	44"	8.75 lbs.
654	54"	10.75 lbs.
658	58"	11.75 lbs.

649 49" 9.75 LBS.

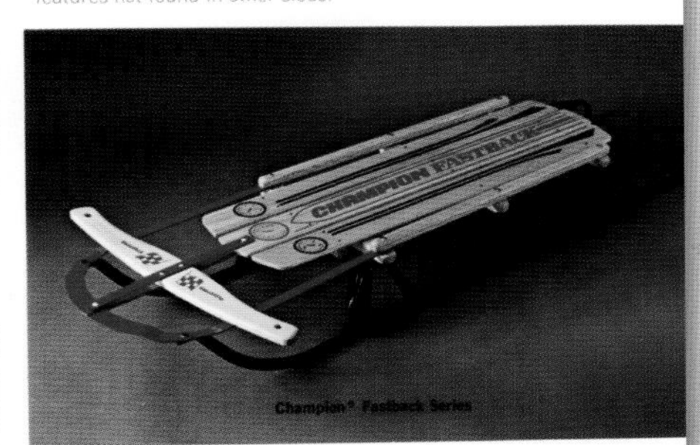

Champion® Fastback Series

Rocket Racer Series

Profit Producing Specialties

A. Champion® Pull-Handle Baby Sled
Strong pull handle. Quality sled guard. Individually cartoned and parcel postable. Model 500, size 28". App. Shpg. Wgt. 9¼ lbs.

B. Champion® Convertible Stroller Sled
When sidewalks are snowy, retract the wheels. When clear—let them down with easy wheel positioner. Gleaming tubular steel push handle, decorated sled guard. Individually cartoned and parcel postable. Model 503A, length 30", width 13". App. Shpg. Wgt. 13¼ lbs.

C. New Champion® Push-Type Baby Sled
Same great features as model A. above, BUT with bright metal push handle. Individually cartoned and parcel postable. Model 504, size 28". App. Shpg. Wgt. 11 lbs.

D. Champion® Sno-Tryke
The thrill of sledding with the excitement of skiing on three wood-enamelled wood skis. Sparkling handle bars, colored handle tassles, "see-through" windshield, sturdy seat. Individually packaged in self-selling carton that's parcel postable. Model 300, length 41". App. Shpg. Wgt. 10 lbs.

E. & F. Champion® Sled Guard and Tow Rope
Both great items in self-selling, display cartons. Sled guard clamps easily onto sled for tot's protection. Hardwood with weather-resistant varnish. Sno-flake decorated back support. (F) Adjustable tow rope fits all sleds. Tow rope of durable, colorful polyethylene. Both items parcel postable.
Sled Guard—
Model 501, packed 6 to carton. App. Shpg. Wgt. 15 lbs./carton.
Tow Rope—
Model 505, packed 24 to carton, size 9". App. Shpg. Wgt. 4 lbs./carton.

G. Champion Flying Disc® Deluxe
Snow sliding, turning and twisting fun on the original, Fiberglass "Flying Disc." Permanent Sno-Brite orange, light weight, safety handle straps, point-of-purchase label. Also available as Snow Disc, slightly smaller, without label. A real volume mover.

Model 100	26" dia.	6/ctn.	24 lbs./ctn.
Model 95	25" dia.	12/ctn.	38 lbs./ctn.

Champion® Sled Display Rack
The only sled display rack on the market—a real merchandising tool for increasing your sled sales. Holds eight sleds. Folds flat for storage. Model 600, height 52", width 21". Packed one to a carton. App. Shpg. Wgt. 24 lbs.

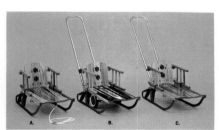

A. B. C.

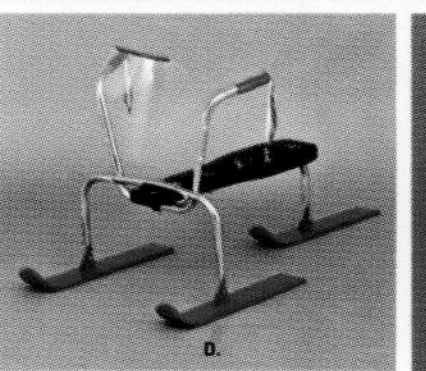

D.

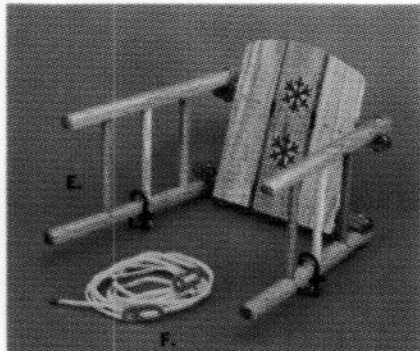

E. F.

G.

GLADDING-KALAMAZOO
Sled and Toy, Inc.
World's Largest Maker of Children's Sleds
844 Crosstown Parkway, East,
Kalamazoo, Michigan 49001
Phone: (616) 344-6114

Division of GLADDING Corporation
America's First Name in Recreation Products

GLADDING-KALAMAZOO
Division of GLADDING Corporation

1971
Champion® Quality Sleds
OUR "77th" ANNIVERSARY

Gladding-Kalamazoo 1971 catalog. Note "Blue Racers'" Champion GTO. Gladding appears on the deck. American Clipper Sled introduced. $15-20.

GLADDING-KALAMAZOO
Champion® "Blue Racers"
Nationally Advertised. They won't go out without Gladding-Champion. We're seeing to it.

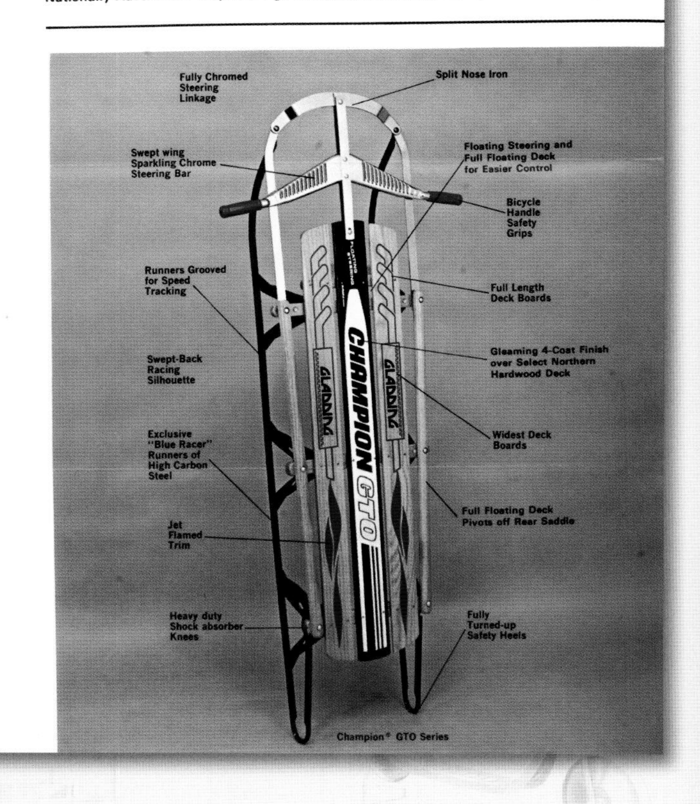

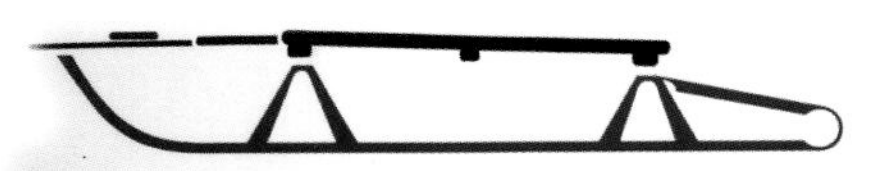

Galdding-Kalamazoo sleds are the World's most wanted, backed by a major national advertising campaign and by our famous "Geared-to-the Weather" service. They're part of the famous Gladding family of quality recreation products, known for great value, profitability and appeal. This year, we offer more than ever before. More value, more advertising, more sales appeal.

Gladding Champion® GTO Series

More exciting features than any other sled. (See features at left.) Our exclusive "Blue Racers" have established a standard of performance and prestige that generates top dollar sled sales volume. The Champion GTO's modern design and styling combine with 77 years of sled-making know-how to produce America's finest sled line. Six popular sizes.

Style Number	Size	Shipping Weight (Each)
102	41"	9.0 lbs.
103	45"	10.0 lbs.
104	50"	11.5 lbs.
105	55"	13.0 lbs.
106	59"	14.5 lbs.

Champion® Fastback Series

America's best-selling sled line, now with **exclusive "Blue Racer"** runners and racing styling. "AIR-SPEED INDICATOR," "TACHOMETER" and "FUEL GAUGE" with "RACING FLAG" select hardwood steering bar. All the special features of the GTO, except chrome. Guaranteed to build record sales with more features than any other sled. Every Champion is backed by the famous Gladding-Kalamazoo reputation for quality construction and fastest service — geared to the weather.

Style Number	Size	Shipping Weight (Each)
01	37"	7.5 lbs.
02	41"	8.5 lbs.
03	45"	9.5 lbs.
04	50"	11.0 lbs.
05	55"	13.0 lbs.
06	59"	14.0 lbs.

New American Clipper T.M.

Like its nautical forerunner, it offers smooth, fast "sale-ing." Loaded with value. Fully decorated deck. Grooved, steel runners with full safety heels. Sturdy deck boards and steering. It's "Sales Ahoy!" with this volume item designed to offer every competitive edge.

Style Number	Size	Shipping Weight (Each)
636	36"	7.0 lbs.
640	40"	7.75 lbs.
644	44"	8.75 lbs.
649	49"	9.75 lbs.
654	54"	10.75 lbs.
658	58"	11.75 lbs.

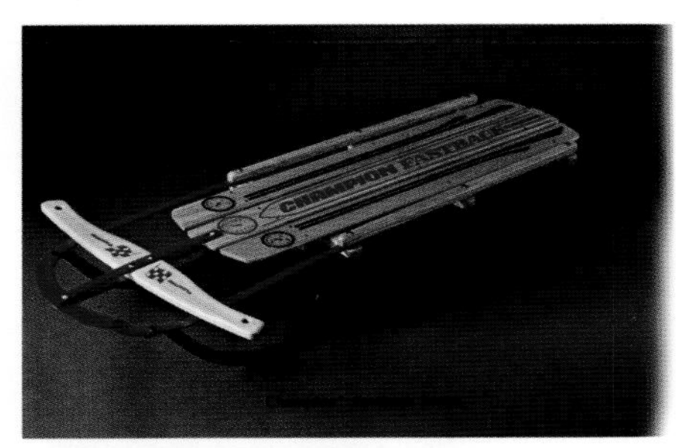

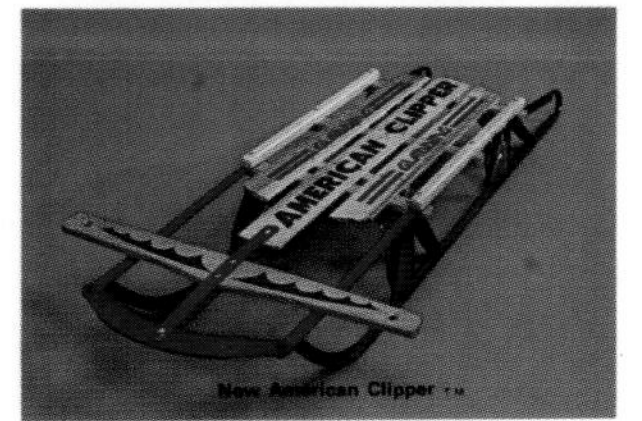

New American Clipper T.M.

Profit Producing Specialties

A. Champion® Flying Disc Deluxe
Snow sliding, turning and twisting fun on the original, Fiberglass "Flying Disc." Permanent Sno-Brite orange, light weight, safety handle straps, point-of-purchase label. Also available as Snow Disc, slightly smaller, without label. A real volume mover.
Model 100 26" dia. 6/ctn. 24 lbs./ctn.
Model 95 25" dia. 12/ctn. 38 lbs./ctn.

B. Champion® Push/Pull Baby Sled
Sturdy baby sled with guard has handle that pivots front or back so sled can be pushed or pulled. Each cartoned and parcel postable. Model 507. Size 30". Weight 11 lbs.

C. Champion® Convertible Stroller Sled
When sidewalks are snowy, retract the wheels. When clear — let them down with easy wheel positioner. Gleaming tubular steel push handle, decorated sled guard. Individually cartoned and parcel postable. Model 503A, length 30", width 13". App. Shpg. Wgt. 13¼ lbs.

D. Champion® Sled Guard and Tow Rope
Both great items in self-selling, display cartons. Sled guard clamps easily onto sled for tot's protection. Hardwood with weather-resistant varnish. Sno-flake decorated back support. (D) Adjustable tow rope fits all sleds. Tow rope of durable, colorful polyethylene. Both items parcel postable.
Sled Guard—
Model 501, packed 6 to carton. App. Shpg. Wgt. 15 lbs./carton.
Tow Rope—
Model 505, packed 24 to carton, size 9". App. Shpg. Wgt. 4 lbs./carton.

E. Champion® Sled Display Rack
The only sled display rack on the market — a real merchandising tool for increasing your sled sales. Holds eight sleds. Folds flat for storage. Model 600, height 52", width 21". Packed one to a carton. App. Shpg. Wgt. 24 lbs.

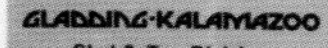

Sled & Toy Division
GLADDING Corporation
844 Crosstown Parkway East
Kalamazoo, Michigan 49001
Phone: (616) 344-6114
Gladding--Pioneers in outdoor recreation, since 1816

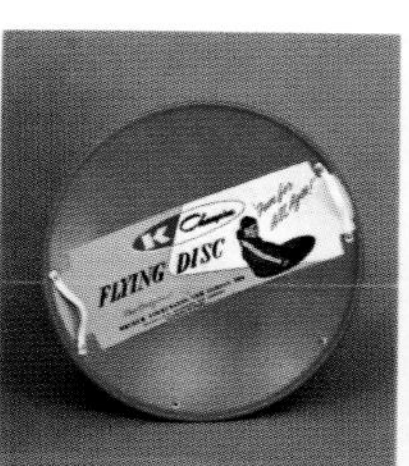

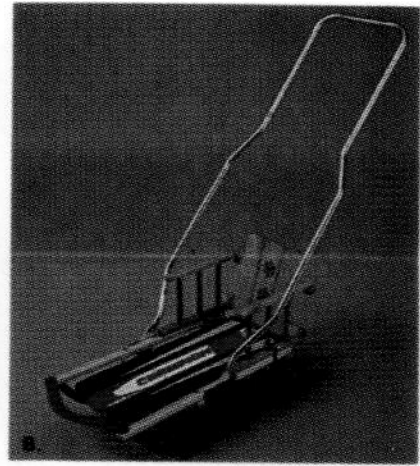

GTO, the one to show

Setting sales and performance standards for 75 years.

Gladding Champion® "Blue Racers" have set a performance and sales standard for 78 years. Now, these years of sled-making and marketing expertise bring you the most modern design and styling in America's finest sled line. Gladding still sets the goal for the competition to attempt to meet.

All Champion® GTOs feature: fully-chromed steering linkage, swept wing sparkling chrome steering bars, grooved runners for speed tracking, swept-back racing silhouette, exclusive "Blue Racer" high carbon steel runners, jet flamed trim, heavy-duty shock absorber knees, split nose iron, floating steering and full floating deck for easier control, heavy-duty bicycle handle safety grips, full length deck boards, fully turned-up safety heels.

Gladding still puts a gleaming four-coat finish over the Select Northern Hardwood Deck—the final touch a standard setter like the Champion® GTO merits. That's the sort of care America's oldest leisure products manufacturer puts into a sled.

GLADDING
Champion®
GTO

GLADDING

Gladding-Champion GTO 1972 company advertisement. $5-10.

Champion, because it is

Winning styling, old fashioned quality.

Gladding's Champion® Fastback isn't America's best-selling sled line by accident. We designed it that way. With 112 years in the sled-making business, we know that parents buy safe sleds. We also know that parents are value-minded. So we give them both value and safety in Fastback.

But we also know kids. Fastback appeals to their sense of adventure with exclusive "Blue Racer" runners, these feature call-outs on the deck: "Air Speed Indicator," "Tachometer," "Fuel Gauge" plus the racing flags on the steering bar. No kid ever returned a Fastback the day after Christmas—he was too busy using it.

Fastbacks are safer because of their careful design and improved steering control. Tell that to buying moms and dads. Show them the floating steering, grooved runners for speed tracking, full-length deck-boards, four-coat finish over select Northern Hardwood and the widest deckboards available. That's the kind of value you expect—and get—from Gladding, America's oldest continuous leisure products manufacturer. Turn the page over for more details.

GLADDING
Champion®
Fastback

GLADDING

Gladding-Champion Fastback 1972 company advertisement. $5-10.

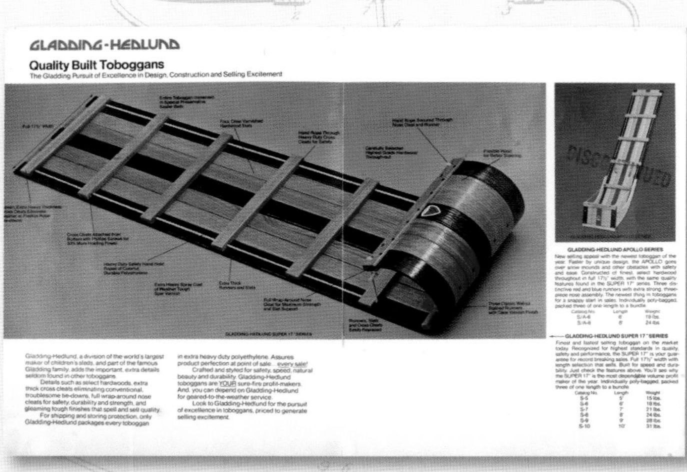

Gladding-Hedlund Quality Toboggans, Division of Gladding Kalamazoo, 1969 catalog. $5-7.

Gladding Snow-Flake Baby Sleds 1972 company advertisement. Note that it is the same as Champion Baby Sleds with changed graphics. $5-10.

Gladding-Hedlund Quality Toboggans, Division of Gladding Kalamazoo, 1970 catalog. $5-7.

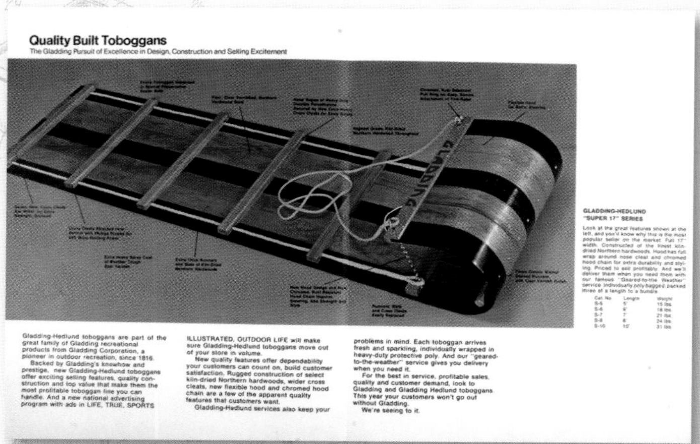

Gladding-Hedlund Quality Toboggans, Gladding Corporation, 1971 catalog. $5-7.

Paris Manufacturing Company

With the purchase of Paris by Gladding in 1970, many changes occurred. In April 1970, Paris Manufacturing released their latest addition to the Paris line of sleds, the "Speedy," joining the already popular Speedaways and Speed-Flex models. Pay close attention to the minor changes when dating these sleds. The 1970 Gladding Speedaway Series, at first glance, does not appear to be any different than the Paris Speedaway, but take a closer look and you will see the letters "Glad ding" in black on top of the red steering bar decoration, with a space in the middle. By 1972 the familiar Speedaway had only the letters "Glad ding" in a wavy pattern on the steering bar, still with the space. The most dramatic change happened the following year. The runners were converted to the distinctive safety heels of the Kalamazoo Line. "Gladding" now appeared on the steering bar in a straight line with the space eliminated. After purchasing Kalamazoo, Hedlund, & Paris, Gladding moved their sled operation to South Paris, Maine, placing more than two centuries of combined experience in sled making under one roof. It was here, in 1972, that the new "Gladding American Flyer" was made. The steering bar was taken from the Hedlund "American," the deck from the old Kalamazoo "Flying Arrow," the grooved runners and double-ribbed knees from the Gladding Speedaway, the safety heel design was that of the Gladding Champion, and, last but not least, the nose bumper of the Gladding Speedflex. The deck was adorned with Old Glory, an arrow and the words Gladding American Flyer, repeated on the steering bar. The Gladding American Flyer was offered as "a great new line from our great old sled makers," available in six sizes from 40" to 60". Gladding also introduced a Gladding Bob Sled called a "Glad-O-Bob" (formerly the "Bob-O-Link," by Withington) before disappearing from the scene. The Gladding Corporation sold to Mr. Henry Morton in 1978 ending decades of making "sno-fun."

Updated Chronology of the Paris Manufacturing Company

1970 • March, Gladding Corporation purchases the Paris Mfg.Co.

1970 • April, Speedy Models introduced, joining Speedaways and Speed Flex Models. Gladding adds their name to all Models by the close of 1970

1971 • Gladding moves all Sled manufacturing to South Paris Me.

1972 • Gladding American Flyer, along with the "Glad-O-Bob" introduced

1978 • Gladding Corporation sold and disappeared from the market

Paris Manufacturing Company, South Paris, Maine. Store photo, circa 1900. C.D. Kidder, proprietor.

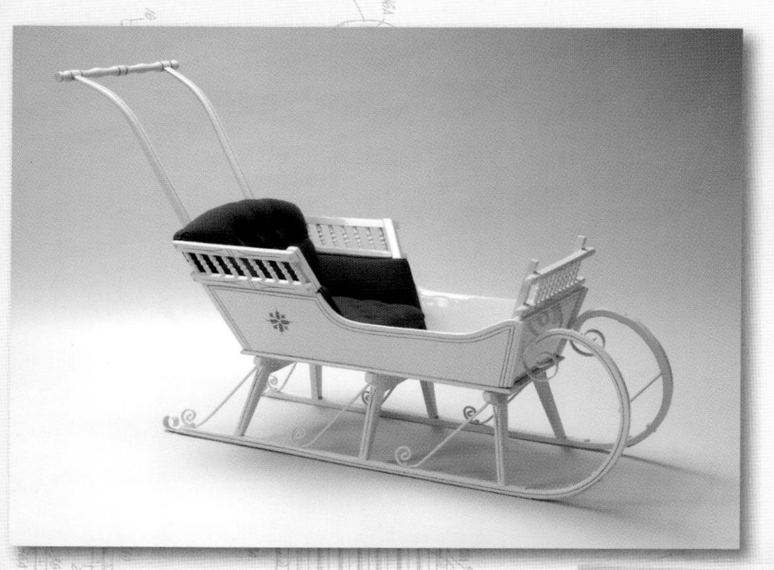

Paris Mfg. Co., c. 1880. White enamel push sleigh with green stenciling and upholstery. L 40", W 20", H 20". $1500-3200. *Courtesy of Carole and Lou Scudillo.*

Paris Mfg. Co., c. late 1860s. # 63 Bow Runner. Red deck with daisies. L 33" W 12", H 7-1/2". $1400-2900. Courtesy of Carole and Lou Scudillo.

Paris Mfg. Co., c. late 1860s. #63 Bow Runner, green deck with yellow rose. L 33", W 12" h 7 ½". $1250-2950. *Courtesy of Carole and Lou Scudillo.*

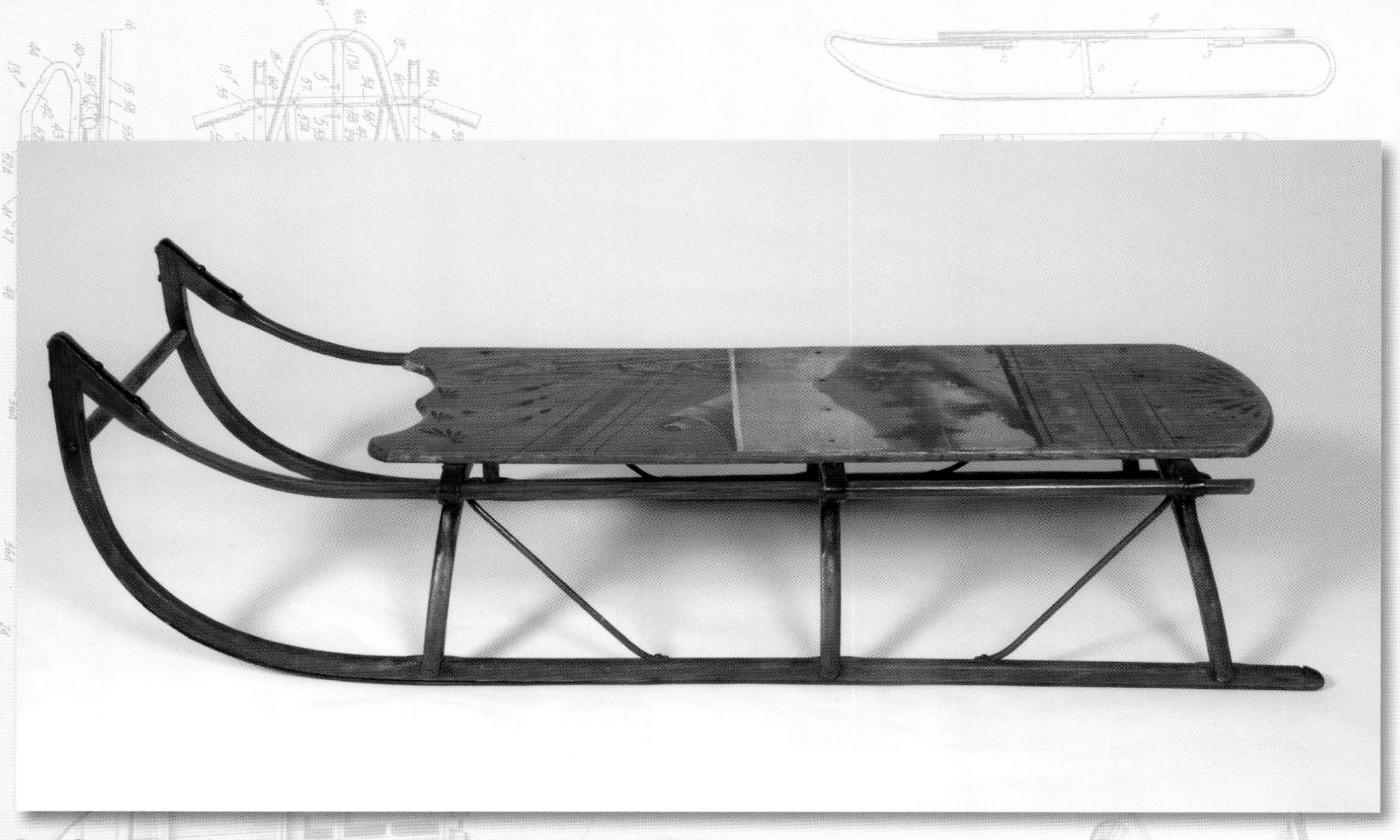

Paris Cutter #23A. L: 36"; W: 12"; H: 7-1/2". Scenic deck painted by William Morton. $1500-2500. *Courtesy of Paul Cote.*

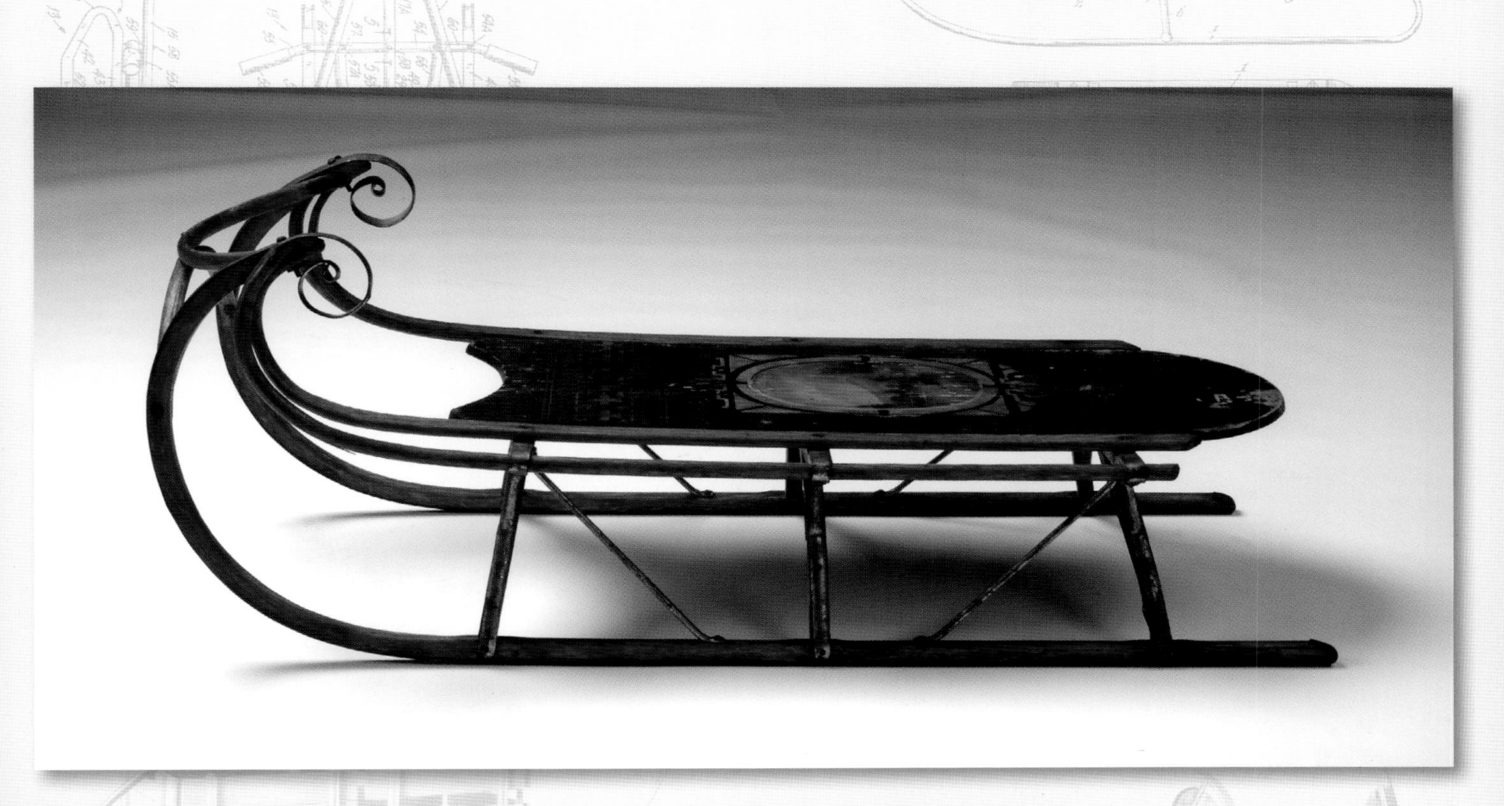

Paris Mfg. Co. Cutter, c. 1870.Green deck with moutain scene. L: 36", W: 15-1/2" H: 8". $1500-3500.
Courtesy of Carol & Lou Scudillo.

Paris Snow Fairy #82, c. 1900s. L:40"; w: 14"; H: 8-1/2". Described in the catalog as "a thing of beauty." Note two small bells attached to the runners. $3000-6000. . *Courtesy of Paul Cote.*

Paris Snow Fairy #82. L: 40"; W: 14"; H: 8-1/2". Painted by William Morton. $3000-6000. *Courtesy of Paul Cote.*

Paris Snow Fairy #85 with extremely rare Holly Deck. L: 36"; w: 13"; H: 8-1/2". Same as Snow Fairy #82 with the exception of having the seat upholstered in duck and finished in imitation lether. $2000-3000.
Courtesy of Paul Cote.

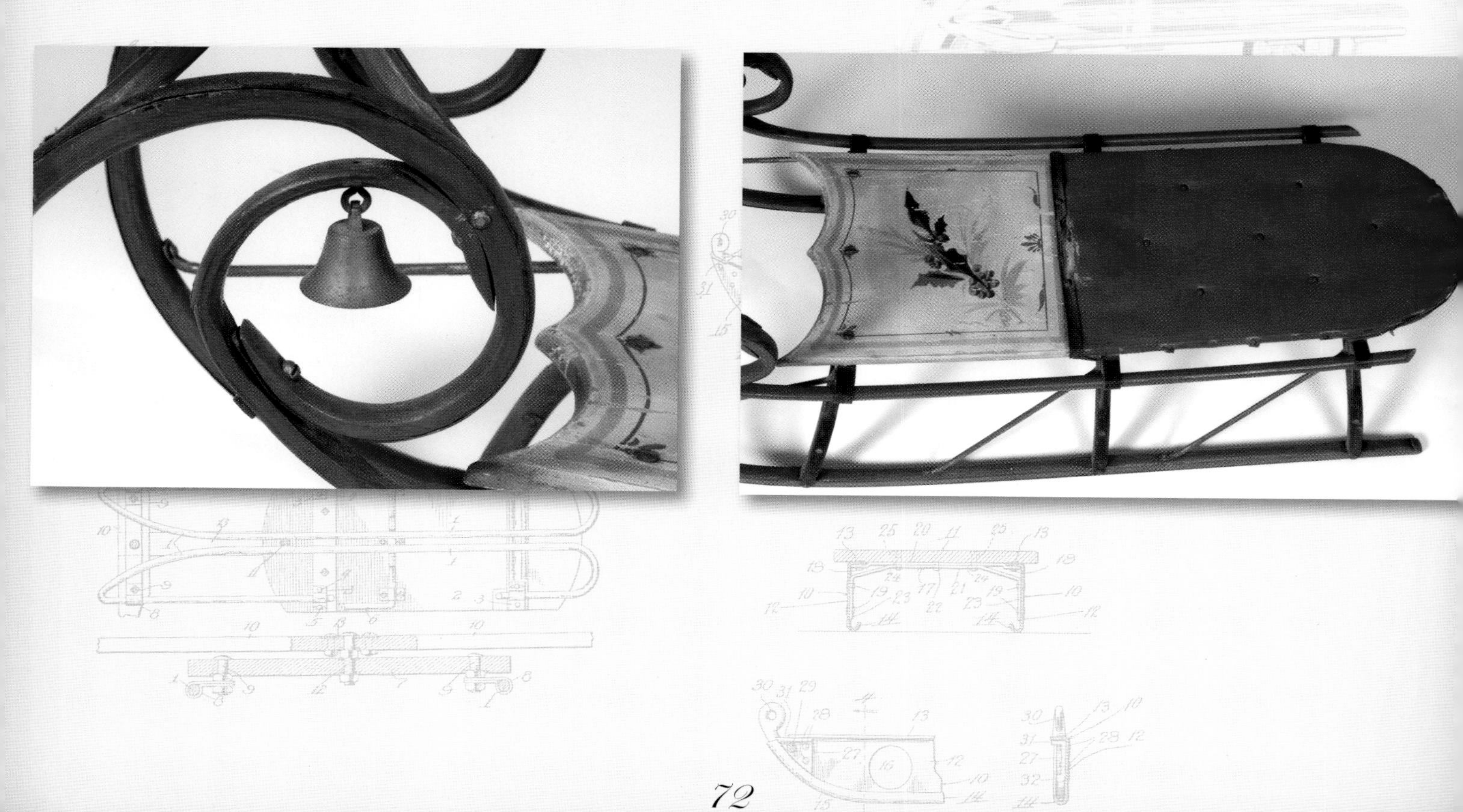

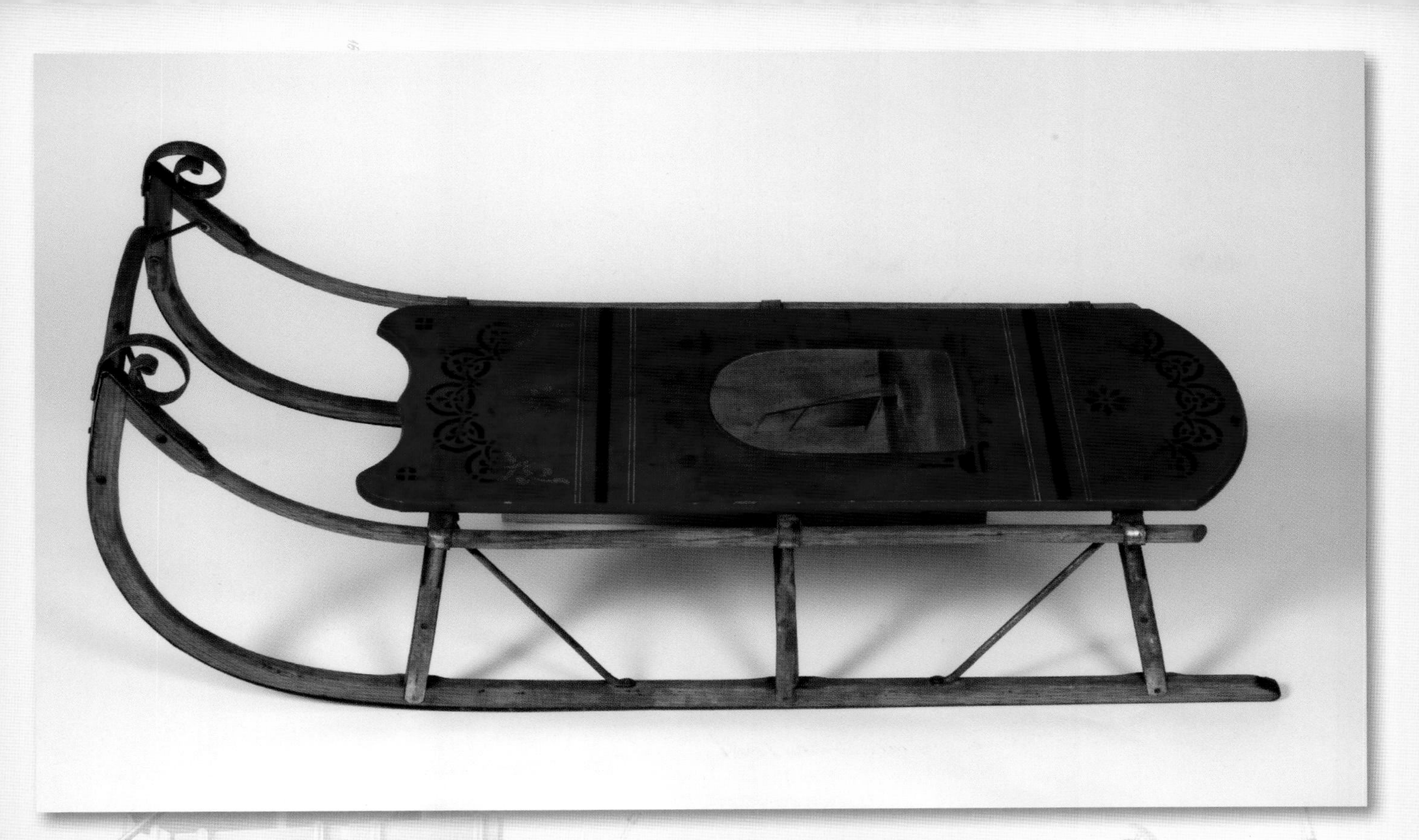

Paris Cutter, c. 1890. Painted with a sailboat by William Morton. $1500-2500. *Courtesy of Paul Cote.*

Paris Cutter #68. L: 36"; W: 12"; H: 7-1/2". Note rare background color (white) deck with morning glories and finches. $1500-2500. *Courtesy of Paul Cote.*

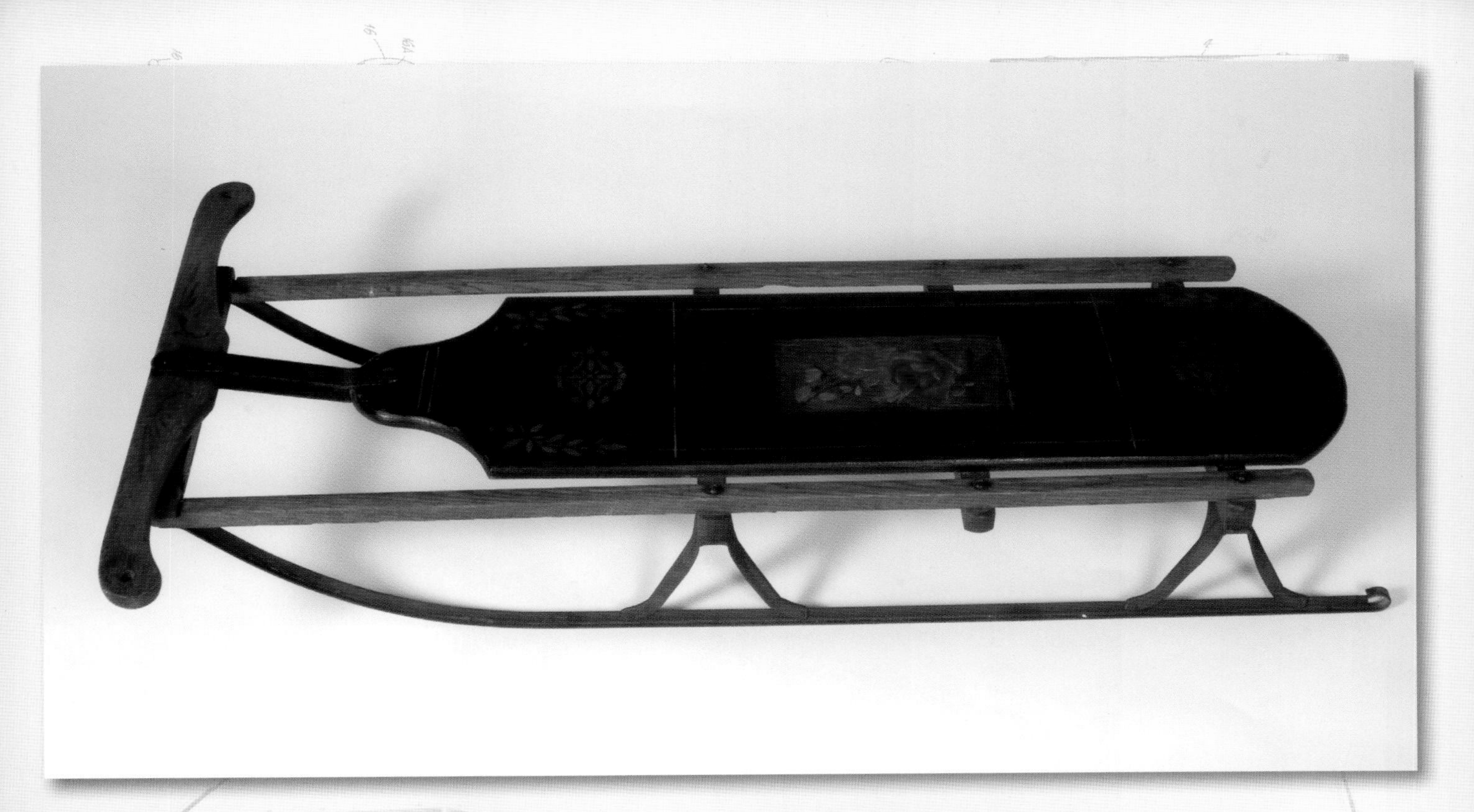

Paris Flyer #92, c. 1912. Floral design solid deck. Hall & Knight distributor. $900-1200. . *Courtesy of Paul Cote.*

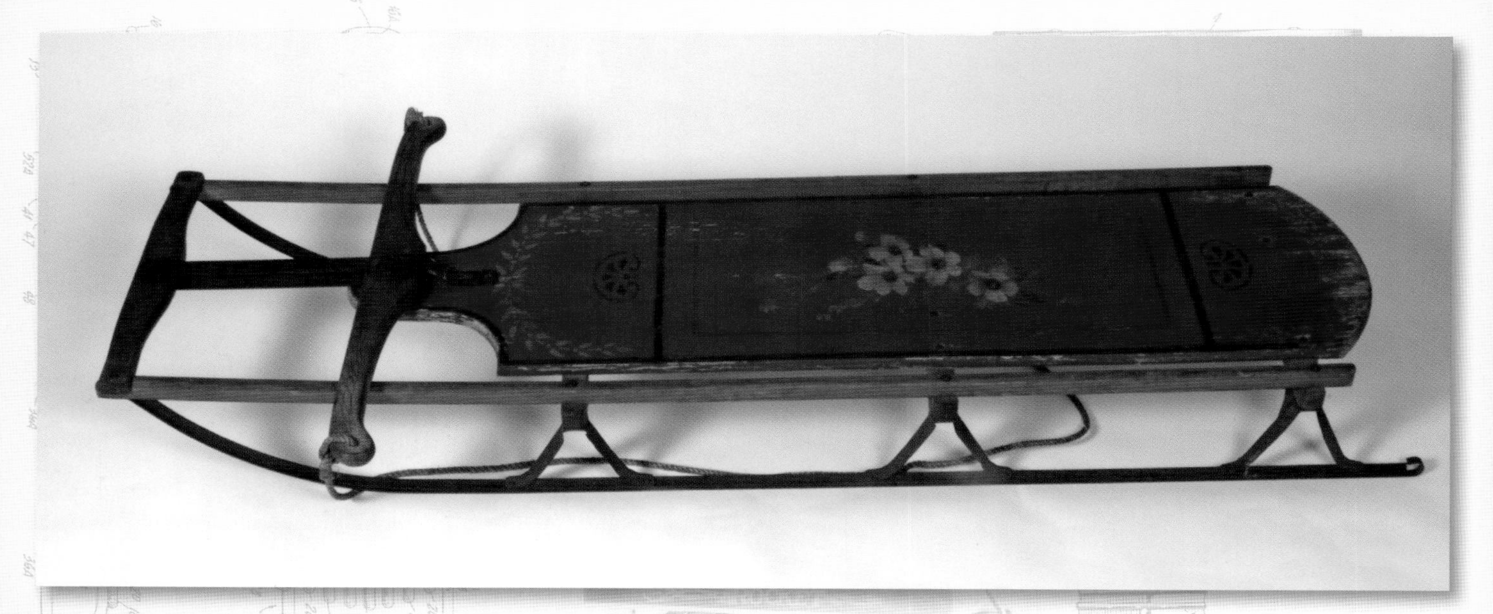

Paris Racer, c. 1912. Floral design solid deck. $900-1200. . *Courtesy of Paul Cote.*

Paris Racer #124, c. 1912. Wooden bumper. $1200-1500.
Courtesy of Paul Cote.

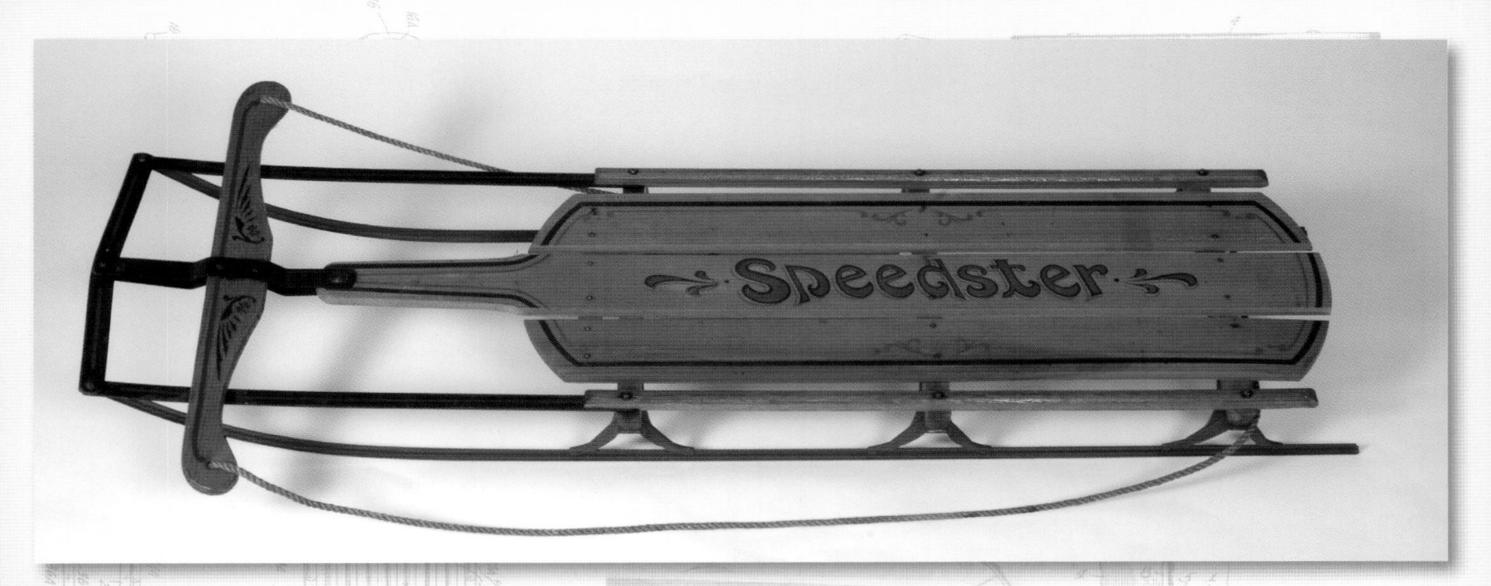

Paris Speedster #355, c. 1920. $150-$300. . *Courtesy of Paul Cote.*

Paris Racer #553, c. 1930. Round bumper. $200-250. . *Courtesy of Paul Cote.*

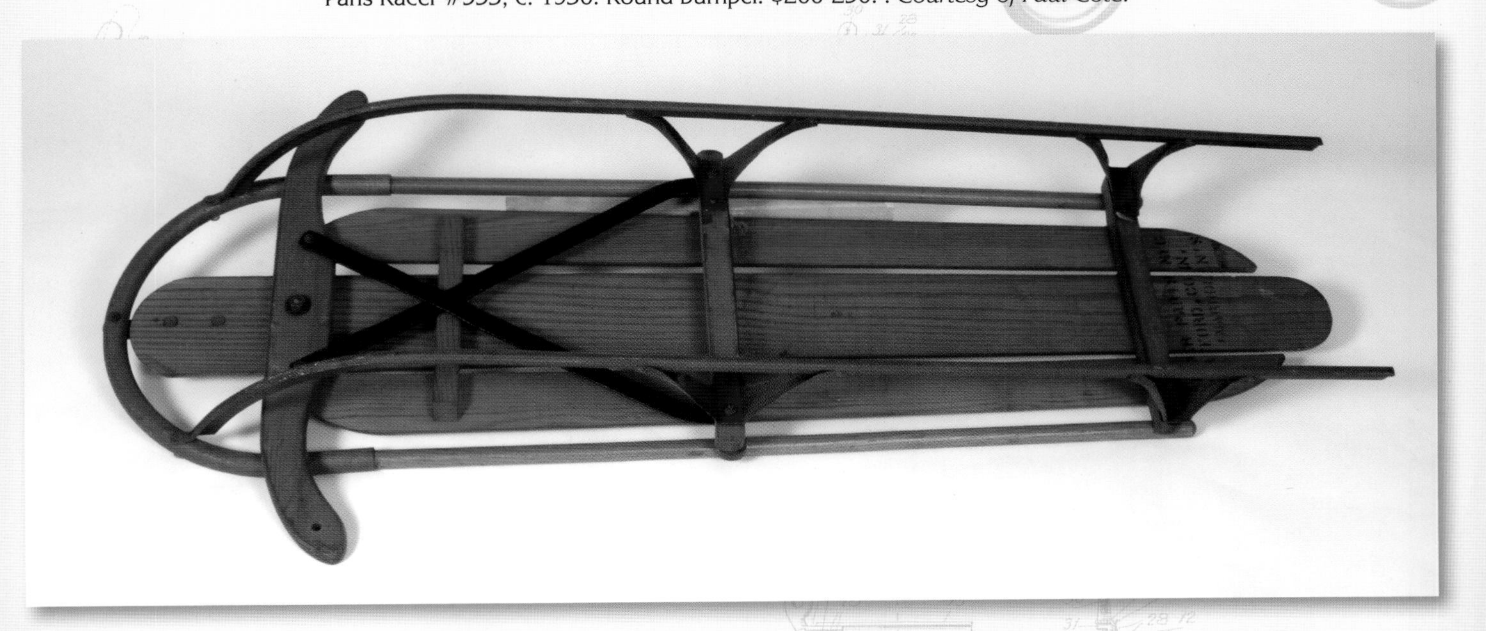

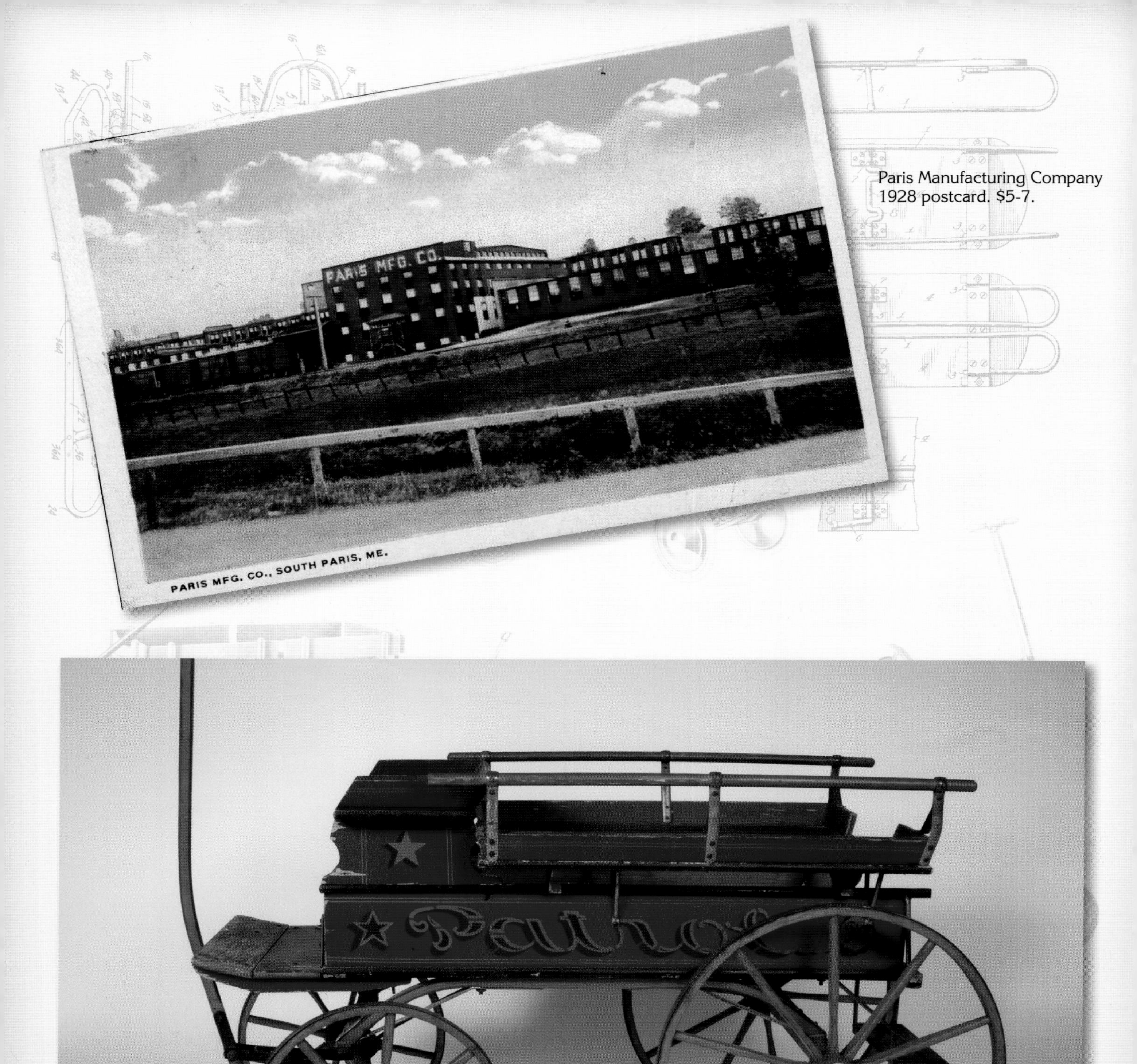

Paris Manufacturing Company 1928 postcard. $5-7.

Paris Patrol Wagon #89, C. 1890. Body size: 16" x 36". Wheels 15" & 21". $3000-5000.
Courtesy of Paul Cote.

Paris Oak Frame Ribbed Body Wagon #74, c. 1890. body size: 24" x 44". Wheels 18" & 24". Note horse on the dasher. $2500-4000. *Courtesy of Paul Cote.*

Paris Daisy Wagon #29-30, c. 1900. Body size: 13" x 26"; wheels 11" & 16-1/2"
Note PMCo logo. $1500-2000. *Courtesy of Paul Cote.*

Paris Farm Wagon with buggy seat, c. 1900. Body size: 18" x 36"; wheels 15" & 21". $1500-2500. *Courtesy of Paul Cote.*

Paris Express Wagon #26. Body size: 18" x 19"; wheels 7" & 10" $350-500. . *Courtesy of Paul Cote.*

Paris Express Wagon #100, c. 1900.
Body size: 18" x 36"; wheels 10" & 15".
$1000-1500. *Courtesy of Paul Cote.*

Paris Express Wagon #112, c. 1900. Body size: 18" x 36"; wheels 15" & 21". Complete with sideboards and dasher. $1500-2000. *Courtesy of Paul Cote.*

Paris Heavy Duty (Delivery) Wagon, c. 1920. Body size: 16" x 40"; 11" artillery wheels. $750-1250. *Courtesy of Paul Cote.*

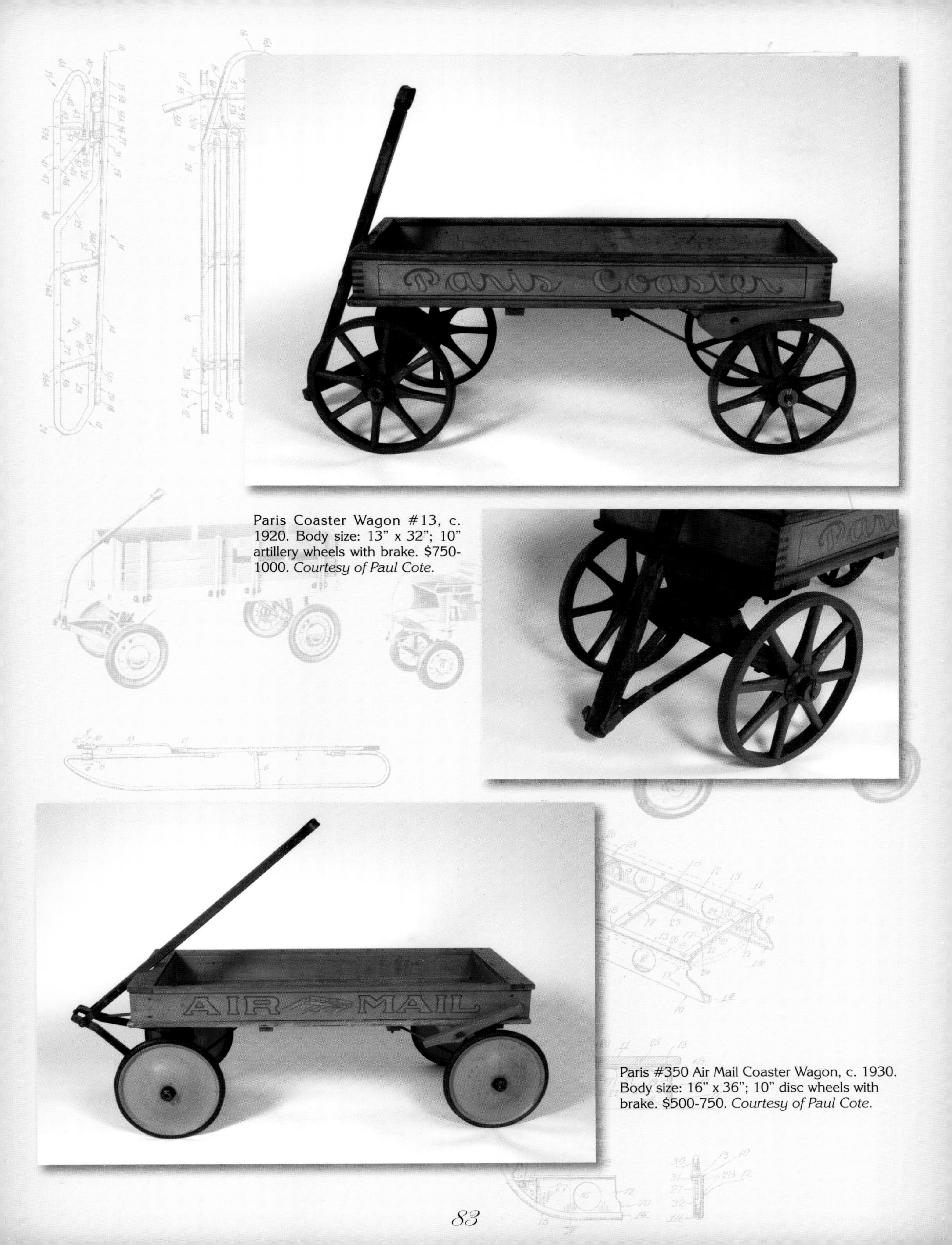

Paris Coaster Wagon #13, c. 1920. Body size: 13" x 32"; 10" artillery wheels with brake. $750-1000. *Courtesy of Paul Cote.*

Paris #350 Air Mail Coaster Wagon, c. 1930. Body size: 16" x 36"; 10" disc wheels with brake. $500-750. *Courtesy of Paul Cote.*

Paris Junior Coaster #10, c. 1930. Body size: 10" x 20"; wheels 5-1/2". $350-450. *Courtesy of Paul Cote.*

Paris Junior Coaster #12, c. 1930. Body size: 12" x 28"; wheels 5-1/2". $350-450. *Courtesy of Paul Cote.*

Paris Coaster Wagon #16, c. 1930. Body size: 18" x 40"; 12" wheels. Note foot pedal brake. $1000-1500. *Courtesy of Paul Cote.*

Buddy #5, c. 1930. Body size: 8-3/4" x 19"; 3-1/2" wheels. $350-450. . *Courtesy of Paul Cote.*

Paris Buddy Coaster Wagon #400, c. 1930. Body size: 15" x 36"; 10" wheel. Note Indian head on the front. $750-1000. *Courtesy of Paul Cote.*

Paris Olympic Coaster Wagon, c. 1930. Body size: 16" x 32"; 10" disc wheels. $3000-5000. *Courtesy of Paul Cote.*

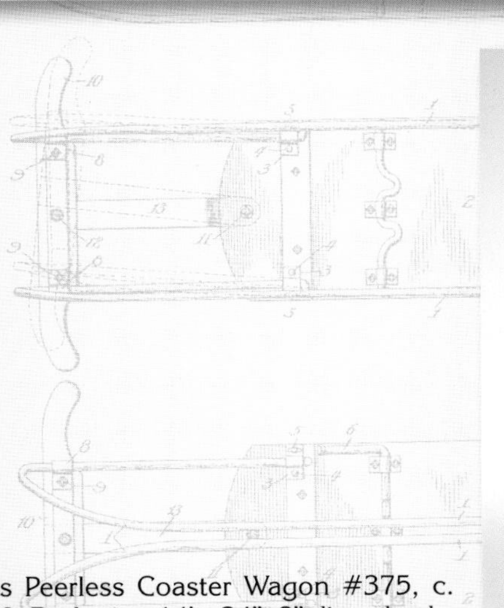

Paris Peerless Coaster Wagon #375, c. 1930. Body size: 14' x 34"; 8" disc wheels. $300-500. *Courtesy of Paul Cote.*

Paris Manufacturing Company 1970 catalog & April price list. $5-7.(continued on opposite page)

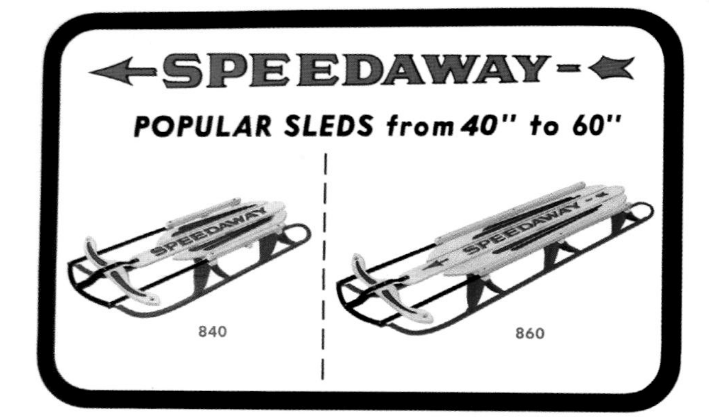

840 Length: 40 inches; 4 Knee; Weight: 8 1/2 lbs. per sled
844 Length: 44 inches; 4 Knee; Weight: 9 lbs. per sled
848 Length: 48 inches; 4 Knee; Weight: 10 lbs. per sled
852 Length: 52 inches; 6 Knee; Weight: 11 lbs. per sled
856 Length: 56 inches; 6 Knee; Weight: 12 lbs. per sled
860 Length: 60 inches; 6 Knee; Weight: 12 1/2 lbs. per sled

PARIS MANUFACTURING CO.
SOUTH PARIS • MAINE

More dealers choose this Best Seller

"SPEEDAWAY" is a household word

To generations of Americans, "Speedaway" is synonymous with "sled". Production of Paris sleds began in 1861—the year the Civil War began. Since then, millions have tasted the thrill of winter sports on Paris sleds. That's why they have confidence in buying Paris sleds for their own youngsters. They know Paris sleds are strong, sturdy and built to last. They are familiar with the skilled Maine craftsmanship of America's oldest sled manufacturer. Paris Speedaways are sleds that sell on their reputation.

all "SPEEDAWAY" sleds have:

- Two-piece nose irons which double runner flexibility and allow twice the steerability.
- Full turned-up runners on back of sled to eliminate sharp ends.
- An extra heavy, full-sized top.
- Oversize bars and side rails.
- A center bar.
- A cleat-reinforced top.
- New streamlined Full-Tapered design.
- All wooden parts are selected Northern White Ash with two coats of varnish.
- Runners are spring steel and all metal parts are finished in high grade baked-on enamel.

SLED GUARD

A sturdy Sled Guard for small children. Folds flat for storage. Fits any Speedaway Sled. Packed 1 or 6 per carton. Weight: 4 or 24 lbs.

SLEIGHETTE

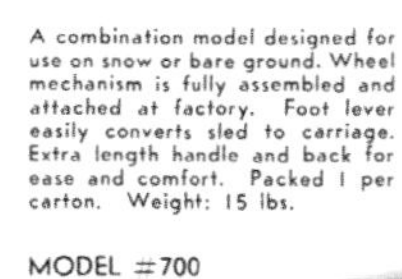

A combination model designed for use on snow or bare ground. Wheel mechanism is fully assembled and attached at factory. Foot lever easily converts sled to carriage. Extra length handle and back for ease and comfort. Packed 1 per carton. Weight: 15 lbs.

MODEL #700

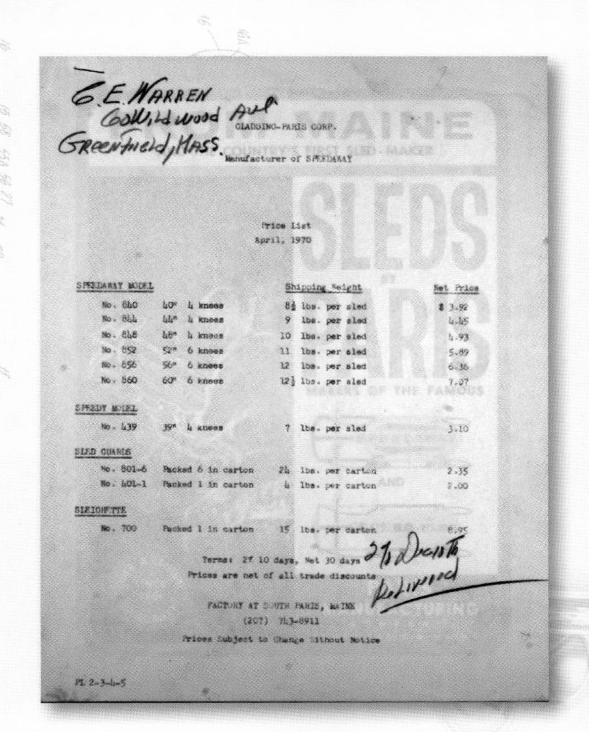

G.E. Warren
Collinwood Ave
Greenfield, Mass.

GLADDING-PARIS CORP.

Manufacturer of SPEEDAWAY

Price List
April, 1970

SPEEDAWAY MODEL			Shipping Weight	Net Price
No. 840	40"	4 knees	8½ lbs. per sled	$ 3.92
No. 844	44"	4 knees	9 lbs. per sled	4.45
No. 848	48"	4 knees	10 lbs. per sled	4.93
No. 852	52"	6 knees	11 lbs. per sled	5.89
No. 856	56"	6 knees	12 lbs. per sled	6.36
No. 860	60"	6 knees	12½ lbs. per sled	7.07
SPEEDY MODEL				
No. 439	39"	4 knees	7 lbs. per sled	3.10
SLED GUARDS				
No. 801-6	Packed 6 in carton		24 lbs. per carton	2.35
No. 401-1	Packed 1 in carton		4 lbs. per carton	2.00
SLEIGHETTE				
No. 700	Packed 1 in carton		15 lbs. per carton	8.95

Terms: 2% 10 days, Net 30 days 2% Discount
Prices are net of all trade discounts Delivered

FACTORY AT SOUTH PARIS, MAINE
(207) 743-8911

Prices Subject to Change Without Notice

PL 2-3-4-5

NEW
SPEEDY
MODEL #436 439
436
4 knee 36" long
Weight 7 lbs. per sled
LATEST ADDITION TO THE
PARIS FAMILY OF QUALITY SLEDS
FEATURING:
• All wooden parts are selected Northern White Ash.
• Oversize bars and side rails.
• A cleat-reinforced top.
• Two-piece nose irons which allow double runner flexibility and twice the steerability.
• New streamlined Full-Tapered design.
• Runners are spring steel and all metal parts are finished in high grade baked-on enamel.

MODEL #401
POPULAR SLED GUARD
A sturdy Sled Guard for small children . . . especially designed for SPEEDY. Folds flat for storage. Packed 1 per carton only. Weight: 4 lbs.
PARIS MANUFACTURING CO.
SOUTH PARIS, MAINE 04281
PHONE: (207) 743-8911

Speedaway Gladding Corporation 1970 catalog. $15-25.

CORPORATION

GLADDING-PARIS

Terms, Dating, Specifications & Price List

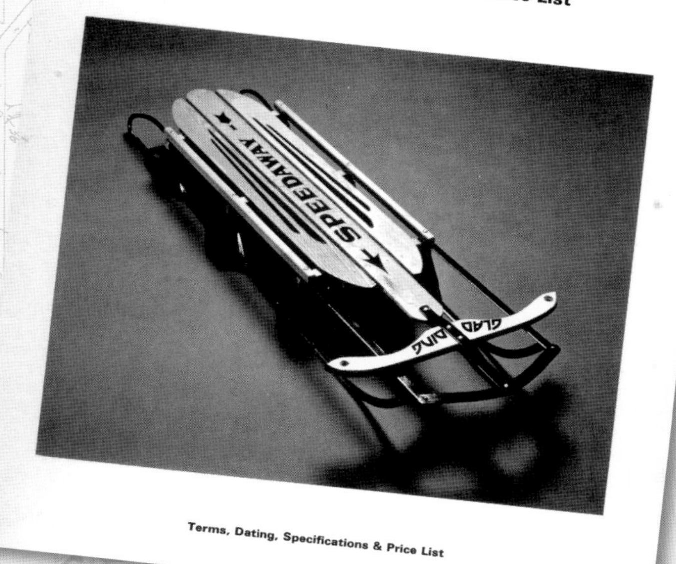

Terms, Dating, Specifications & Price List

Gladding-Paris, March 1972 catalog and price list. $5-7.

Here's a great sled from the great old sled makers.

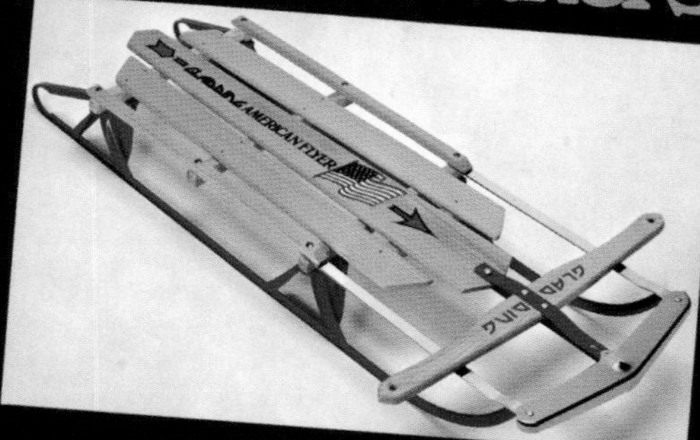

GLADDING
American Flyer

Combine the features developed and proven over the years by America's oldest sled maker (Paris Manufacturing Company, founded in 1861) together with Kalamazoo Sled & Toys Company, founded in 1893, the Hedlund Manufacturing Company, successor to the American Sled Company founded in 1946, put them all together in a great new plant and the Gladding American Flyer is what you get.

We placed the more than two centuries of combined experience of our great old sled makers under one modern roof—the 80-year-old Gladding Kalamazoo Division, the 27-year-old Gladding-Hedlund Division and the 112-year-old Gladding-Paris Division—and they've built a winner. It's a sled with the rugged steering bar of the famous old Hedlund "American", the sturdy deck of the old Kalamazoo "Flying Arrow", the proven grooved tracking runners and double-ribbed knees used for years on the old reliable Gladding "Speedaway®", the safety heel design and strength of the Gladding "Champion®" and the nose bumper of the Gladding "Speedflex."

We're making these great sleds at the world's most modern sled making facility, including an addition of 70,000 square feet of fully conveyorized floor space completed in March, 1972. This headquarters plant of the 157-year-old Gladding Wood and Youth Products Division, has an unsurpassed record of shipping controlled quality products on schedule with its "geared-to-the-weather" service. This, plus shipping from several strategic locations, means we can guarantee delivery of sleds in quantity when you need them. Let our great old sled makers give you a great new line of profitable merchandise. Turn the page for all the details.

GLADDING
Corporation
Wood and Youth Products Division
P.O. Box 250, South Paris, Maine 04281
First in Outdoor Recreation, Since 1816

Gladding American Flyer, 1972 company advertisement. $10-15.

Speedaway, doesn't coast on reputation

Traditional quality construction, popular styling.

"Speedaway" ® is synonymous with sled to millions of Americans. Production began in 1861 and has never stopped because parents have confidence in Speedaways.

Gladding, America's oldest continuous leisure products manufacturer, combines streamlined full-tapered design with traditional Speedaway® quality and safety. Just as sturdy as the originals because Gladding's quality control experts are traditionalists in the strictest sense of the word.

All wooden parts are still kiln-dried selected Northern Hardwoods with two coats of varnish. Speedaway® runners are spring steel. All metal parts are finished in high grade baked-on enamel.

These quality features are standard on all Speedaways: extra-heavy full-sized tops; oversize bar and side rails; center bar for floating steering; cleat-reinforced tops; two-piece nose irons which increase runner flexibility 100% and allow twice the steering ability; full turned-up safety runners at back to eliminate sharp ends.

Compare and you'll know Gladding isn't just coasting on past reputation. We live up to it . . . all the way.

GLADDING
Speedaway®

Gladding Speedaway, 1973 company advertisement. $10-15.

Standard Novelty Company

Established in 1904, the home of the "Lightning Guider," has truly stood the test of time. After eighty six years of sled production it closed its doors and emerged as The Old Sled Works. A unique antique mall, complete with an authentic 1950s soda fountain, penny arcade, and a museum dedicated to all products manufactured by the Standard Novelty Works.

When we hear the words Lightning Guider we immediately think of sleds, however they also manufactured wagons. It is not clear as to when wagon production began. A 1917 company catalog does not include wagons, however a company brochure circa 1920 illustrates two models: the Lightning Wheel Coaster in two sizes # 200 & #300 and the Lightning Speedster, a durable Stake Wagon. The sides and end gates were removable making it easy for coasting. In addition to wagons they produced a scooter not to be confused with a wheeled version. The Lightning Scooter has three runners, circa 1920. Last but not least, they manufactured a Lightning Wheel Guider. In all my years of collecting I had never seen one nor did I know of anyone who had one, until recently, when Skip Palmer contacted me. Thanks to Skip for sharing his mint condition Lightning Wheel Guider so everyone can see this wonderful example. Pictured in a 1911 catalog the ad reads: "Runs on the level with very little energy, turns around on an eight foot sidewalk, besides all these pleasant features, it can be used in many ways as an express wagon" Today thanks to Mr. Rosen we can enjoy the wonderful collection of sleds and sledding memorabilia preserved for generations to come at The Old Sled Works in Duncannon, Pennsylvania, a must see!

Standard Novelty Works, 1917 company catalog page featuring the Lightning Wheel Guider. $3-5 seen on page 90.

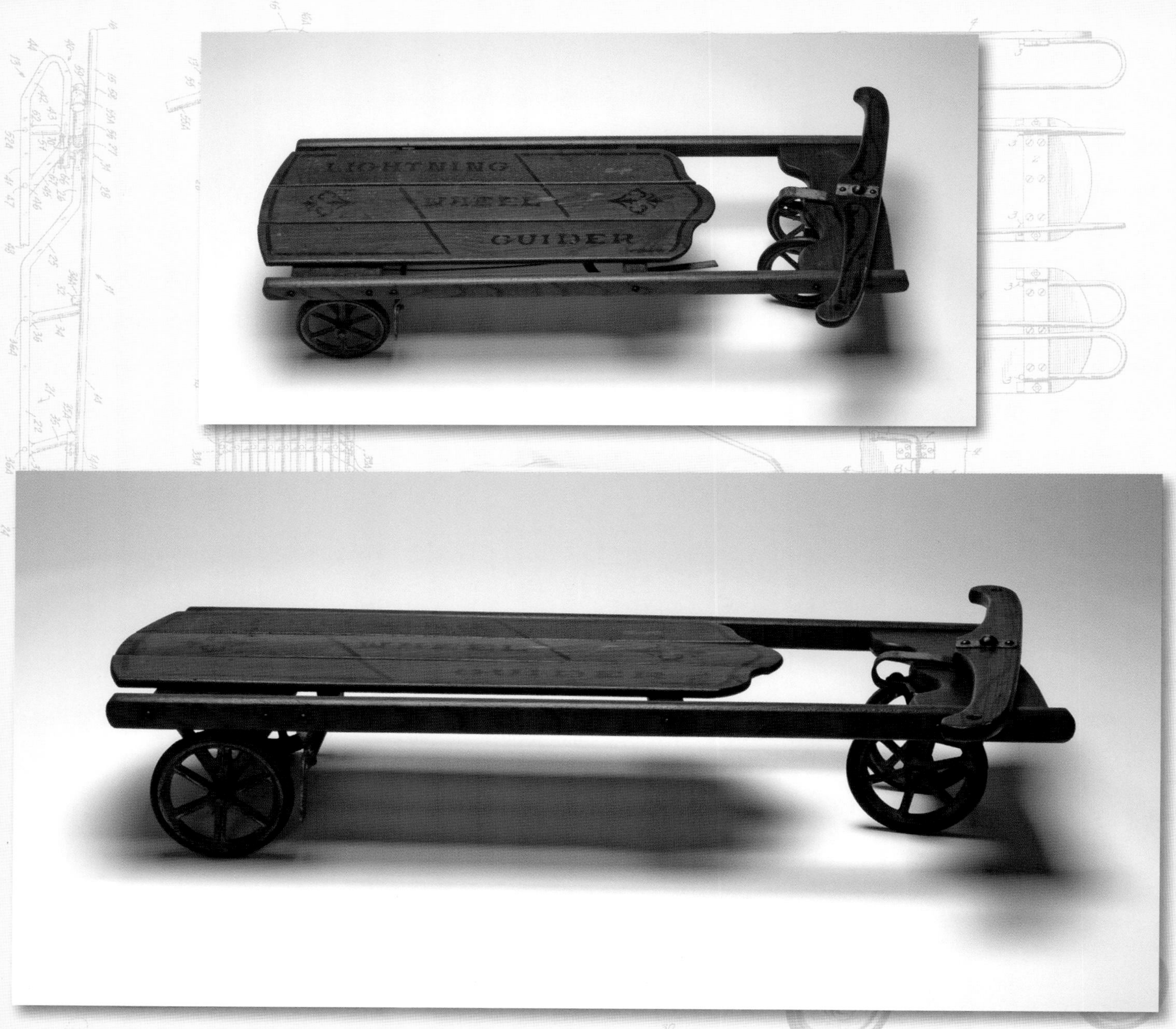

Lightning Wheel Guider, c. 1911. L: 33-3/4"; W: 12-3/4"; HL: 6-1/2". $350-500.
Courtesy Carroll Palmer, Jr.

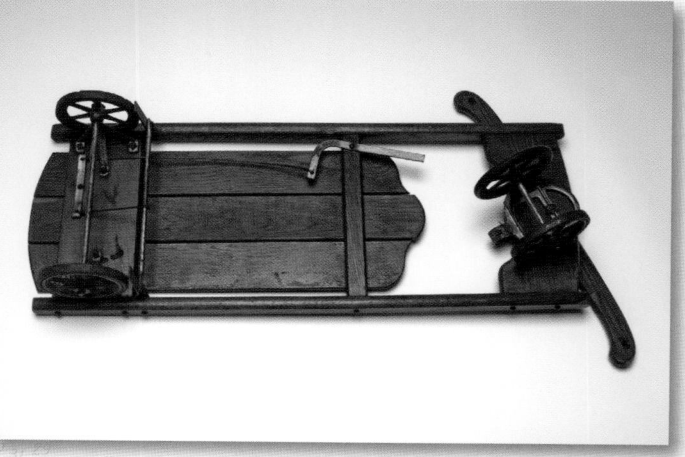

Standard Novelty Works Lightning Guider Scooter, c. 1917. L: 40"; W: 7 1/2"; H: 3 1/2" deck, 28" handle. $250-450.

Nothing else like the
LIGHTNING SCOOTER

AND you will find it easy to sell. It's instant appeal to the kiddies assures you quick sales and big profits.

A thoroughly reliable, rigidly constructed toy that will last indefinitely.

Made of air-dried white ash—spring steel runners—riveted throughout—and finished with two coats of weather varnish.

Rear of front runner acts as a POSITIVE BRAKE when handle is pulled back

Handle-bar is furnished to insert in loop for foot steering

Made by the Makers of the Famous

LIGHTNING GUIDER
SLEDS

Non-slip Rubber Mat

STANDARD NOVELTY WORKS DUNCANNON, PENNA.

EST. 1905

Write for Sample and Price

Lightning Wheel Guider advertisement, c. 1911. *Courtesy of the Old Sled Works.*

Lightning Guider Sleds

STRONGLY BUILT

of the Best Materials and Workmanship

OH, HOW THEY STEER ! !

Ask Any Boy or Girl Who Owns One

The Boy Loves Speed

HE GETS IT IN A

Lightning Guider Sled

One for Every BOY AND GIRL

Standard Novelty Works

Duncannon, Pa.

The New All Steel Front "CHALLENGER"

The Strongest Steering Sled ever Built

No. 36

Length .... 36 in.
Height, Front .... 6½ in.
Height, Rear .... 4½ in.
Width .... 13 in.
Weight .... 98 lbs. to doz.
Packed two in a bundle

No. 45

Length .... 45 in.
Height, Front .... 6½ in.
Height, Rear .... 4½ in.
Width .... 13 in.
Weight .... 156 lbs. to doz.
Packed two in a bundle

This construction gives the sled the proper and graceful stream lines to make it up to the minute. It looks well and has lots of speed.

SPECIAL FEATURES

Heavy inch channel steel benches or knees. You must closely examine the construction of the knee to appreciate how the knee is attached to the runner giving strength and a perfect fit.

Spring steel high carbon grooved runners. Look at the gauge of the steel used in these runners. They have the weight to stand up to any test. Air dried WHITE ASH lumber. Tops are finished with two coats of high grade weather varnish. Nicely decorated.

Ask for Prices on our Complete Line of All Steel Front Sleds

Standard Novelty Works, company brochure, c. 1920. $25-50.

Lightning Guider Sleds

THE Lightning Guider Sled has proven its own quality by the enormous increase in sales from year to year. We have maintained this same high standard of quality and solid construction throughout by using Air-dried Selected White Ash Lumber; the very best High-Carbon Spring Steel Runners (these runners are concaved or grooved which promotes speed and easy guiding, prevents skidding and holds sled to its true course over snow or ice) Heavy Gauge Stamped-Steel Parts; and last but not least, the finish—three coats of High Grade Coach Varnish which gives the sled a nice lasting finish from year to year.

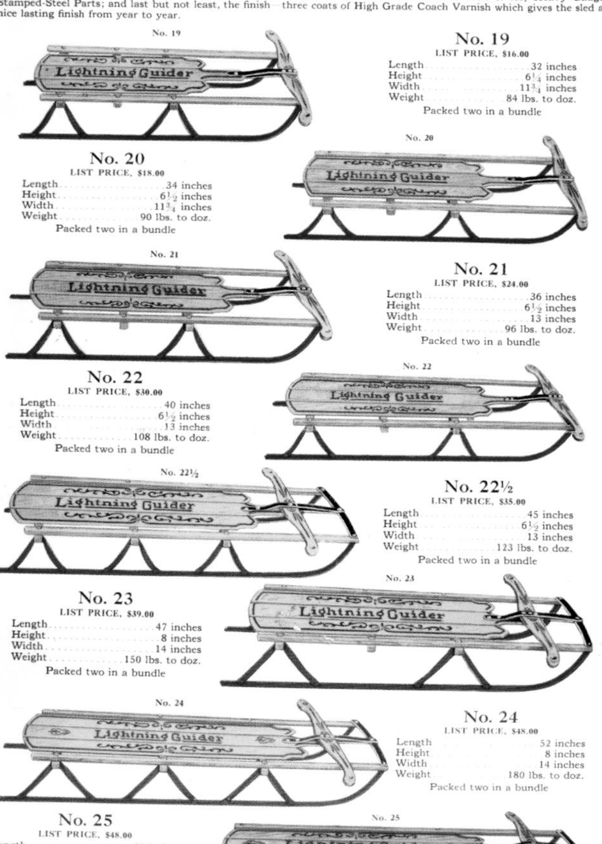

No. 19
LIST PRICE, $16.00
Length .... 32 inches
Height .... 6¼ inches
Width .... 11¾ inches
Weight .... 84 lbs. to doz.
Packed two in a bundle

No. 20
LIST PRICE, $18.00
Length .... 34 inches
Height .... 6½ inches
Width .... 11¾ inches
Weight .... 90 lbs. to doz.
Packed two in a bundle

No. 21
LIST PRICE, $24.00
Length .... 36 inches
Height .... 6½ inches
Width .... 13 inches
Weight .... 96 lbs. to doz.
Packed two in a bundle

No. 22
LIST PRICE, $30.00
Length .... 40 inches
Height .... 6½ inches
Width .... 13 inches
Weight .... 108 lbs. to doz.
Packed two in a bundle

No. 22½
LIST PRICE, $35.00
Length .... 45 inches
Height .... 6½ inches
Width .... 13 inches
Weight .... 123 lbs. to doz.
Packed two in a bundle

No. 23
LIST PRICE, $39.00
Length .... 47 inches
Height .... 8 inches
Width .... 14 inches
Weight .... 150 lbs. to doz.
Packed two in a bundle

No. 24
LIST PRICE, $48.00
Length .... 52 inches
Height .... 8 inches
Width .... 14 inches
Weight .... 180 lbs. to doz.
Packed two in a bundle

No. 25
LIST PRICE, $48.00
Length .... 58 inches
Height .... 6½ inches
Width .... 13 inches
Weight .... 162 lbs. to doz.
Packed two in a bundle.

They are built to last

SLED AND WAGON GUARD

A New Safety Attachment

Specifications

THIS Guard is rigidly constructed of white ash lumber, with two coats of weather varnish, which gives it a beautiful natural finish.

The back is decorated in red. All metal parts are rust proof.

Slotted, adjustable parts held in position by stove bolts and wing-nuts. It straps right on any size or make of sled, no screws or nails are needed. The back can be reclined to any comfortable position.

Made in one size only. Size of back 14", length of side 24", height of side 7". Shipping weight 5 lbs.

Packed in one-half and one dozen lots in large cartons. Also each guard is neatly packed in seperate container ready for the customer.

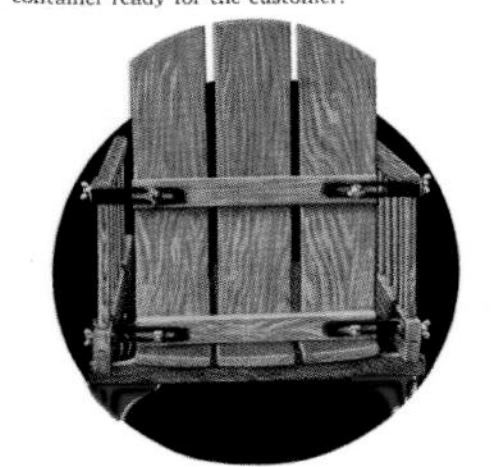

The above illustration shows how the Guard can be used on any width sled or wagon. The adjustable parts have much room for expansion as shown on the illustration.

The Guard is just as serviceable for wagons as for sleds. It is used on wagons by adjusting with four wood screws. It makes the wagon entirely safe for the kiddie.

Built by the makers of the famous Lightning Guider Sleds.

The LIGHTNING WHEEL COASTER

A Guaranteed Roller-Bearing Coaster with these Special Selling Features

Made in Two Body-Sizes
No. 200 Wagon-Body-size 14 x 34 in.
No. 300 Wagon-Body-size 16 x 38 in.

SPECIFICATIONS

BODY STYLE—Removable wood body, made of the best second-growth air-dried white ash, worked to one-half in full. Cross-end pieces mortised into side pieces.
TONGUE—1 inch square. White ash. Round corners.
WHEELS—10-in. Double-disc, dust proof. Roller bearings.
BRAKE—Steel Roller.
TIRES—¾-inch, best grade rubber.
BOLSTERS—Reinforced channel steel, ⅜-inch.
AXLES—½-inch cold-drawn steel.
FINISH—Wheels are red enamel. Body, best grade paint, finished with two coats of Spar Varnish.
PACKED—One in wooden crate, with all parts complete.
WEIGHT—No. 200 Wagon, 47 pounds, including crate. No. 300 Wagon, 48½ pounds, including crate.

The LIGHTNING SPEEDSTER

A strong and durable Stake Wagon. The only wagon of its kind on the market today. The removable sides and end gate make it practical for easy coasting or bulky hauling.

Patented Fifth Wheel. 10-in. Double Disc Wheels. Roller Bearing. ¾-Inch Rubber Tires. Air-Dried Ash Lumber, Natural Finish. Weather Varnish. Size of Bed 17 x 40 Inches. Strongly built and will carry 1000 pounds. Steel Roller Positive Brake.

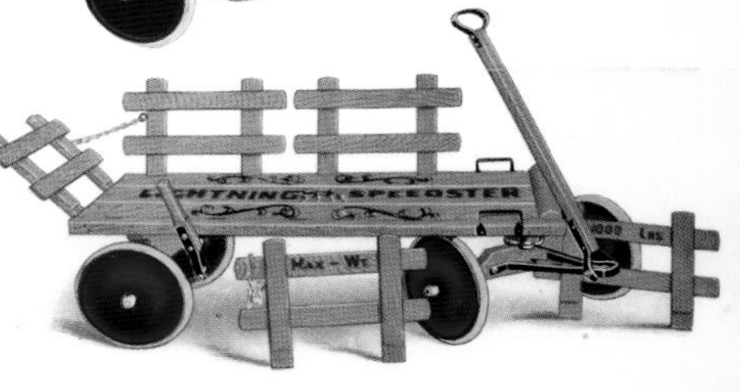

STANDARD NOVELTY W

Standard Novelty Works, company brochure, c. 1920. $25-50.

New Companies

Crosby, Gilzinger & Company

A patent was issued to Sebastian Gilzinger on March 13th. 1877, for an improvement in sleds, the runners and cross braces being constructed of one continuous piece of metal. One half of the patent was assigned to Abel Crosby. Their new Sled was called "The Ulster" The company was located in Rondout, New York. Unable to find Rondout, New York, on any map, I focused on Ulster County. After many months and phone calls, I discovered Rondout was now Kingston New York. Traveling to Kingston, I found that somewhere between the Catskill Mountains and the Hudson River lies a tiny hamlet called Esopus. A small plaque on the wall of the local Klyne Esopus Museum read "Sleighs made in Sleightburg by Crosby & Gilzinger. The factory established just before the Civil War and was destroyed by fire in 1901. Every child in town had an 'Ulster' sleigh. Many enterprising boys made bobsleighs using two Ulsters. On Colonel Payne's estate in West Park, children would ride the long hill of Rt. 9W, one on horse back would haul the bobsled back to the top of the hill after each run" (author unknown). Little else is known about the company and none of the story can be validated because a huge fire destroyed all newspapers and records for the years between 1900 and 1904. These sleds are few and far between. I can honestly say in the past thirty five years I have only seen four or five in good to mint condition. If you own an "Ulster" you are lucky!

Crosby, Gilzinger & Co.,

"THE ULSTER" SLED.

Trade Mark registered Oct. 23, 1876. Patented March 13, 1877. Patented in Canada Feb'y 19, 1878.

Patentees and Manufacturers of

THE ULSTER SLED

AND

The Sheridan Velocipede.

"THE SHERIDAN" VELOCIPEDE.

Patented in U. S. August 22, 1876.
Patented in England August 19, 1876.
Patented in France November 6, 1876.
Patented in Germany.

Centennial Carriage, Sulkey and Seat Springs, Centennial Seat Fasts, *The Dexter Shifting Bar for Sleighs*—can be changed from centre to side draft instantly, requiring no wrench, and without unhitching horse. *The Ulster Side Bar Wagon Spring*, the latest and best side bar spring ever made. We also make a very superior Trotting Sulkey, with our patent Centennial Spring and Brace Attachment.

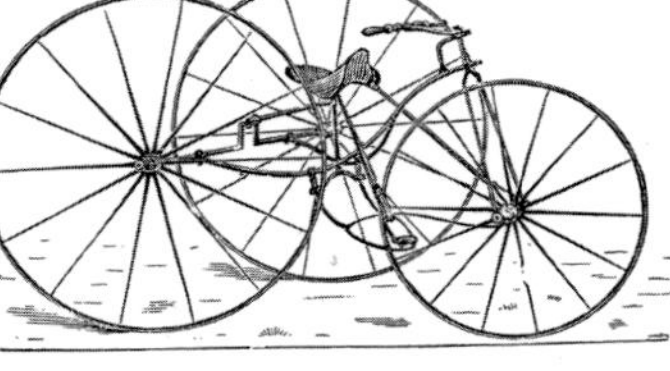

Send for Illustrated Circular. RONDOUT, N. Y.

Crosby Gilzinger & Company "The Ulster" Sled advertisement, December 1877, *Century Magazine*. $5-10.

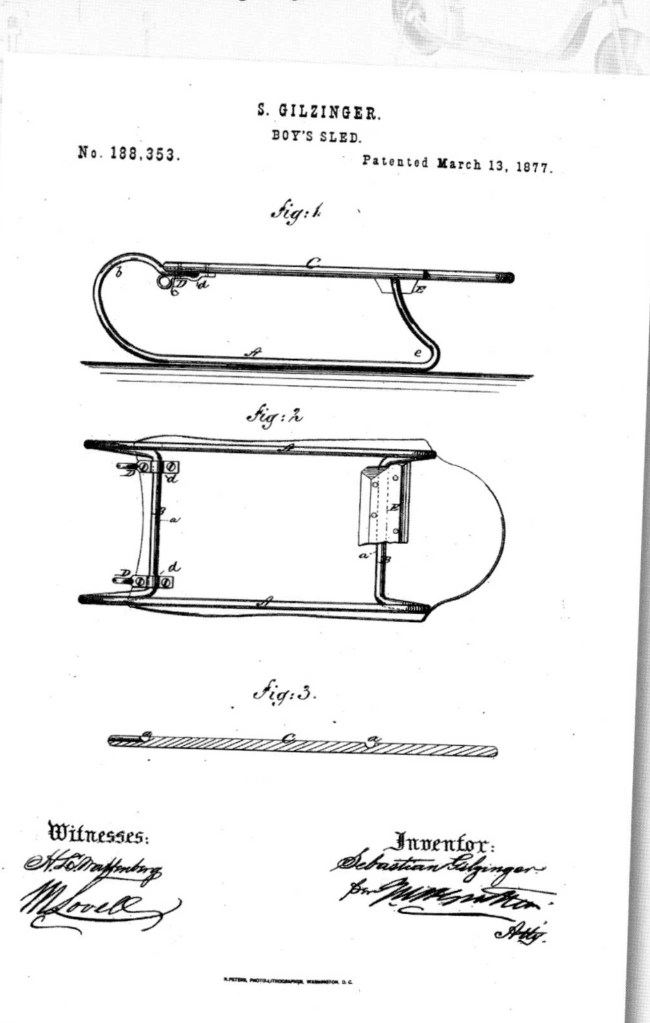

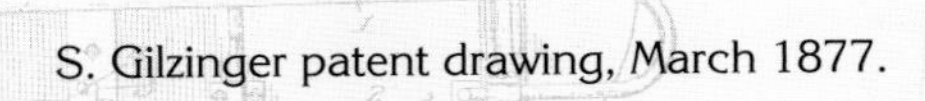
S. Gilzinger patent drawing, March 1877.

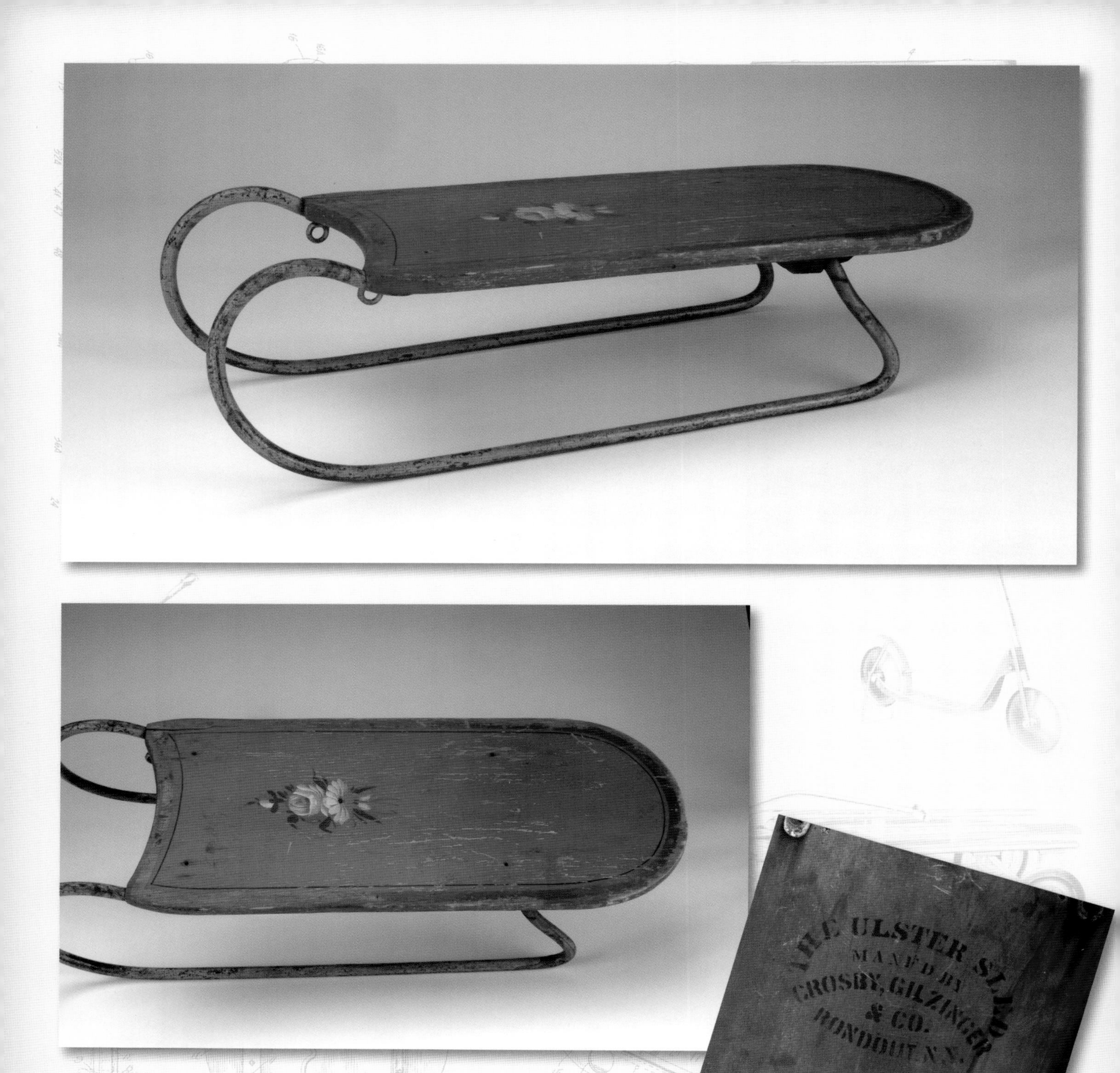

Crosby Gilzinger & Co. "The Ulster" sled, c. 1876. L: 25"; W: 9"; H: 6". $150-200.

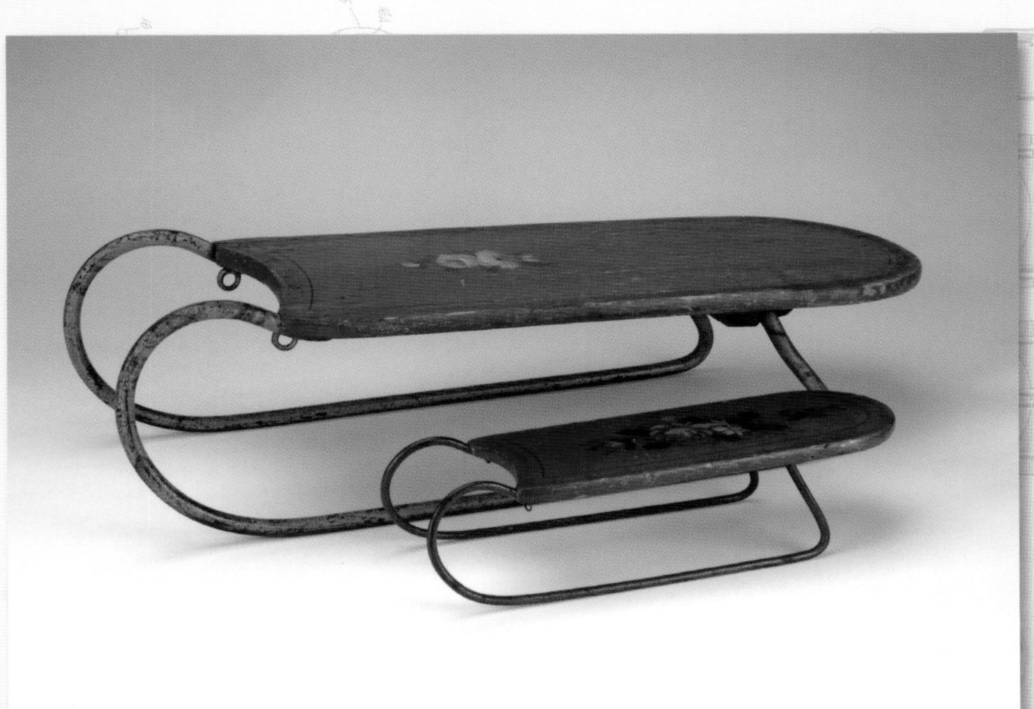

"The Ulster" sled shown beside its U.S. Patent model.

Crosby Gilzinger & Co. Ulster Sled U.S. patent model, October 23, 1876. L: 12"; W: 4 1/2"; H: 3". $1,000-1,200.

The Duralite Corporation

In 1949, Morton Thomas and Bertrand Lesser of New York, formed the Duralite Corporation to manufacture light weight aluminum furniture. In 1965 they moved operations to Passaic, New Jersey, where they began to manufacture sleds, using light weight tubular metal to form the decks and runners, there by minimizing the weight of the sleds. In addition to the steerable sled they also manufactured bobsleds using the same construction. The popularity of the Duralite Sled was regional as was the advertising and distribution. When asking the locals about the Duralite Corporation, most did not recall the sleds but did remember the "Art Deco" aluminum furniture. By 1969 financial difficulties led to the deterioration of the partnership between Morton and Lesser. Production ceased in the early 1970s.

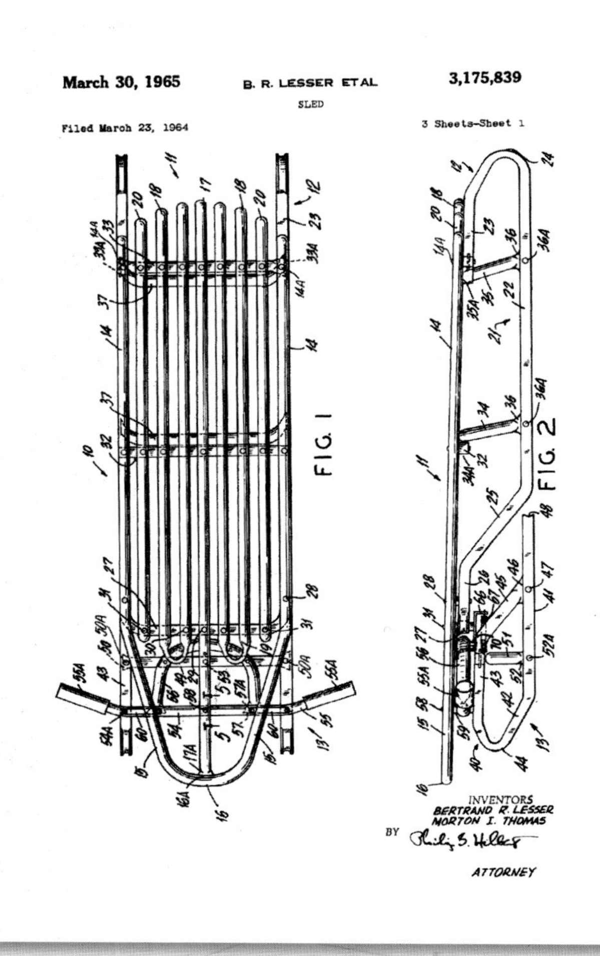

Patent drawing, 1965, for the Duralite Racer.

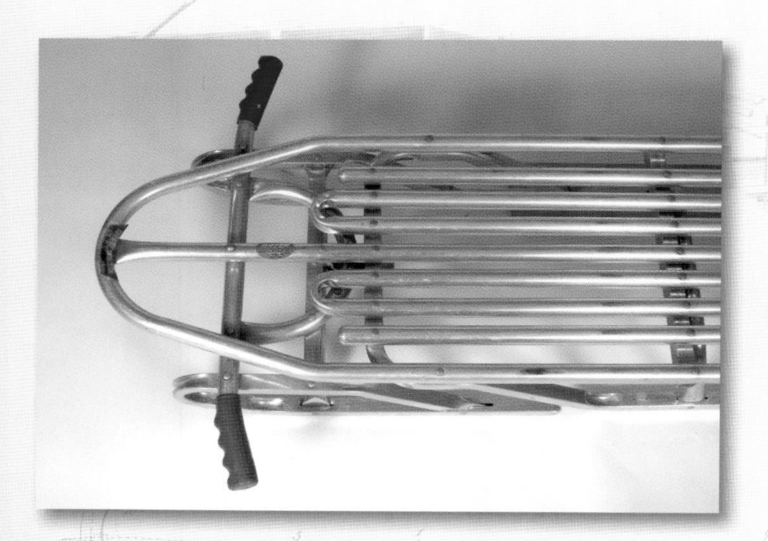

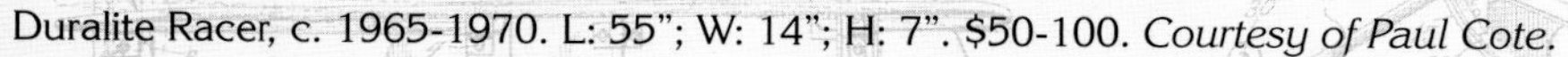

Duralite Racer, c. 1965-1970. L: 55"; W: 14"; H: 7". $50-100. *Courtesy of Paul Cote.*

The Ellingwood Turning Company

Little is known about the Oxford Sled manufactured by the Ellingwood Turning Company, West Paris, Maine. The company incorporated on February 11, 1913 and was famous for snowshoes, pick poles, and hockey sticks. Early catalogs (1913-1914) do not mention sleds or toboggans. The Oxford Company was incorporated in 1919 or 1920. In 1923 it became the selling agent for the Ellingwood Turning Company. Their trademark is a diamond embellished with the word Oxford and pierced with an arrow. (Oxford is the county where the Ellingwood Turning Co. was located.) The sleds had the word Oxford in a diamond on their decks, and were marked on the underside "Manufactured by the Ellingwood Turning Company." The Oxford Sled was available in three sizes, #36, #43, & #50, in the 1926 Catalog. The 1926 catalog and the 1931 catalogs are identical with the exception that on the 1931 catalog it reads, "Manufactured by the Oxford Corporation" no mention of the Ellingwood Turning Company.

I have been asked many times to date Oxford sleds and based on the information I have, any sled with the Diamond, embellished with the word Oxford, and pierced with an arrow, sled dates from 1923-1931. For any sled with the word Oxford on the deck and no diamond, your guess is as good as mine: before 1919 or after 1931?

Top: Ellingwood Turning Co., Oxford Sled #150. $250-350
Bottom: Ellingwood, Oxford salesman's sample sled, L: 21"; W: 5-1/2"; H: 2-1/2". $1200-1500.
Courtesy of Paul Cote.

CATALOGUE OF

SNOWSHOES · SKIS · TOBOGGANS · SLEDS
HOCKEY STICKS · TENNIS RACKETS · ETC.

Manufactured by
ELLINGWOOD TURNING CO.

Selling Agents
THE OXFORD CORPORATION
WEST PARIS, MAINE, U. S. A.

Ellingwood Turning Company Catalog, 1926. Note manufactured by Ellingwood Turning Company, Selling agent The Oxford Corporation. Courtesy of Paul Cote.

Ellingwood Turning Company logo, courtesy Paul Cote.

CATALOGUE OF

SNOWSHOES · SKIS
TOBOGGANS · SLEDS

Manufactured by
THE OXFORD CORPORATION
WEST PARIS, MAINE, U. S. A.

PRINTED IN U.S.A.

Oxford catalog with price list for 1931. Identical to the 1926 Ellingwood Catalog except manufactured by the Oxford Corporation with no mention of the Ellingwood Turning Company.

OXFORD TOBOGGANS

Model A

	weight	each
6' Rope Rails	24 lbs.	$12.00 each
7' Rope Rails	27 "	14.00 "
8' Rope Rails	29 "	16.00 "
9' Rope Rails	34 "	20.00 "
10' Rope Rails	37 "	25.00 "

Model B

	weight	each
4' Rope Rails	16 lbs.	$ 5.00 each
5' Rope Rails	18 "	7.00 "
6' Rope Rails	21 "	8.50 "
7' Rope Rails	24 "	10.00 "
8' Rope Rails	26 "	12.00 "
9' Rope Rails	28 "	13.50 "
Cushions		$1.20 per foot
Cushions, Fancy Top		1.30 " "

OXFORD SLED

	weight	each
No. 36—36 inch	9 lbs.	$4.00 each
No. 43—43 inch	11 "	5.00 "
No. 50—50 inch	14 "	6.00 "

NASH PACK CARRIER

Men's size	3½ lbs.	$3.50 each

[5]

TOYS

	weight dozen	list
No. 30 Sidewalk Skis	15 lbs.	$8.40 Doz.
No. 40 " "	30 "	9.60 "
No. 48 " "	36 "	10.80 "
No. 28 Ski Set	24 "	12.00 "
No. 3 Ski Pole	6 "	3.00 "
No. 36 Toboggan	96 "	48.00 "
No. 120 Doll Sleigh	33 "	27.00 "
	weight	
No. 111 Kiddie-Slyde	20 lbs.	$6.00 Each
Par-9 Golf Game	9 "	6.00 "

Main Office and Factory, West Paris, Maine
New York Sales Office, 200 Fifth Ave.
Freight Office, Bates, Maine
Canadian National Railways

Terms 2% ten days, net 30 days. F.O.B. Factory

Printed in U. S. A.

PRICE LIST No. 31

(Effective February 1st, 1931)

SNOWSHOES
SKIS, TOBOGGANS
SLEDS AND
TOYS

THE
OXFORD CORPORATION
WEST PARIS, MAINE, U. S. A.

Maplewood Craft Industries,

The allure of research is the thrill of discovery, and thrill it was when I realized the "Steermaster Bob-Ski" may have been influenced by the Arts and Crafts Movement. This started me looking for anything that had Maplewood in its name, in Minnesota. It was not until I purchased a third Bob-Ski which had a complete paper label indicating manufactured in Hutchinson, Minnesota, that I was able to track the origin of this unusual sled to the Maplewood Academy, a Christian-based Seventh-Day- Adventist, Prep School. The Minneapolis Preparatory School, as it was called when it opened in 1888, went through many changes, and emerged as the Maplewood Academy in 1904 when they purchased and moved to a ninety-four acre farm near Maple Plain, Minnesota. The school operated a farm with dairy cows, chickens, and horses for transportation. From 1910 to 1928 Maplewood operated in conjunction with the Danish-Norwegian Seminary of Hutchinson. The two schools merged onto the Hutchinson campus in 1928 and Maplewood Academy has remained there until the present time.

As the Academy grew, in addition to the farm they operated a bookbindery, a print shop, and a craft shop. In addition to redwood furniture, they manufactured toys. One of these toys was the Steermaster Bob Ski. Unfortunately a patent was never applied for so dating is left up to the "senior alumnae." The consensus is that it was sometime between 1954-1962.

As it turned out, the only similarities to the Arts & Crafts Movement is the time frame and the fact that a craft shop was maintained and furniture was also hand made. Unlike the Arts & Crafts Movement, whose "Mission Furniture," created between 1890 and 1914, is very well known and sought after by many collectors, the redwood furniture, once produced by the Maplewood Academy is long forgotten. It is the "Steermaster Bob Ski" that is sought after by today's collector.

Maplewood Academy Steermaster Bob Ski, c. 1954-1962. Note the rudder. L: 37" deck, 41" overall; W: 12 1/2"; H: 6 1/2". $150-250.

Maynard Miller Manufacturing Company

The maker of Sea Shells and Camels, previously listed as unknown, has been identified! Who would have ever thought that a sled with such an odd duo on the deck would not have been one of a kind or a special order? Much to my surprise upon entering an Antique Shop in upstate New York, my eyes fell upon an identical sled with one exception; stamped on the underside was Manufactured by Maynard Miller, Seneca Falls, N.Y. Patented Jan. 4, 1876.

Thanks to the patent date I was able to track a patent issued to William Wentworth, of Seneca Falls N. Y., for an improvement in knees for children's sleighs, and the witness was none other than Maynard Miller. The knees are unique in design, two semi-circles back to back, joined in the center.

According to the Seneca Falls Village Directory 1874, Maynard Miller's business was established in 1870, for the purpose of manufacturing wooden goods, prominently on the patents of William Wentworth, which included snow shovels, hand sleighs, water drawers, and pumps. In the 1884-85 catalog it is written: "Recently we have obtained a patent for our device of Malleable Iron Sled Knees, and have also the exclusive manufacture of Hand Sleds using this style of knee. In them we have combined the excellent qualities of a frame runner with an indestructible knee, making a sled that is light, graceful and very durable. We have unusual facilities for manufacturing each Sled in its style, and purpose making the popular clipper a specialty hereafter." Unfortunately there are no clear illustrations in the catalog or any mention of how the decks were decorated. It is not clear when Sled production ceased or when the company name changed to the "National Advertising Company—Manufactures of Wooden Specialties for Advertising Purposes." Letter heads dated 1909 show the name change. With advertising in mind one can only speculate about what exotic vacation land Camels & Sea Shells may have been advertising. Maynard Miller fell on hard times and was no longer associated with the business in 1909. Maynard Miller Sleds are rare, and easily identified by the unusual knees. Keep your eyes out, and you may be fortunate enough to locate one.

Maynard Miller Catalog, 1884-1885. Courtesy of the Historic Society of Seneca Falls, New York.(continued on opposite page)

TERMS.

Water Drawers, Wooden Specialties, etc., Thirty Days.

Shovel and Sled bills, payable Jan. 1st, following sale.

☞ *All bills not settled at maturity, will be subject to Sight Draft.*

FREIGHTS.

We have special contracts with leading Freight lines, for the Transportation of our goods, whereby very low rates are ensured.

On shipments amounting to $50.00 and upwards, via Boston, New York or Philadelphia, Freights will be allowed to those points

PACKING.

Our goods are securely packed, and when required strong cases are used, for which no extra charge is made.

CATALOGUE AND PRICE-LIST

—OF—

WATER DRAWERS,

PATENT MALLEABLE IRON KNEED

HAND SLEDS,

CLIPPER SLEDS, SLED BOXES,

Patent Malleable Iron Tip and Boys'

SNOW SHOVELS,

—AND—

WOODEN SPECIALTIES,

MANUFACTURED BY

DAY, TOWNS & MILLER,

SENECA FALLS, N. Y.

1884-5.

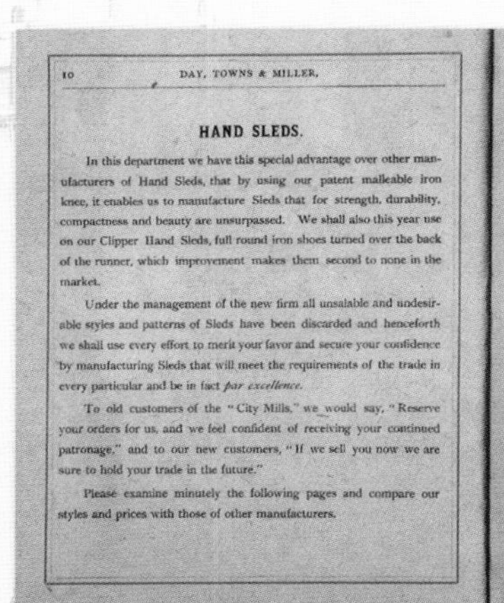

10 DAY, TOWNS & MILLER,

HAND SLEDS.

In this department we have this special advantage over other manufacturers of Hand Sleds, that by using our patent malleable iron knee, it enables us to manufacture Sleds that for strength, durability, compactness and beauty are unsurpassed. We shall also this year use on our Clipper Hand Sleds, full round iron shoes turned over the back of the runner, which improvement makes them second to none in the market.

Under the management of the new firm all unsalable and undesirable styles and patterns of Sleds have been discarded and henceforth we shall use every effort to merit your favor and secure your confidence by manufacturing Sleds that will meet the requirements of the trade in every particular and be in fact *par excellence*.

To old customers of the "City Mills," we would say, "Reserve your orders for us, and we feel confident of receiving your continued patronage," and to our new customers, "If we sell you now we are sure to hold your trade in the future."

Please examine minutely the following pages and compare our styles and prices with those of other manufacturers.

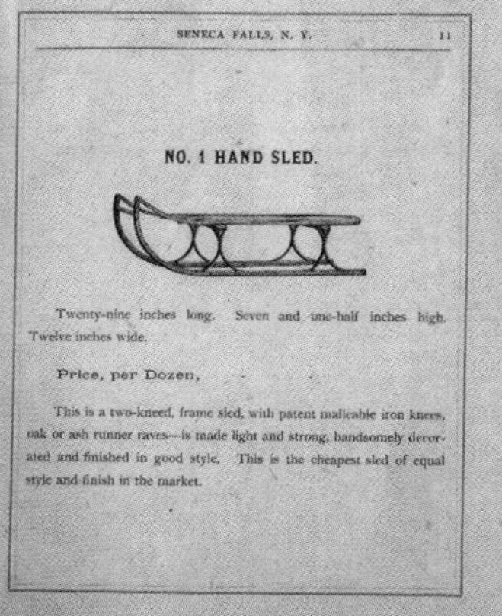

SENECA FALLS, N. Y. 11

NO. 1 HAND SLED.

Twenty-nine inches long. Seven and one-half inches high. Twelve inches wide.

Price, per Dozen, - -

This is a two-kneed, frame sled, with patent malleable iron knees, oak or ash runner raves—is made light and strong, handsomely decorated and finished in good style. This is the cheapest sled of equal style and finish in the market.

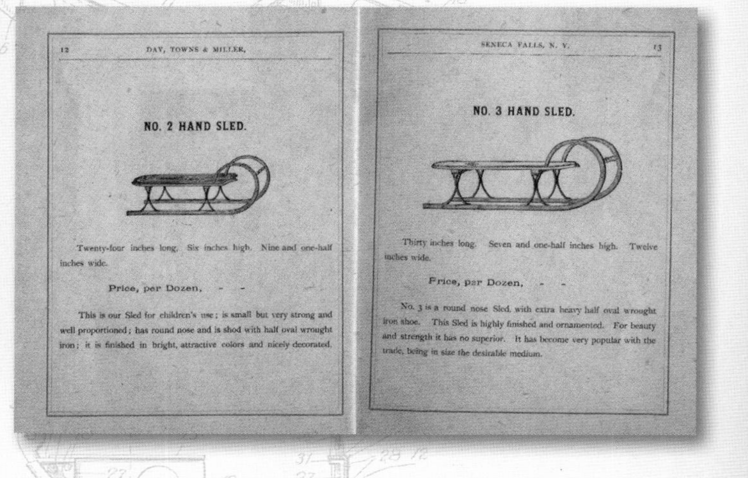

12 DAY, TOWNS & MILLER,

NO. 2 HAND SLED.

Twenty-four inches long. Six inches high. Nine and one-half inches wide.

Price, per Dozen, - -

This is our Sled for children's use; is small but very strong and well proportioned; has round nose and is shod with half oval wrought iron; it is finished in bright, attractive colors and nicely decorated.

SENECA FALLS, N. Y. 13

NO. 3 HAND SLED.

Thirty inches long. Seven and one-half inches high. Twelve inches wide.

Price, per Dozen, - -

No. 3 is a round nose Sled, with extra heavy half oval wrought iron shoe. This Sled is highly finished and ornamented. For beauty and strength it has no superior. It has become very popular with the trade, being in size the desirable medium.

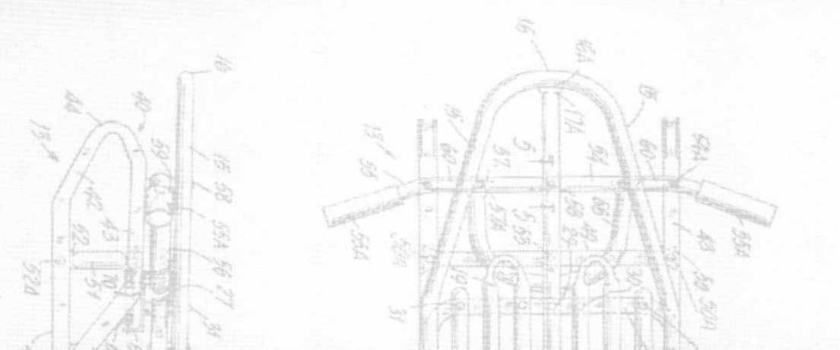

NO. 4 HAND SLED.

Thirty-four inches long. Eight and one-half inches high. Twelve inches wide.

Price, per Dozen, - -

This Sled is the coaster of its class. Its construction is similar to No. 3, having the round nose and half oval wrought iron shoe, extra heavy. It is, however, considerably larger. The finish is equal to that of No. 3.

Besides being a noble coaster, it is quite desirable to use with a box and perambulator handles for children.

NO. 5 HAND SLED.

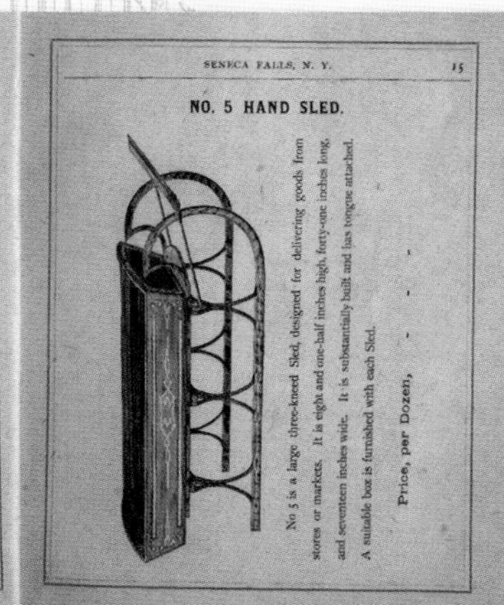

No 5 is a large three-kneed Sled, designed for delivering goods from stores or markets. It is eight and one-half inches high, forty-one inches long, and seventeen inches wide. It is substantially built and has tongue attached. A suitable box is furnished with each Sled.

Price, per Dozen, - - -

PERAMBULATOR.

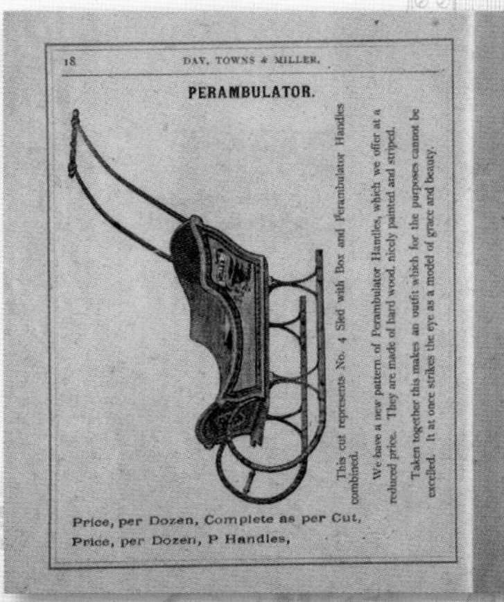

This cut represents No. 4 Sled with Box and Perambulator Handles combined.

We have a new pattern of Perambulator Handles, which we offer at a reduced price. They are made of hard wood, nicely painted and striped.

Taken together this makes an outfit which for the purposes cannot be excelled. It at once strikes the eye as a model of grace and beauty.

Price, per Dozen, Complete as per Cut,
Price, per Dozen, P Handles,

NO. 10 CLIPPER.

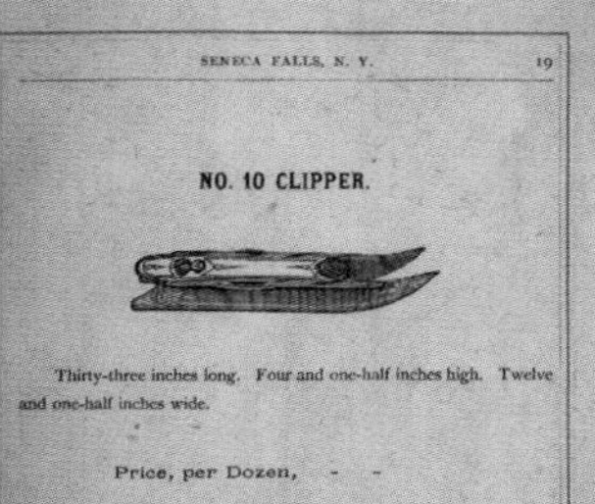

Thirty-three inches long. Four and one-half inches high. Twelve and one-half inches wide.

Price, per Dozen, - -

No. 10 has full round iron spring shoes, highly polished; is made of well-seasoned oak and ash; has eliptical holes in raves, making good handles; malleable iron rings in end of runner to draw by, and is put together solely with a view to durability, having wrought iron braces riveted in. The finish is oil and varnish, finely striped and ornamented.

NO. 6 HAND SLED.

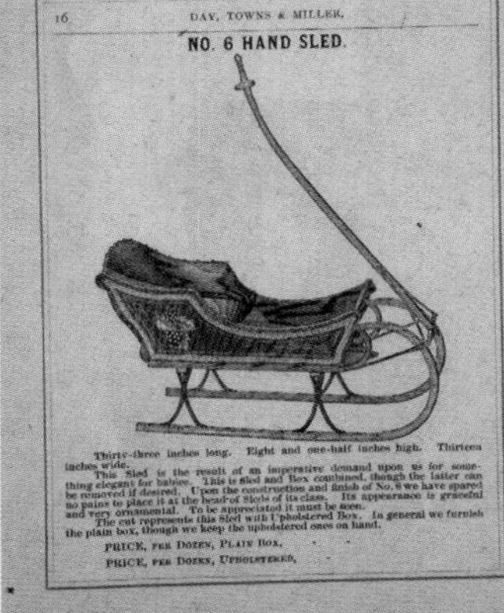

Thirty-three inches long. Eight and one-half inches high. Thirteen inches wide.

This Sled is the result of an imperative demand upon us for something elegant for babies. This is Sled and Box combined, though the latter can be removed if desired. Upon the construction and finish of No. 6 we have spared no pains to place it at the head of Sleds of its class. Its appearance is graceful and very ornamental. To be appreciated it must be seen.

The cut represents this Sled with Upholstered Box. In general we furnish the plain box, though we keep the upholstered ones on hand.

PRICE, PER DOZEN, PLAIN BOX, - -
PRICE, PER DOZEN, UPHOLSTERED, -

SLED BOX.

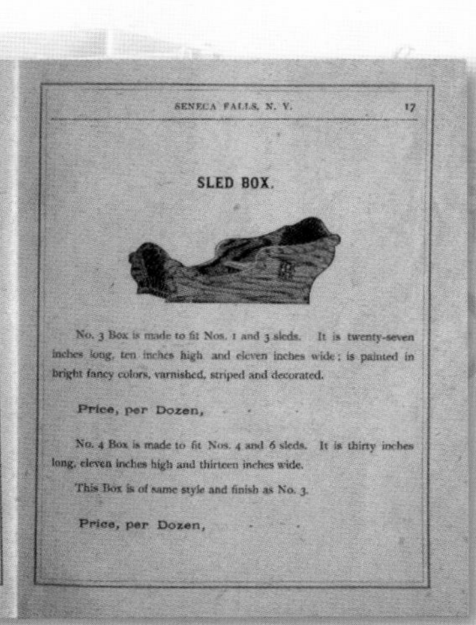

No. 3 Box is made to fit Nos. 1 and 3 sleds. It is twenty-seven inches long, ten inches high and eleven inches wide; is painted in bright fancy colors, varnished, striped and decorated.

Price, per Dozen, -

No. 4 Box is made to fit Nos. 4 and 6 sleds. It is thirty inches long, eleven inches high and thirteen inches wide.

This Box is of same style and finish as No. 3.

Price, per Dozen, - -

NO. 11 CLIPPER.

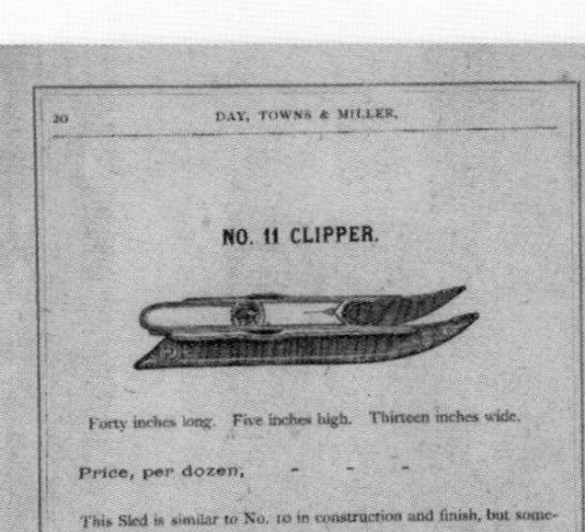

Forty inches long. Five inches high. Thirteen inches wide.

Price, per dozen, - - -

This Sled is similar to No. 10 in construction and finish, but somewhat larger.

This being our medium sized hardwood Clipper, it has all the points the ordinary boy appreciates so well. All Clippers of this class are built with a view to running qualities, and for this reason we use the full round shoe, which presents so little resistance, passes over obstructions easily and runs with great speed.

NO. 12 CLIPPER.

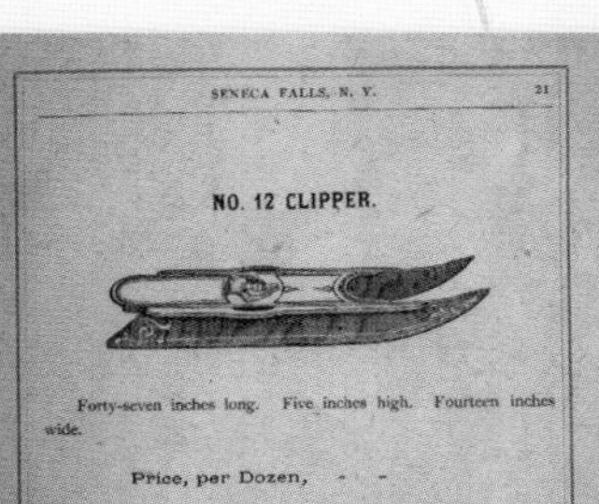

Forty-seven inches long. Five inches high. Fourteen inches wide.

Price, per Dozen, - -

Our No. 12 is the *ne plus ultra* of coasting Sleds. Its construction, style and finish are like the other two sizes of our best Clipper.

It is considerably larger than No. 11, and we offer it particularly for the edification of the larger boys. They will find in it all that can be asked of any coaster.

NO. 19 CLIPPER.

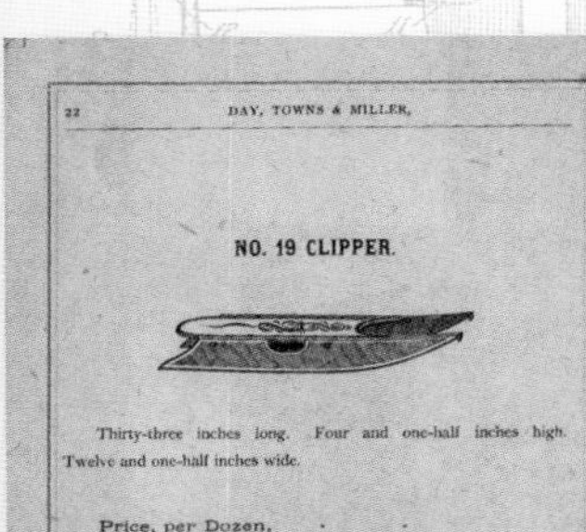

Thirty-three inches long. Four and one-half inches high. Twelve and one-half inches wide.

Price, per Dozen,

This Sled has round iron spring shoes, highly polished. It is built of the best seasoned oak or ash; has hand holes cut in sides. It is finished in oil and varnish, very handsomely ornamented and striped.

NO. 20 CLIPPER.

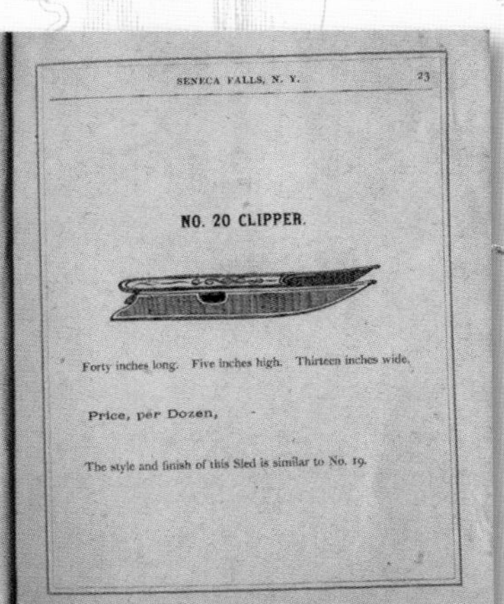

Forty inches long. Five inches high. Thirteen inches wide.

Price, per Dozen,

The style and finish of this Sled is similar to No. 19.

NO. 21 CLIPPER.

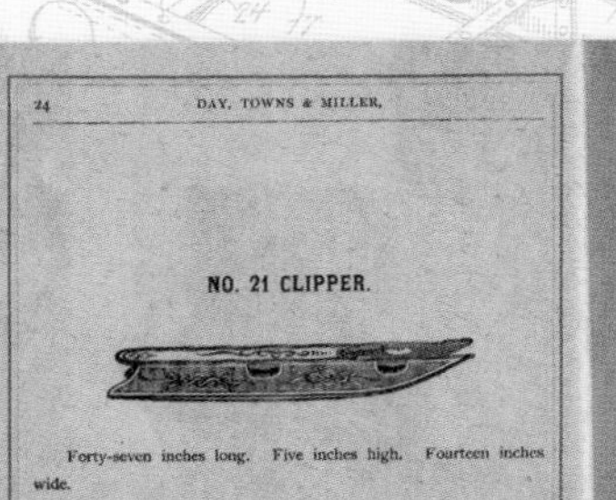

Forty-seven inches long. Five inches high. Fourteen inches wide.

Price, per Dozen, -

The construction, style and finish of this Sled is the same as No. 20.

NO. 24 HAND SLED.

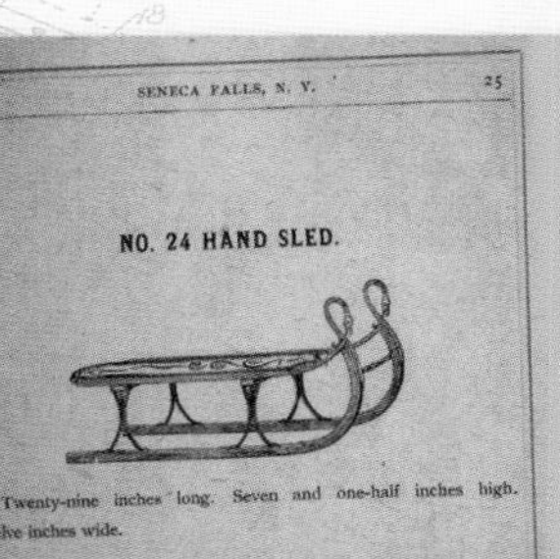

Twenty-nine inches long. Seven and one-half inches high. Twelve inches wide.

Price, per Dozen, - -

No. 24 is a two-kneed frame Sled, with our patent malleable iron knees; is made light and very strong, and has swan necks. It is painted very finely, and finished and decorated in a handsome manner.

Maynard Miller, Seneca Falls, N.Y., January 4, 1876. Seashells & Camels, unique knees. L 31", W 11 ½", H 7 ½". $650-1450. *Courtesy of Carole and Lou Scudillo.*

Pratt Manufacturing Co.:

Pratt Manufacturing Company was founded in the 1870s in Coldwater, Michigan, by John Buggy and the father of the late Allen J. Pratt. During the early years they manufactured ladders and horse-drawn cutters, hence they were known to the town's people as "The Cutter Factory." Later, in the 1890s they added non-steerable sleds to their line. Between 1910 and 1920 the old style wooden sleds and cutters were replaced by the new modern steering sled called the "Flyaway." However, the most well known children's item, sold all over the United States, was the rocking horse known to the trade as "Shooflys." With the outbreak of World War II, the company began to manufacture folding camp cots, stools, camp tables, and mosquito nets for the Navy Department. During this time, the company was reorganized as a partnership and operated as such until November 1952, when sled production ceased and the machinery and patents were sold. The old factory was used as a warehouse by the L.A. Darling Company until it was deeded to the city in 1966. The building was demolished in 1967, leaving us with only fond memories of the "Flyaway" sled

Chronology of the Pratt Manufacturing Company

1870s • Founded by John Buggy and the late father of Allen J. Pratt manufactured ladders and horse-drawn cutters. Known as the "Cutter Factory"

1882 • Building acquired at the corner of Division St. and Park Ave. Coldwater, Mich. by the Pratt family

1905 • Company incorporated and employed 125 men

1910-1920 • Flyaway, steerable sled introduced along with Rocking Horses better known as "Shooflys"

1941-1945 • Awarded the coveted Navy "E" for their outstanding war production record, operating 24hrs./day building folding canvas cots and mosquito nets for the navy

1945 • Company reorganized, sled and shoofly production resumed

1952 • November, partnership dissolved, patents and machinery sold. Sled and shoofly production ceased

1966 • Building deeded to city

1967 • Building demolished in March and the site cleared for new industrial use

Pratt Mfg. Co., Cold Spring, Michigan, oak fender sled, c. 1880.. Black deck with rose center. Note the extra-long swan heads and that the fender passes through the knees. L: 38", W: 15" H: 8-1/2". $1450-2950. *Courtesy of Carole and Lou Scudillo*. (continued on next page)

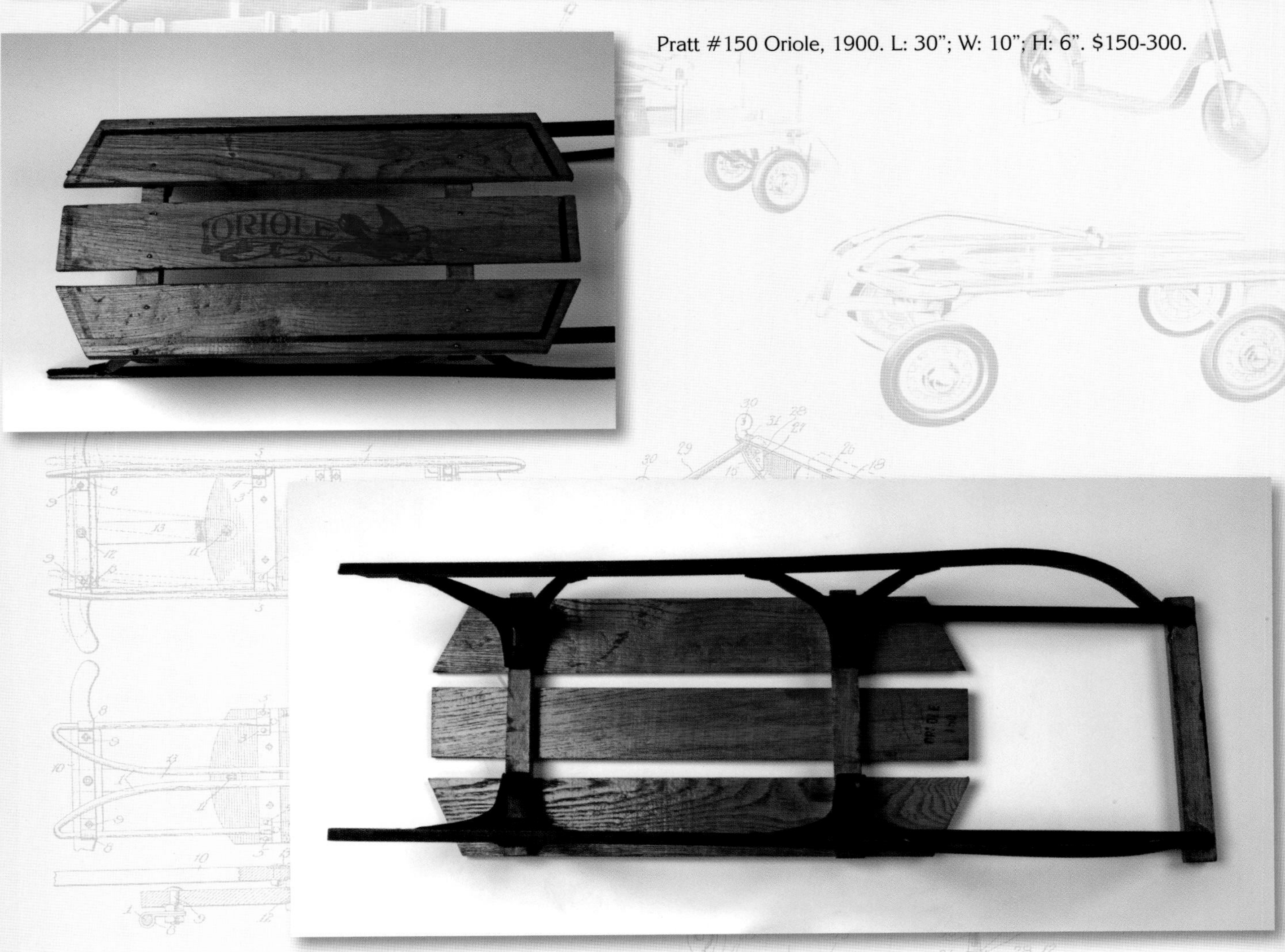

Pratt #150 Oriole, 1900. L: 30"; W: 10"; H: 6". $150-300.

SLEDS

PRATT MFG. CO.
COLDWATER, MICH.

Pratt Mfg. Co. catalog circa 1910. $75-125. (continued on the next four pages)

Steering Coasters

No. 150

Size: Width 10 inches, Length 30 inches

The great popularity of the STEERING SLEDS—shown on the opposite page—led us to include this clever little number. It is made with the same runners and knees that are used in our "FLYAWAY" Steering Sleds, and it makes an attractive number for the boy who has not the price of a steering coaster. The iron work is painted red, and the board is of ash in three pieces, neatly printed and striped, and the whole sled is varnished.

Tied two together. Weight 60 pounds **per dozen**

No. 00

Size: Width 10 inches, Length 31 inches

This sled is made by converting our No. 150 into a steering sled. It has the same style of runners and knees as the other numbers in the line, but has no wooden side rails, the seat boards covering the entire width of the top.

Tied two together. Weight 72 pounds **per dozen**

"Flyaway" Steering Coasters

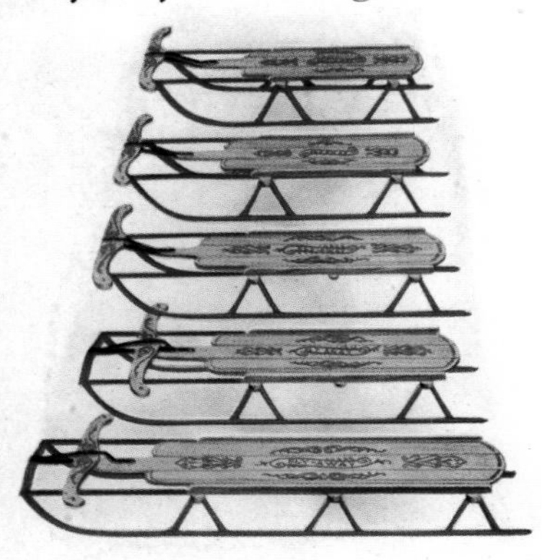

The runners are especially rolled "⊥" shaped steel of high carbon and the knees are pressed from sheet steel in a single piece. The steel parts are painted red, and the board, made of three pieces of selected ash, is finished on the natural wood, and is finely scrolled and striped.

	Width inches	Length inches	Height inches	Weight **per dozen**
No. 0	12	32	6	84
No. 1	12	36	6	90
No. 2	13	40	6	108
No. 2½	13	45	6	120
No. 3	15	46	8	162
No. 4	17	51	8	192
No. 5*	14½	46	....	150
No. 6*	14½	52	....	175

* Nos. 5 and 6 have three 6-inch knees on each runner.

Girl's Big Value Sleds

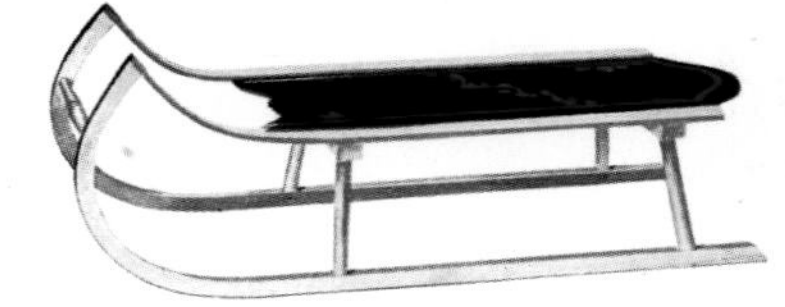

No. 122

Size: Width 12½ inches, Length 33 inches

This, we believe, is the best cheap sled made. It has round post knees mortised into beams the full size of the knees. The two benches stand bracing. Top is full ⅜ inch stock which is exceptional for this class. Gear finished natural. Has flat shoes. Seat board painted red and has printed ornaments in fine assortment.

Six in crate. Weight 46 pounds **per dozen**

No. 162

Size: Width 12½ inches, Length 33 inches

Bent knees, made of rock elm, finished natural wood color varnished. Seat board is painted red and has variety of printed ornaments. Flat shoes.

Six in crate. Weight 46 pounds **per dozen**

Sleds

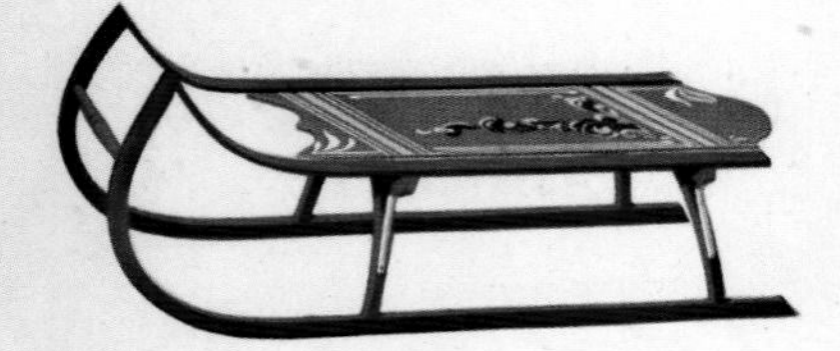

No. 142

Size: Width 12½ inches, Length 33 inches

A sturdy, handsome sled. Bent knees of rock elm. Gear painted red and striped. Braced with half-length tinned braces. Seat board decorations are hand painted flowers and scrolls (on a variety of colored boards). Flat iron shoes.

Six in crate. Weight 48 pounds **per dozen**

No. 243

Size: Width 12½ inches, Length 36 inches

This is our painted leader. Gear is red. Seat boards are handsomely ornamented by hand on various colored seat boards. Rock elm, bent knees with six tinned braces. This number has flat shoes and tinned swan-necks.

Six in crate. Weight 70 pounds **per dozen**

PRATT MFG. CO., COLDWATER, MICH.

Girl's Quality Sleds

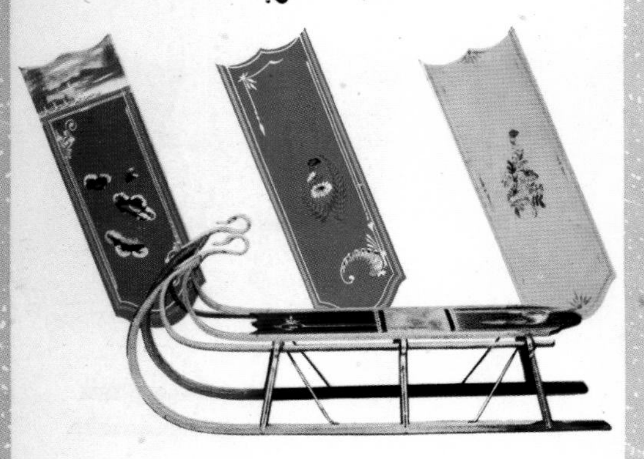

No. 445

Size: Width 15 inches, Length 40 inches

Bent knees made of rock elm, and braced with six tinned knee braces and four tinned runner braces. The runners and knees are chamfered, and the whole gear is painted red and finely striped. It has half oval shoes and extra large swan-necks. The seat board is decorated with hand-painted designs, and is expensively striped.

Two in crate. Weight 102 pounds **per dozen**

Page Seven

PRATT MFG. CO., COLDWATER, MICH.

Oak Fender Sleds

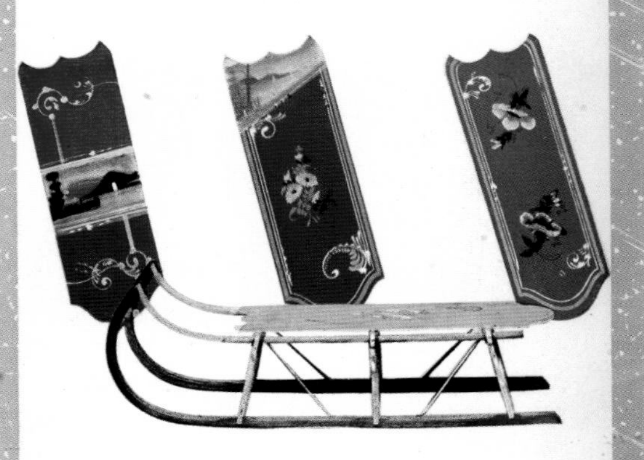

No. 393

Size: Width 12 inches, Length 35 inches

This sled is made only of selected oak, with the highest grade of natural wood finish. It is light and graceful and still strong and serviceable, being well braced with six knee braces, and has four side braces. It has no raves, but has round fenders, and is shod with half oval iron. The decorations are hand-painted flowers and landscapes.

Two in crate. Weight 66 pounds **per dozen**

Page Eight

PRATT MFG. CO., COLDWATER, MICH.

Bow Runner Sleds

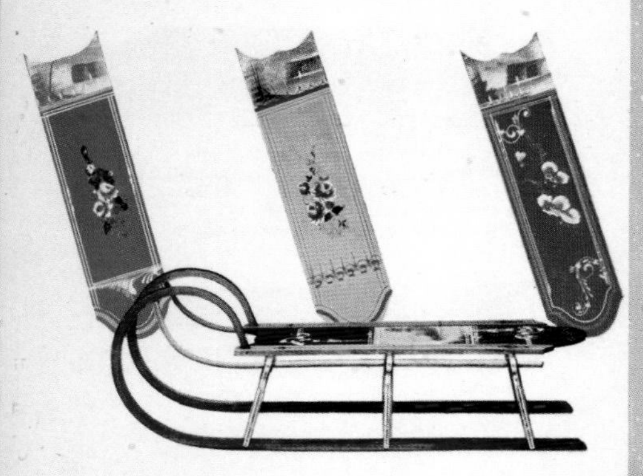

No. 334

Size: Width 14 inches, Length 38 inches

Has bow runners carefully fastened to the raves, and also has fenders. The three benches are braced with tinned braces, the gear is painted and neatly striped, and the seat board finely decorated by hand with flowers and landscapes. Shod with half-oval iron.

Two in crate. Weight 72 pounds **per dozen**

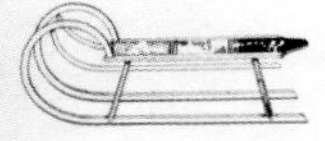

No. 332—Same general description as No. 334, but different size.

Size: Width 13 inches, Length 35 inches
Six in crate. Weight 64 pounds **per dozen**

Page Nine

PRATT MFG. CO., COLDWATER, MICH.

Coasters

No. 719

Size: Width 10 inches, Length 32 inches

This is the first of our coaster line. Finished natural color with seat board painted red showing printed ornaments. Shod with ¼ inch round spring steel.

Six in crate. Weight 62 pounds **per dozen**

No. 713

Size: Width 11 inches, Length 36 inches

Made of hardwood, varnished natural color. Seat board painted red and has printed ornaments. Two hand holes on each side. Shod with spring steel.

Six in crate. Weight 90 pounds **per dozen**

No. 714

Size: Width 12 inches, Length 42 inches

Made of hardwood, finished on the natural wood, and shod with round spring steel. The seat board finely decorated by hand, with flowers and scrolls.

Four in crate. Weight 117 pounds **per dozen**

Page Ten

PRATT MFG. CO., COLDWATER, MICH.

Coasters

No. 701

Size: Width 12 inches, Length 42 inches

Made of maple or other close grained, hard woods, with a high pointed runner. Shod with round spring steel, and seat board decorated by hand with flowers and scrolls.

Four in crate. Weight 120 pounds **per dozen**

Miniature Bob Sleigh

No. 500

Size: Width 17 inches, Length 48 inches

Made in exact imitation of farmers' bob sleighs. The runners are shod with heavy bevel-edged steel and are bolted to the runners. The hounds and bolsters are made of two-inch stock, and the reach is adjustable so the bobs can be stretched apart. The bottom is loose and can be quickly removed. The gear is painted red and the bottom painted green and both are varnished.

Packed singly. Weight 30 pounds **each**

Page Eleven

PRATT MFG. CO., COLDWATER, MICH.

Guards

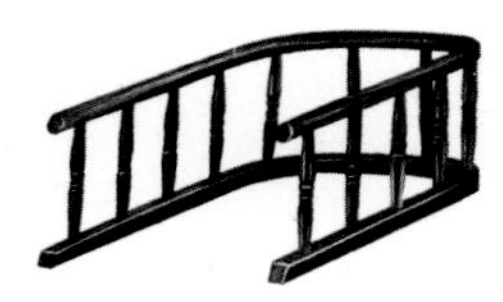

This guard can be attached to any of our sleds. Guard is six inches high, finished natural, strongly made, and easily attached by drilling four ⅜-inch holes through the raves of any sled, and using the four bolts furnished.

Two in crate. Weight 21 pounds **per dozen**

No. 334 G

Size: Width 14 inches, Length 38 inches

This shows the guard attached to our No. 334 sled as described on page 9.

Two in crate. Weight 93 pounds **per dozen**

No. 334 GW, same as No. 334 G, except that both sled and guard are finished by hand in white enamel. The board is decorated with flowers and the gear striped in attractive colors.

Two in crate. Weight 93 pounds **per dozen**

Page Twelve

PRATT MFG. CO., COLDWATER, MICH.

Guards

No. 142 G

This shows the guard attached to our No. 142 sled. See page 5 for description of No. 142.

Two in crate. Weight 70 pounds **per dozen**

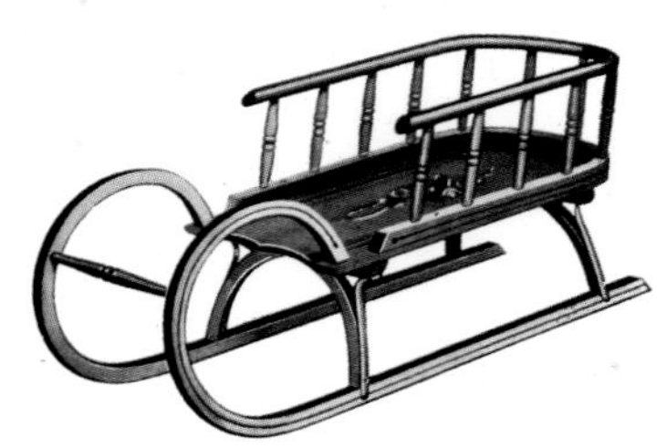

No. 332 G

This guard also looks well on this sled, as shown on page 9. In fact, it can be used on any sled and makes a very satisfactory substitute for a baby cutter.

Two in crate. Weight 85 pounds **per dozen**

Page Thirteen

PRATT MFG. CO., COLDWATER, MICH.

Doll Cutters

No. 13

No description of these Doll Cutters is necessary, except to say that they are painted red and neatly striped and ornamented. The runners are shod with flat iron. The cutter on the left is No. 13, the one on the right is discontinued.

Six in crate. Weight 60 pounds **per dozen**

Page Fourteen

PRATT MFG. CO., COLDWATER, MICH.

Baby Cutters

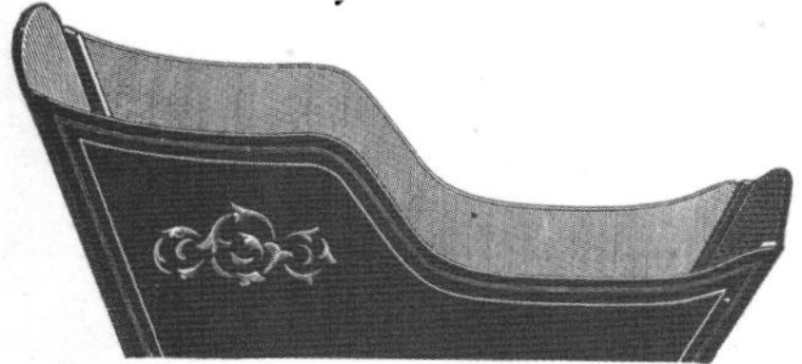

SINGLE BOXES

The illustration shows this box so plainly that it is only necessary to say that it will fit any of our sleds.

We also make the same style of the box in double form, for two children. In this the children face each other. Both these boxes are finished in red and are finely striped.

Six in crate. Weight 82 pounds **per dozen**

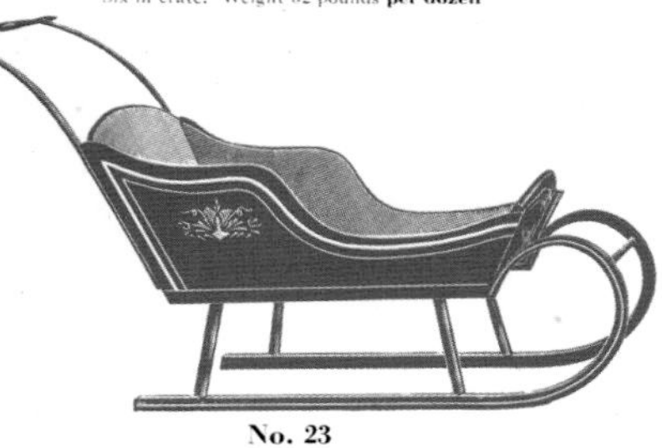

No. 23

Round knees, mortised into runner and rave. Brightly painted and striped outside, plain inside and shod with flat iron. Push handles will be furnished without extra charge. Can be furnished in white enamel at an extra charge.

Two in crate. Weight 36 pounds **per crate**

Page Fifteen

PRATT MFG. CO., COLDWATER, MICH.

Baby Cutters

No. 20

Size: Width 16 inches, Height 21 inches, Length 35 inches

This baby cutter has become so well known during the past twelve years that an extended description of it seems unnecessary except for our new trade, and to say that some slight improvements are made in its shape each season.

The bottom and dash is one piece of solid hardwood, the grain running lengthwise, the dash being bent to the required shape. A man can rest his full weight on the dash without in any way injuring it. The gear has two benches, with bent knees braced with tinned braces and shod with flat iron. The box is brightly finished outside and painted inside, with painted seat. Push handles are furnished without extra charge.

Furnished plain or upholstered in damask or plush. In ordering be particular to say whether plain or upholstered, and how upholstered. See price list.

Two in crate. Weight 37 pounds **per crate**

Page Sixteen

PRATT MFG. CO., COLDWATER, MICH.

Baby Cutters

No. 21

Size: Width 16 inches, Height 21 inches, Length 35 inches

This is built on the same principle as No. 20, but is more desirable, and of course, costs more. The gear has three benches and fenders, is braced with tinned braces and shod with half oval iron. The bodies are painted red, dark green or white, and have painted ornaments on the sides, back and dash.

The inside of the body is painted and the seat is upholstered. Reversible push handles are sent without extra charge.

Furnished plain or upholstered in damask or plush. In ordering be particular to say whether plain or upholstered, and how upholstered. See price list.

Two in crate. Weight 39 pounds **per crate**

Page Seventeen

PRATT MFG. CO., COLDWATER, MICH.

Baby Cutters

No. 80. Russian Portland

Size: Width 16 inches, Height 22 inches, Length 41 inches

This is one of the best of our line of baby cutters, the box being of extra length, and made of extra thick lumber. The cutter is long enough for a second child to be placed in the front end, and the high sides keep the robes and wraps well up around the child. It is finished in white enamel. The gear is shod with half oval iron, and braced with tinned braces. The body is upholstered in velvetta, and has wool plumes. Push handles are furnished without extra charge. Plumes and upholstered.

Crated singly. Weight 33 pounds **each**

Page Eighteen

Richards & Wilcox Manufacturing Company

It is not always the historic societies, research libraries, or museums that uncover clues to a company's past. That was the case with Richards & Wilcox Manufacturing Company in Aurora, Illinois. Thanks to Jim Pauzé and his printed ephemera, I have been able to put together a partial time line. The Wilcox Mfg. Co. dates back to 1880. They produced a sundry of items, such as carpet sweepers, roller shades, and barn door track and hangers, however there is no mention of sleds. According to historical records, the Richards Mfg. Co. was organized in 1903 to manufacture parlor and barn door hangers. This is where the facts are confusing. An advertisement dated 1901, courtesy of the Aurora Historical Society, shows a Richards 600 Little Giant Folding Sled with the caption "Made by Richards & Wilcox." This would not have been possible because Richards & Wilcox did not merge until 1910.

In the 1904 Richards Door Hanger Catalog there is a detailed description of the "Little Giant" sled with patent pending. Here is where the plot thickens. The actual patent for the "Little Giant " #600 was granted December 22, 1908 to Otis L. Beardsley of Chicago and another patent was also granted to Mr. Beardsley on July 1, 1913 showing the application of a steering bar. To date there is no information about the identity of Otis L. Beardsley other than that he was the patent holder of the folding sleds (Little Giant & Flex-O-Fold). By 1910 the Richards Flex-O-Fold #610 and #610, models 1 & 2, appear with a steering bar and in 1914 the Flex-O-Fold #610, model #3, was added to the line with a logo on the sled deck. In a response to a request by The Winter Sports Club of America for Richards & Wilcox to build a Sled for them, Richards & Wilcox respectfully declined in a letter dated January 29, 1919, stating that they had discontinued sled production several years before (1915 -1916). Despite sled production being discontinued, the Richards & Wilcox Mfg. Company continued to flourish and still exists today.

THE RICHARDS MANUFACTURING COMPANY : AURORA, ILLINOIS, U.S.A.

No. 600

Richards Little Giant Folding Sled

NOVEL: PRACTICAL: SIMPLE: STRONG

Size steel runners................ $\frac{3}{8}$-inch
Brace .... $\frac{7}{16}$-inch
Length .... 35 inches

Height .... $4\frac{3}{4}$ inches
Width .... $11\frac{1}{2}$ inches
Weight .... 8 pounds

List, per dozen...$15.00

Discount..............................

SHOWING THE SLED FOLDED

¶ The folding feature permits of economy in handling, transportation and storage. The user can carry the sled under his arm when climbing hills or riding on street cars. These sleds can be hung on the wall, out of the way, when not in use. *Runners and Braces*—Made of Bessemer Spring Steel. Securely attached with corrugated clips bolted to wood top with nickle-plated bolts. *Top*—Made of good quality bone dry hardwood. *Finish*—Steel parts finished in finest black baked enamel. Wood top finished in red, blue and green, well varnished and handsomely decorated. *Speed*—There are no braces to obstruct the speed of the sled, the only contact with the snow being the round steel runners which offer greater speed than any other type of runner. *Capacity*—We guarantee Little Giant and Flexofold Sleds to be stronger and more durable than any other sled of the same size or weight on the market. They are practically indestructible and will carry a load up to 1,000 pounds. Each sled wrapped in heavy express paper, retaining the lustre of finish. So crated as to entirely prevent loss by breakage.

PAGE 124 *Discount*..............................

Richards Manufacturing Product Catalog, circa 1903-1910. #600 Richards Little Giant folding sled. Courtesy of Jim Pauzé.

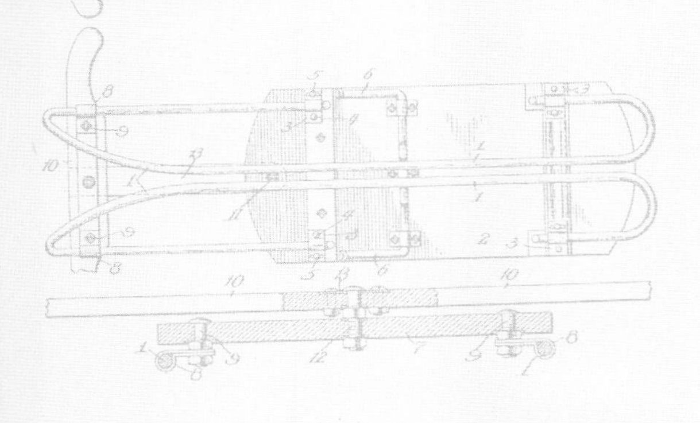

600—Richards Little Giant Folding Sled
The folding feature of the "Little Giant" appeals instantly to the buying public.
FOLDED—PATENT PENDING

600 Richards Little Giant folding sled advertisement. *From the collections of the Aurora Historical Society, Aurora, Illinois.*

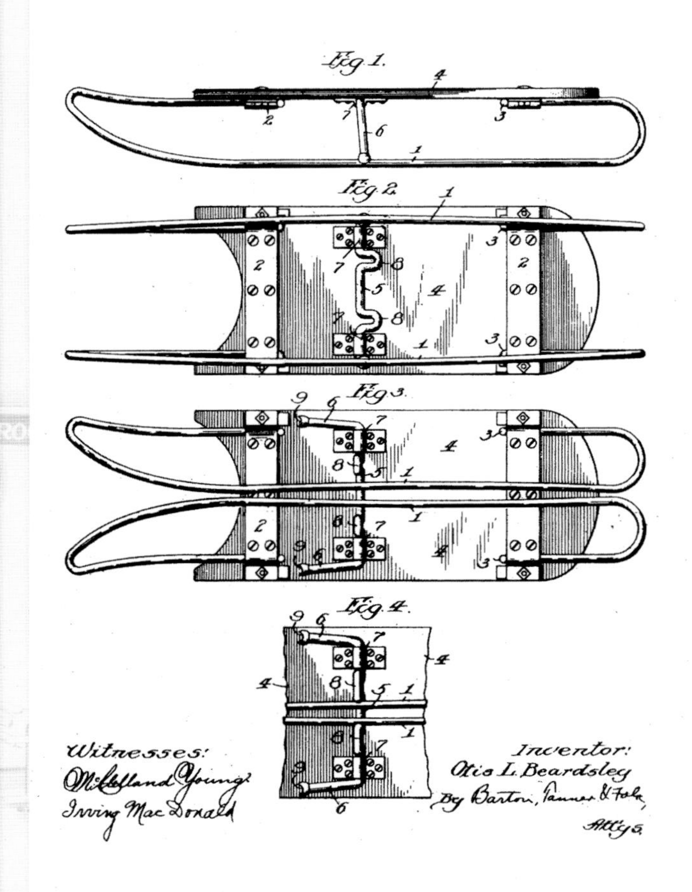

Patent drawing, 1908, for the Richard Folding Sled

THE RICHARDS MANUFACTURING COMPANY : AURORA, ILLINOIS, U.S.A.

Nos. 610-1 & 610-2
Richards Flexofold Steering Sled

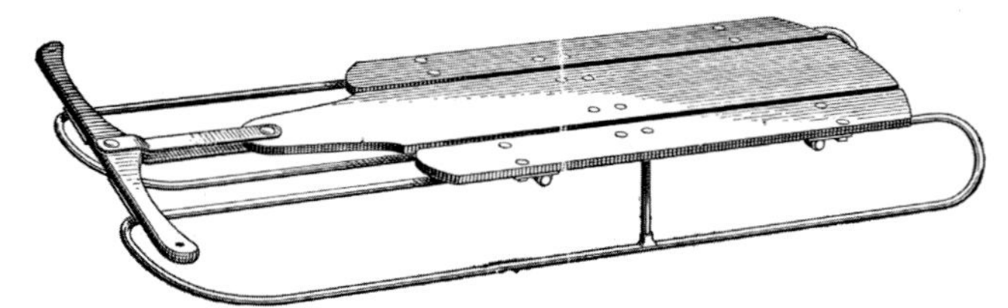

Dimensions, No. 610-1		Dimensions, No. 610-2	
Size steel runners	$\frac{5}{8}$-inch	Size steel runners	$\frac{7}{16}$-inch
Brace	$\frac{7}{16}$-inch	Brace	$\frac{7}{16}$-inch
Length	36 inches	Length	42 inches
Height	5 inches	Height	$5\frac{1}{2}$ inches
Width	$11\frac{1}{2}$ inches	Width	12 inches
Weight	$10\frac{1}{2}$ pounds	Weight	$17\frac{1}{2}$ pounds

No. 610-1, per dozen $30.00
No. 610-2, per dozen 39.00

Richards Flex-O-Fold sled, no logo, circa 1903-1909. *Courtesy of Jim Pauzé.*

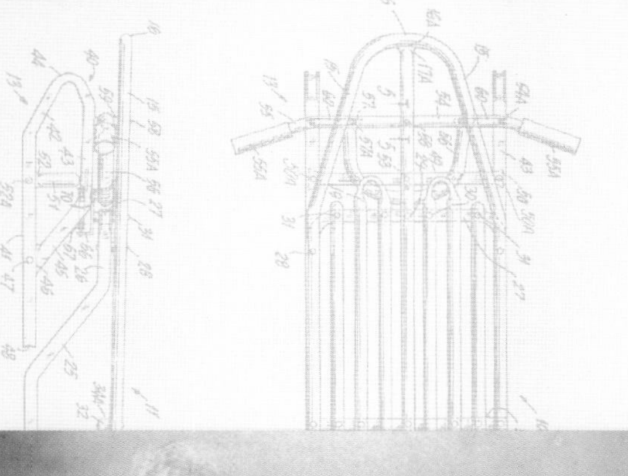

The Flexfold Sled

The Richards-Wilcox Mfg. Co., Aurora, Ill., describe their Flexfold Sled as the child's best winter chum. It folds up, making the carrying of it where it cannot well be drawn, both convenient and easy. It has round steel

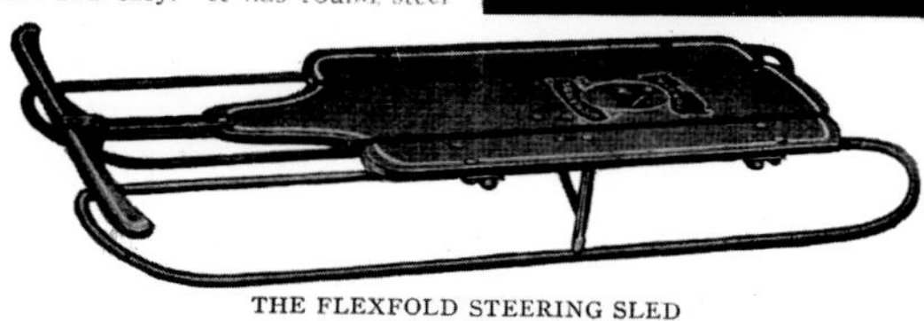

THE FLEXFOLD STEERING SLED

runners which are of course adjuncts of speed; is easy to steer; and is so sturdily constructed as not only to stand hard usage from the lively boy or girl, but will carry a load of one thousand pounds. The manufacturers call especial attention to these added points: "Economy of storage, handling and transportation. Every sled folds, taking up a very small space in your store. No danger of scratching, marring or breakage of sled in shipping, as each one is placed in heavy express paper bag and securely crated."

Richards & Wilcox advertisement, circa 1911-1915. Flex-O-Fold sled. Courtesy of Jim Pauzé.

"IT FAIRLY GLIDES"
JUST LIKE
RICHARD'S DOOR HANGERS.

1909			DECEMBER			1909
SUN	MON	TUE	WED	THR	FRI	SAT
-	-	-	1	2	3	4
5	6	7	8	9	10	11
12	13	14	15	16	17	18
19	20	21	22	23	24	25
26	27	28	29	30	31	-

MFG. CO. TRADE MARK

The Richards Manufacturing Company
AURORA, ILLS. - - - U. S. A.

Richards advertising card, 1909. *Courtesy of Jim Pauzé.*

RICHARDS-WILCOX MFG. CO., AURORA, ILLINOIS, U. S. A.

No. 600
R-W Little Giant Folding Sled
NOVEL, PRACTICAL, SIMPLE, STRONG

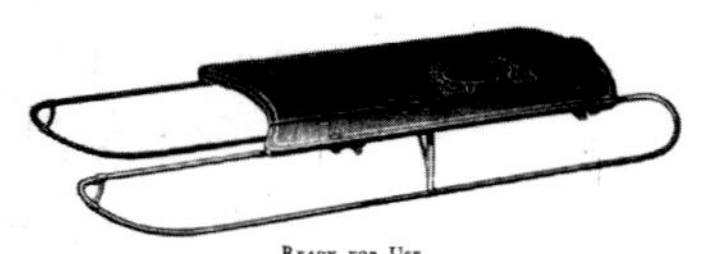

READY FOR USE

FOLDED (PATENTED AUGUST 10, '09)

The folding feature permits of economy in handling, transportation and storage. The user can carry the sled under his arm when climbing hills or riding on street cars. These sleds can be hung on the wall, out of the way, when not in use.

Runners—Bessemer spring steel, $\frac{3}{8}$-inch round. Securely attached with corrugated clips to wood top with bolts.

Braces—Malleable iron, $\frac{7}{16}$-inch round, with special design of ribs which secure minimum deflection or spring under pressure and weight. Securely attached to wood top with bolts.

Top—Made of good quality, bone-dry, hardwood.

Finish—Steel parts finished in finest black baked enamel. Wood top finished in red, blue or green enamel, well varnished and handsomely decorated. Each put in heavy express paper bag, retaining the lustre of finish.

Speed—There are no braces to obstruct the speed of the sled, the only contact with the snow being the round steel runners which offer greater speed than any other type of runner.

Dimensions—$11\frac{1}{2}$ inches wide, $4\frac{3}{4}$ inches high, 35 inches long.

Weight—8 pounds each. Packed one dozen in a crate.

PRICE LIST

No. 600 Little Giant Folding Sled, per dozen $18.00

Discount.........

Richards & Wilcox Catalog, Little Giant #600.
Courtesy of Jim Pauzé.

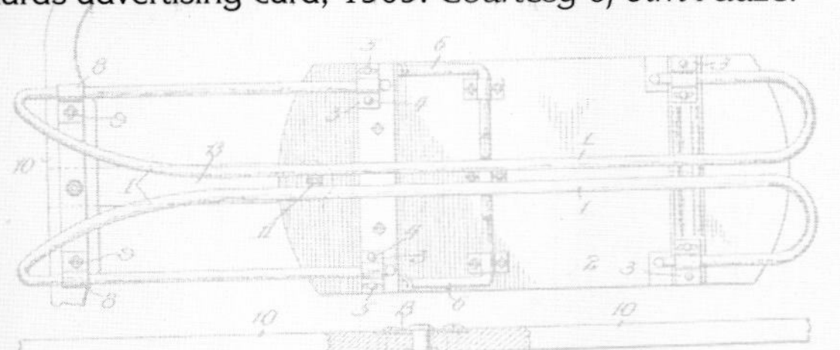

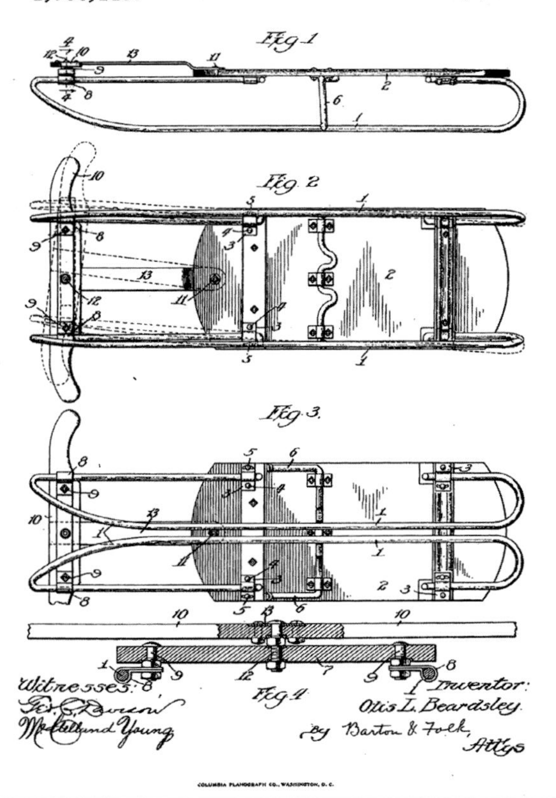

Patent drawing, 1913, for the Richards Folding Sled.

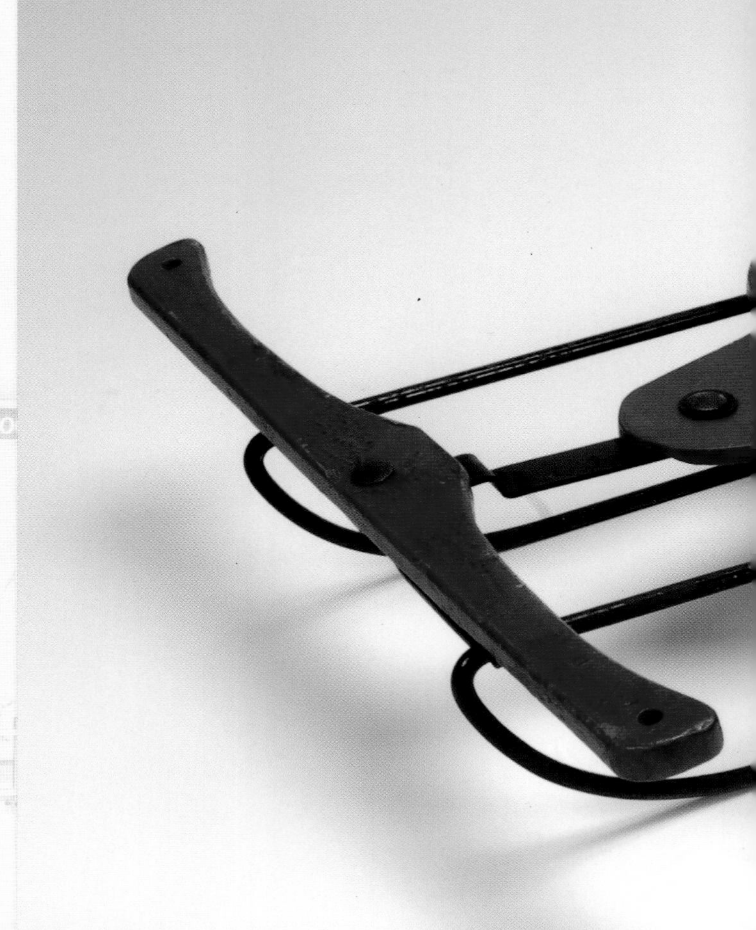

Richards-Wilcox, Flex-O-Fold #1, 1913. $200-400.
Courtesy of Jim Pauze.

No. 600
R-W Little Giant Folding Sled
NOVEL, PRACTICAL, SIMPLE, STRONG

READY FOR USE

FOLDED (PATENTED AUGUST 10, '09)

The folding feature permits of economy in handling, transportation and storage. The user can carry the sled under his arm when climbing hills or riding on street cars. These sleds can be hung on the wall, out of the way, when not in use.

Runners—Bessemer spring steel, $\frac{3}{8}$-inch round. Securely attached with corrugated clips to wood top with bolts.

Braces—Malleable iron, $\frac{7}{16}$-inch round, with special design of ribs which secure minimum deflection or spring under pressure and weight. Securely attached to wood top with bolts.

Top—Made of good quality, bone-dry, hardwood.

Finish—Steel parts finished in finest black baked enamel. Wood top finished in red, blue or green enamel, well varnished and handsomely decorated. Each put in heavy express paper bag, retaining the lustre of finish.

Speed—There are no braces to obstruct the speed of the sled, the only contact with the snow being the round steel runners which offer greater speed than any other type of runner.

Dimensions—$11\frac{1}{2}$ inches wide, $4\frac{3}{4}$ inches high, 35 inches long.

Weight—8 pounds each. Packed one dozen in a crate.

PRICE LIST

No. 600 Little Giant Folding Sled, per dozen..$18.00

Discount..................

Richards & Wicox Catalog, Flex-O-Fold #610. *Courtesy of Jim Pauzé.*

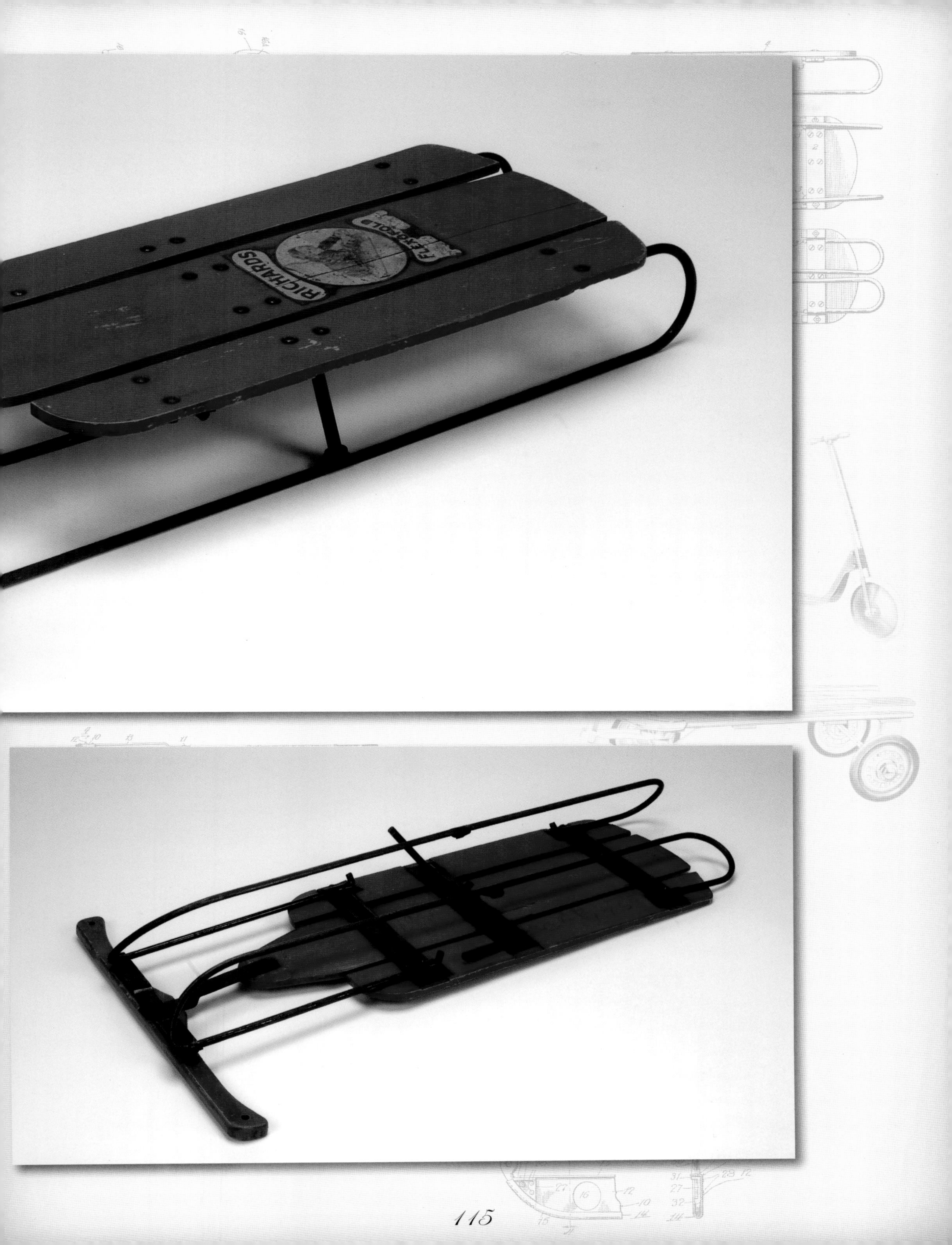
FLEXOFOLD
RICHARDS

The Safety Sled Manufacturing Co.

Frank Hornquist of Kane, Pennsylvania. was granted a patent on September 12, 1911, for an invention related to improvements in the construction of sleds, mainly a jointed runner that would allow for easier steering, making it safer, hence the name Safety Sled Manufacturing Co. It was the maker of the Safety Sled, Race-O-Plane and Easy Guider, not to be confused with the Lightening Guider, manufactured by the Standard Novelty Company, Duncannon, Pennsylvania. In January 1912, Mr. Hornquist and the Mt. Jewett Board of Trade reached an agreement that he would locate his Sled Manufacturing Co. in Mt. Jewett, Pennsylvania.

According to the local newspapers and townspeople, in November 1912, everyone agreed that the Safety Sled was without a doubt, "the lightest, strongest, most serviceable and most boy-pleasing sled they have ever seen, and it's bound to be a winner," and winner it was. With sales in the thousands, forty carloads of sleds were shipped to merchants near and far that same year. With business booming, Mr. Hornquist was granted another patent in 1914, for one-piece construction of benches, runners, and crossbars, a new concept in sled construction, along with non-skid runners, the first on the market in America. They are illustrated on a letterhead dated February 6, 1922, shown in this section. It is not clear when the first "Safety Wagons" were manufactured, but patents were granted to Mr. Hornquist improving on his wagons from the late teens to the mid-twenties. The year 1922 ended with a disastrous fire, claiming the Sled Company. According to an article published December 11, 1922, Mr. Hornquist stated "the plant will be rebuilt if the citizens rebuild it." The company went forward in 1923 and continued to produce sleds and wagons in Mt. Jewett until Mr. Hornquist sold it to C.J. Johnson, a businessman/banker from Johnsonburg, Pennsylvania. The Safety Sled Company relocated to Bradford, Pennsylvania. in May 1928, under the management of Lawrence Heddens and his wife Vera Hornquist Heddens, the new company became known as the C. J. Johnson Sled Company.

Chronology of the Safety Sled Manufacturing Co.

1911 • Frank Hornquist granted a patent to improve sleds

1912 • Safety Sled Manufacturing Co. opens in Mt. Jewitt, Pa.

1914 • Patents granted for one piece bench construction and non-skid runners, first on market in America. Wagon patents granted for improvements.

1922 • Year ends with disastrous fire.

1928 • Safety Sled Co. sold to C.J.Johnson and relocated to Bradford Pa. The Safety Line and Race-O-Plane Sleds and Wagons, continue to be manufactured

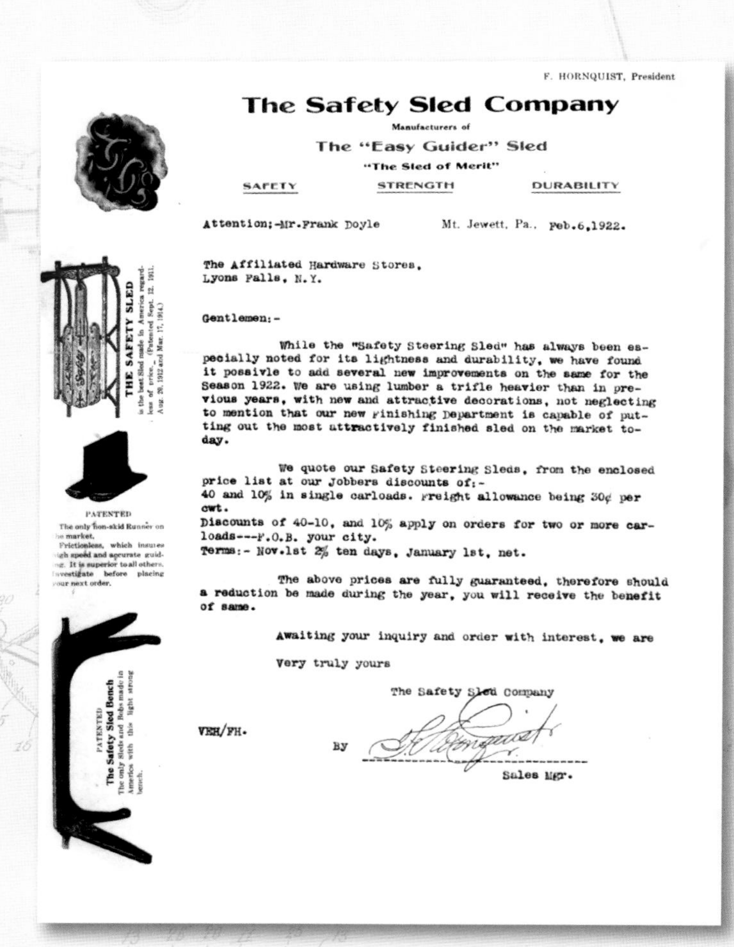

F. HORNQUIST, President

The Safety Sled Company

Manufacturers of

The "Easy Guider" Sled

"The Sled of Merit"

SAFETY STRENGTH DURABILITY

THE SAFETY SLED is the best Sled made in America regardless of price. (Patented Sept. 12, 1911. Aug. 20, 1912 and Mar. 17, 1914.)

PATENTED The only non-skid Runner on the market. Frictionless, which insures high speed and accurate guiding. It is superior to all others. Investigate before placing your next order.

PATENTED The Safety Sled Bench The only Sleds and Bobs made in America with this light strong bench.

Attention:-Mr.Frank Doyle Mt. Jewett, Pa., Feb.6,1922.

The Affiliated Hardware Stores,
Lyons Falls, N.Y.

Gentlemen:-

While the "Safety Steering Sled" has always been especially noted for its lightness and durability, we have found it possivle to add several new improvements on the same for the Season 1922. We are using lumber a trifle heavier than in previous years, with new and attractive decorations, not neglecting to mention that our new Finishing Department is capable of putting out the most attractively finished sled on the market today.

We quote our Safety Steering Sleds, from the enclosed price list at our Jobbers discounts of:-
40 and 10% in single carloads. Freight allowance being 30¢ per cwt.
Discounts of 40-10, and 10% apply on orders for two or more carloads---F.O.B. your city.
Terms:- Nov.1st 2% ten days, January 1st, net.

The above prices are fully guaranteed, therefore should a reduction be made during the year, you will receive the benefit of same.

Awaiting your inquiry and order with interest, we are

Very truly yours

The Safety Sled Company

VEH/FH. By [signature] Sales Mgr.

Safety Sled Company, original letter, February 6, 1916.

Safety Easy Guider Sled advertisement, c. 1914. *Courtesy of Barbara Heddens Hagg.*

C.J. Johnson Sled Company

The new sled company continued to manufacture the Safety Line and Race-o-Plane Sleds and Wagons. New products included lawn furniture, cots, baby gates, folding furniture, and other items. The Johnson Line included Diamond Sleds, skis, wagons, and scooters.

Shooting Star Wagons, the Saint Louis Special Wagon, and,in the late thirties, they introduced the Prize Winner and Streamline Klipper Sled. The last patent granted to Mr. Hornquist was July 1, 1936, for an adjustable guard railing for children's sleds or wagons.

With the outbreak of WWII, sled and wagon production halted. Over 5,000 cots were built at the Bradford plant and supplied to the army throughout the war. Sled and wagon production never resumed. After the death of Lawrence Heddens in 1964, his wife ran the factory until her death the following year. Their daughter Barbara Heddens Hagg managed the company until it closed its doors in 1978.

Chronology of the C.J. Johnson Sled Co.

1929 • Diamond Sleds, skis, wagons and scooters are manufactured by the C.J. Johnson Co. along with Shooting Star Wagons, and the Saint Louis Special Wagon

1935 • The Prize Winner and Streamline Klipper Sleds were introduced

1936 • July 1st, last patent granted to Mr. Hornquist.

1941 • Sled and Wagon production halted and never resumed

1978 • C.J. Johnson closed its doors

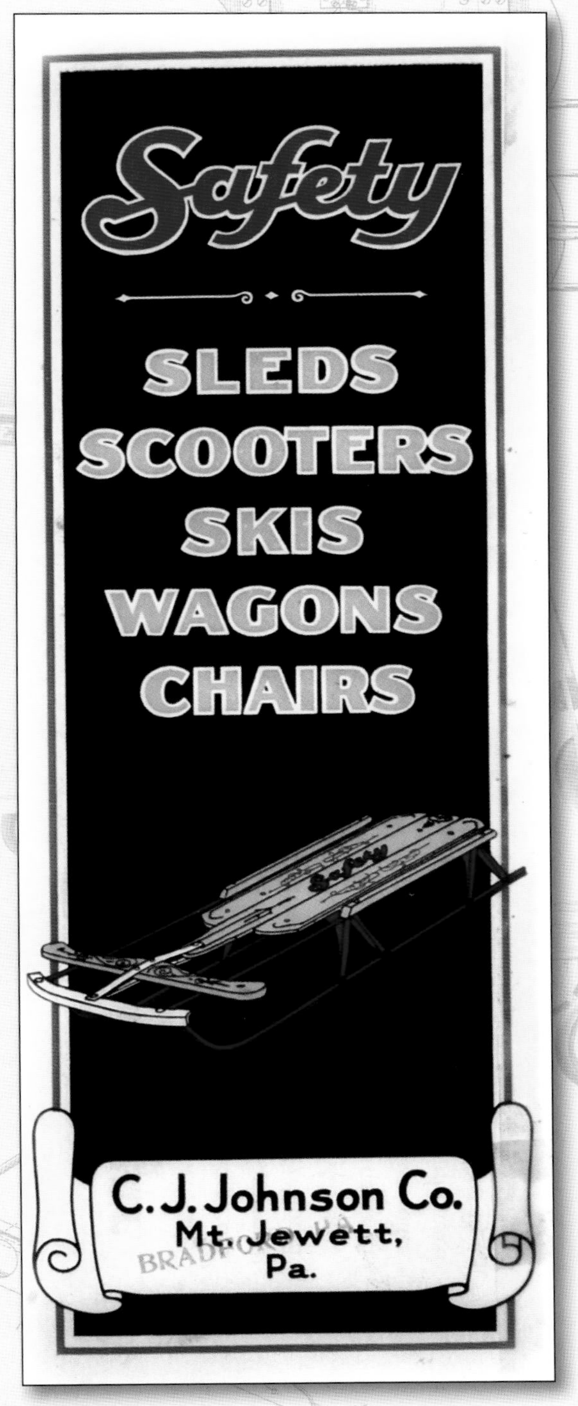

Safety brochure, C.J. Johnson Co., Mt Jewett, Pennsylvania, c. 1920. *Courtesy of Barbara Heddens Hagg.* *(continued on the next page)*

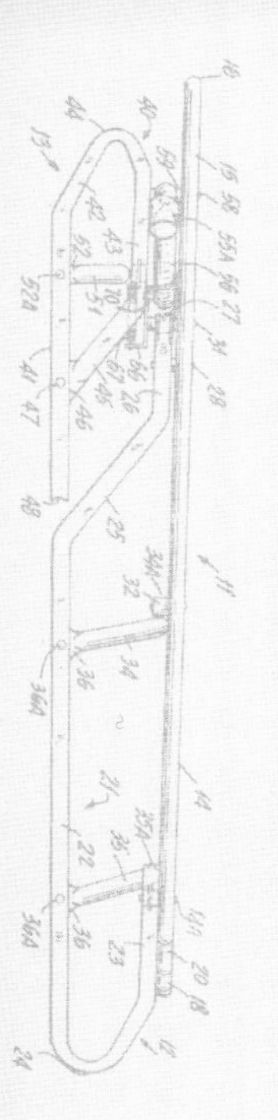

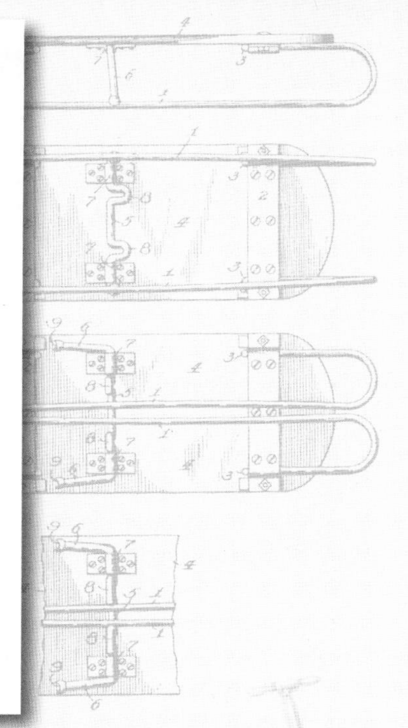

For Speed and Easy Guiding Safety

The King of Self Steering Sleds

The frictionless Non-Skid Runners insure high speed and accurate guiding. (Patented.)

Safety Special	No. 41	No. 51
Length, inches	41	51
Height, inches	6	6
Width, inches	13	13
Weight, pounds	8	13

Packed in Bundles of Two.

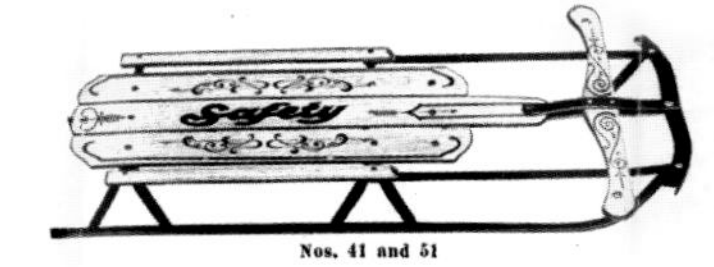

Nos. 41 and 51

THE SAFETY SLED BENCH
(Patented)

The Safety Sled Bench , a solid stamping of wrought steel, is lighter, neater in appearance, and has a strength far beyond any other form of construction. It is only half as heavy as a common sled bench, yet it carried 2200 pounds at the Toronto Exhibition.

The tops are of well seasoned, tough-textured white ash, attractively decorated in three colors and finished with two coats of water resisting varnish. All parts are securely riveted. There are no nails to pull out and no nuts or bolts to lose. The high carbon spring steel runners and Safety Sled benches are finished with a high grade bright red enamel. The new patented steel bumper used on the Nos. 41 and 51 has placed this sled far ahead of all other sleds with steel fronts.

The patented Non-Skid Runner, Safety Sled Bench and Easy Steering Device has made THE SAFETY the outstanding sled line.

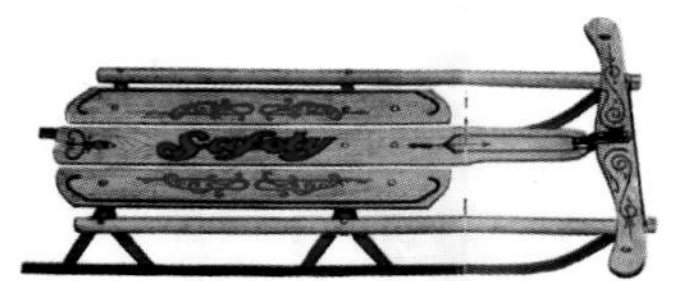

Nos. 27 to 41 Inclusive

	No. 27	No. 30	No. 33	No. 37	No. 41
Length, inches	27	30	33	37	41
Height, inches	5½	5½	5½	6	6
Width, inches	11¼	11¼	11¼	13	13
Weight, pounds	5¾	6	6½	7	7½

Packed in Bundles of Two.

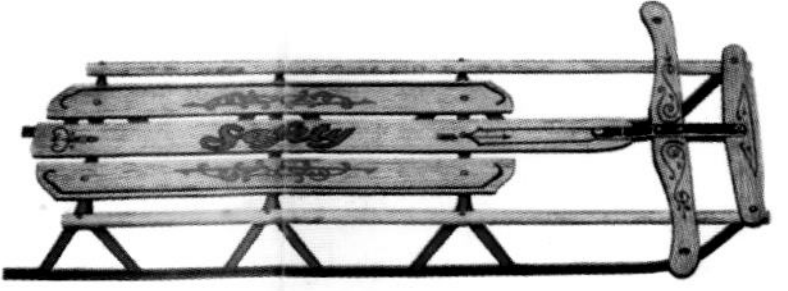

Nos. 46, 51 and 57

	No. 46	No. 51	No. 57
Length, inches	46	51	57
Height, inches	7	7	7
Width, inches	14	14	14
Weight, pounds	11½	13½	14½

Packed in Bundles of Two.

Safety Sled Guard

Patented

Safety Push Guard

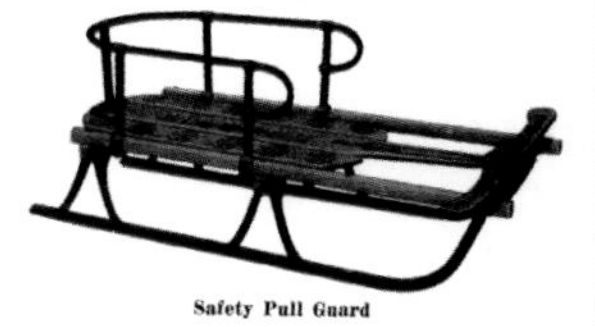

Safety Pull Guard

These guards are so constructed that they can be attached to any make of steering sleds without the drilling of holes. The metal braces clamp around the side rails and hold the guard firmly. Guards are seven inches high, finished in red enamel. Packed six in a carton. This improvement in sled guards has an instant appeal to every parent.

Safety Wheelbarrow

No. 27—Body 8x11 inches, handles 25 inches made of good quality hardwood lumber, finished with a special light oak varnish. The front moulding strip and steel legs are finished in bright red enamel. The six-inch wood wheel is grooved so as to give the appearance of a disc wheel. The edge is finished in red enamel and the center with special light oak varnish.

Packed one dozen to a carton, weighing 45 pounds.

Safety Wheelbarrow

Shooting Star

No. 3—34x14½x4" auto steel bed, sides reinforced with three beads and a large roll top, bottom reinforced with four hardwood cleats. Heavy steel bolsters securely braced to body, steel tongue and loop grip. All parts are finished in bright red enamel, name in golden yellow. Ten inch double disc roller bearing wheels with one inch tires and nickel plated hub caps. They are packed one in a carton, weighing 30 pounds.

Shooting Star

No. 3—34x14½x4" auto steel bed, sides reinforced with three beads and a large roll top, bottom reinforced with four hardwood cleats. Heavy steel bolsters securely braced to body, steel tongue and loop grip. All parts are finished in bright red enamel, name in golden yellow. Ten inch double disc roller bearing wheels with one inch tires and nickel plated hub caps. They are packed one in a carton, weighing 30 pounds.

St. Louis Special

Made from hardwood lumber with 21 inch tongue, and finished with a special light oak varnish. The front and rear mouldings as well as the edges of the wagon bed are finished in bright red enamel.

Our first two numbers 16 and 20 are equipped with wood wheels, which are grooved so as to give the appearance of disc wheels, the edge and running surface is finished in red enamel and the center with special light oak varnish.

The No. 21 wagon is in size, etc., identical to our No. 20 except that it comes equipped with double disc steel wheels with ½" rubber tires.

	Body	Wheels	Wt. Cart.
No. 16	16 x 9 in.	4 in. (wood)	30 lbs.
No. 20	20 x 9 in.	5 in. (wood)	38 lbs.
No. 21	20 x 9 in.	5 in. (disc)	40 lbs.

Packed: 6 in a carton

	Length	Height	Width	Wheels	Tire	Wt. Ea.
No. 1	30 in.	14¾ in.	14 in.	8 in.	¾ in.	26 lbs.
No. 2	34 in.	15¾ in.	15 in.	10 in.	1 in.	33 lbs.

Packed One to a Carton

Shooting Star

All lumber parts used in the construction of these wagons are made of thoroughly seasoned white ash. The sides of the boxes are moulded out of one piece of lumber, eliminating splitting or breaking off of moulding strips. The bottoms are reinforced by four cleats and equipped with our patented steel bolsters. Wood parts are finished with two coats of water resisting varnish, edges trimmed in bright red enamel.

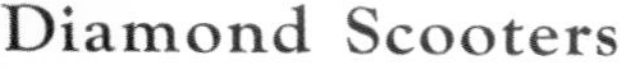

Diamond Scooters

When it comes to scooters, there's nothing like the DIAMOND.

It is equipped with fenders, stand, brake, self-contained roller bearing disc wheels with rubber tires and nickel plated hub caps. Complete in every detail.

The platform is built from a single piece of sheet stamping, heavily flanged in an attractive form with a light corrugation to prevent slipping of the foot.

Finished with a high gloss vermilion enamel, name in golden yellow.

	Length	Height	Wheels	Tire	Wt. Carton
No. 8	34 in.	30 in.	8 in.	¾ in.	22 lbs.
No. 10	36 in.	31 in.	10 in.	1 in.	24 lbs.

Packed two in a carton

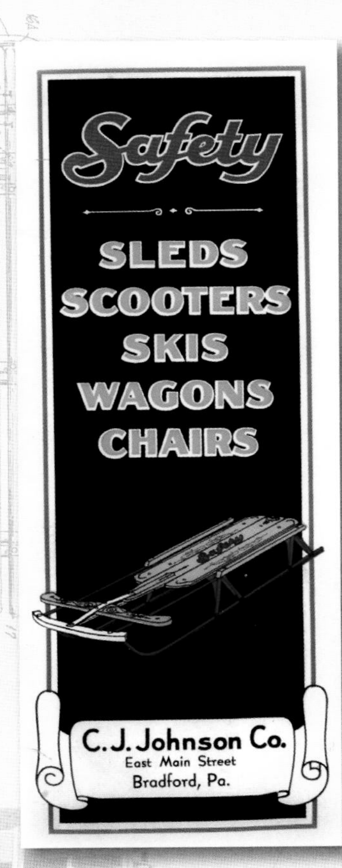

Safety

SLEDS
SCOOTERS
SKIS
WAGONS
CHAIRS

C.J. Johnson Co.
East Main Street
Bradford, Pa.

The frictionless Non-Skid Runners insure high speed and accurate guiding. (Patented.)

Race-O'Plane

For Speed and Easy Guiding

OUR PATENTED
NON-SKID RUNNERS, SLED BENCH, STEEL BUMPER FRONT AND EASY STEERING ARRANGEMENT MAKES THE RACE-O'PLANE,
THE PREMIER SLED LINE

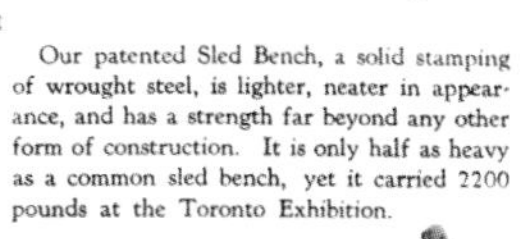

Our patented Sled Bench, a solid stamping of wrought steel, is lighter, neater in appearance, and has a strength far beyond any other form of construction. It is only half as heavy as a common sled bench, yet it carried 2200 pounds at the Toronto Exhibition.

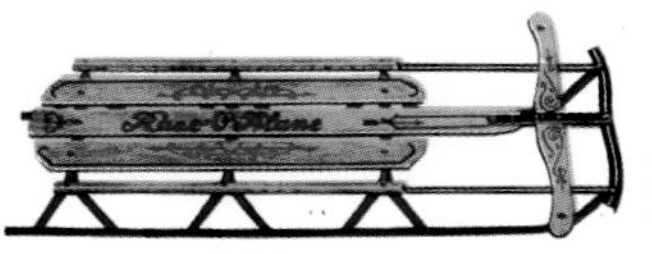

	No. 510	No. 560
Length, inches	51	56
Height, inches	6¾	6¾
Width, inches	14	14
Weight, pounds	12	13

Packed in bundles of two.

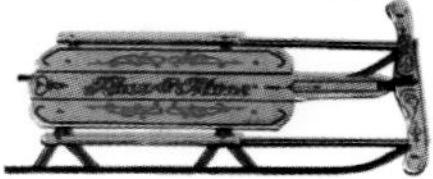

	No. 320	No. 360	No. 400
Length, inches	32	36	40
Height, inches	5½	5½	5½
Width, inches	11¼	12¾	12¾
Weight, pounds	6½	7½	8

Packed in bundles of two.

The tops are of well seasoned, tough-textured white ash, attractively decorated in three colors and finished with two coats of water resisting varnish. All parts are securely riveted. There are no nails to pull out and no nuts or bolts to lose. The high carbon spring steel runners and sled benches are finished with a high grade bright red enamel. The new patented steel bumper used on the Race-O'Plane has placed this sled far ahead of all other sleds with steel fronts.

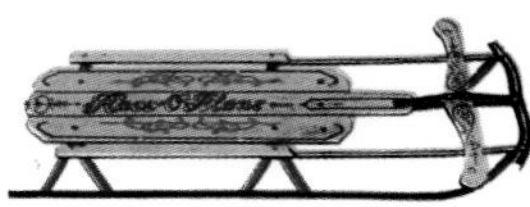

	No. 410	No. 450
Length, inches	40	45
Height, inches	5½	6¾
Width, inches	12¾	14
Weight, pounds	8½	11

Packed in bundles of two.

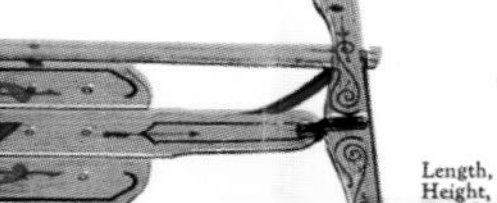

Safety

	No. 30	No. 33	No. 37	No. 41
Length, inches	29	32	36	40
Height, inches	5½	5½	5½	5½
Width, inches	11¼	11¼	12¾	12¾
Weight, pounds	6	6½	7	7½

Packed in bundles of two.

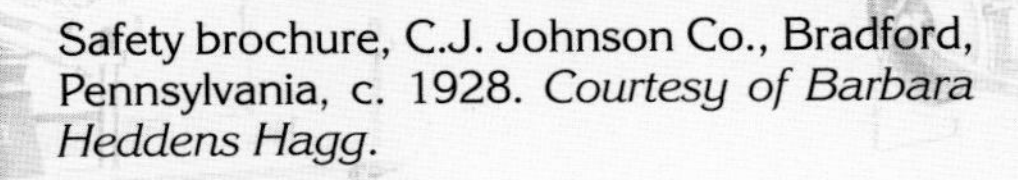

Safety brochure, C.J. Johnson Co., Bradford, Pennsylvania, c. 1928. *Courtesy of Barbara Heddens Hagg.*

Safety Push Guard

Safety Sled Guard

Patented

Safety Pull Guard

These guards are so constructed that they can be attached to any make of steering sleds without the drilling of holes. The metal braces clamp around the side rails and hold the guard firmly. Guards are seven inches high, finished in red enamel. Packed six in a carton. This improvement in sled guards has an instant appeal to every parent.

Diamond Scooters

When it comes to scooters, there's nothing like the DIAMOND.

It is equipped with fenders, stand, brake, shelf-contained roller bearing disc wheels with rubber tires and nickel plated hub caps. Complete in every detail.

The platform is built from a single piece of sheet stamping, heavily flanged in an attractive form with a light corrugation to prevent slipping of the foot. Finished with a high gloss vermilion enamel, name in golden yellow.

	Length	Height	Wheels	Tire	Wt. Carton
No. 8	34 in.	30 in.	8 in.	⅝ in.	22 lbs.
No. 10	36 in.	31 in.	10 in.	¾ in.	24 lbs.

Packed two in a carton

Race-O'Plane

No. 3—34x14x4" auto steel bed, sides reinforced with three beads and a large roll top, bottom reinforced with four hardwood cleats. Heavy steel bolsters securely braced to body, steel tongue and loop grip. All parts are finished in bright red enamel, name in golden yellow. Ten inch double disc roller bearing wheels with ¾ inch tires and nickel plated hub caps. They are packed one in a carton, weighing 30 pounds.

Race-O'Plane

High quality materials and workmanship enter into the construction of these wagons. Wood parts finished with two coats of water resisting varnish. Edge's trimmed in bright red enamel.

	Length	Width	Wheels	Tire	Wt. Ea.
No. 24	24 in.	10½ in.	6 in.	½ in.	9 lbs.
No. 30	30 in.	13 in.	8 in.	⅝ in.	25 lbs.
No. 36	36 in.	16 in.	10 in.	¾ in.	35 lbs.

No. 24 packed two in a carton

Nos. 30 and 36 packed one in a carton

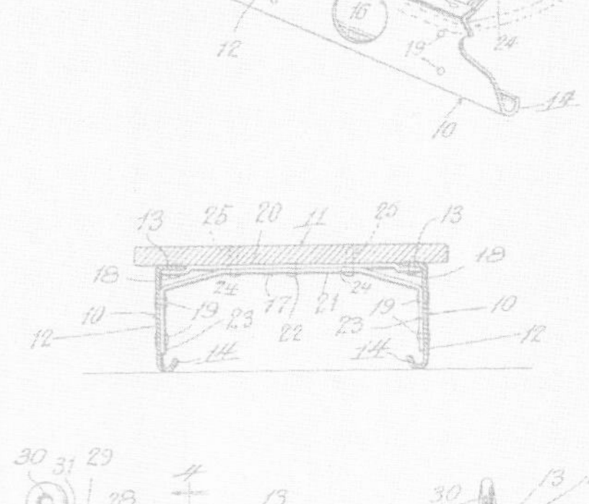

Safety SLED GUARD

PUSH GUARD

Packed, 6 in a carton, weighing 32 pounds

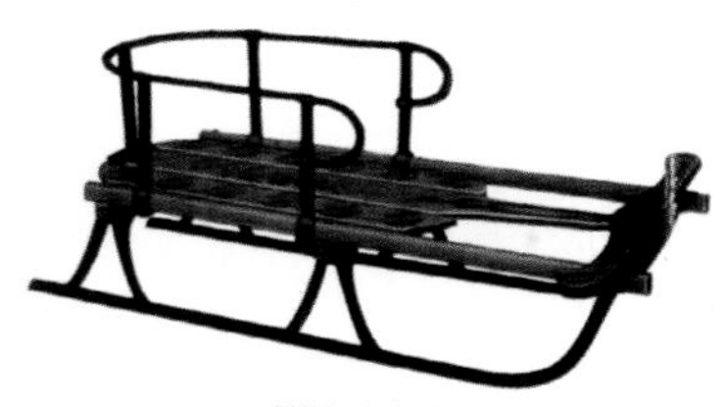

PULL GUARD

Packed, 6 in a carton, weighing 20 pounds

JUNIOR TOBOGGAN

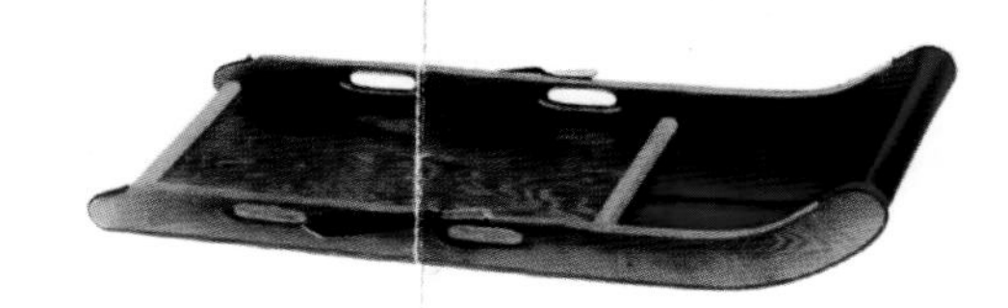

Through it's novel construction affords another source of coasting pleasure; insuring speed and safety.

The running surface is of steel construction with single longitudinal corrugation through the center to provide additional strength and prevent skidding. Edges of which, are re-inforced with strip steel to provide additional wear. Sides are equipped with hand-holes for the rider and two simple but efficient brakes, which operate equally as well for guiding, slowing-up or stopping.

Length	Height	Width	Weight, Per Doz.	Packed
39"	2¾"	14"	140 lbs.	bundles of two

These Guards are so constructed that they can be attached to any make of steering sleds without the drilling of holes, and are adjustable to different widths. The metal braces clamp around the side rails, and hold the guard firmly. This improvement in sled guards has an instant appeal to every parent.

Guards are seven inches high, finished in bright red enamel.

SAFE-MADE LINE

Also Includes

FOLDING-FURNITURE

SCOOTERS

WAGONS

C. J. JOHNSON COMPANY

BRADFORD, PENNSYLVANIA

Safe-Made Line, C.J. Johnson Compnay, c. 1930. *Courtesy of Barbara Heddens Hagg.*

RACE-O'-PLANE

for Speed and Easy Guiding

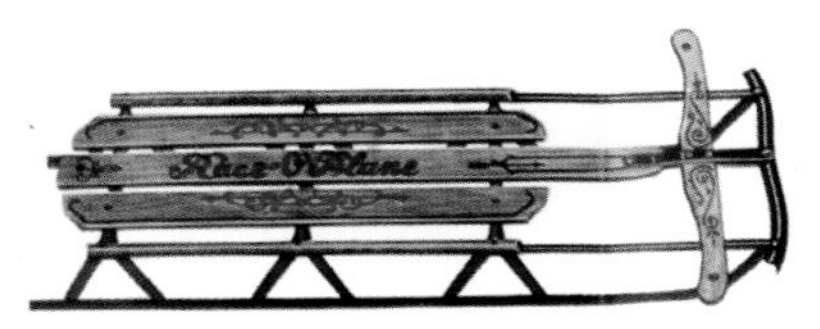

	No. 510	No. 560
Length, inches	51	56
Height, inches	6¾	6¾
Width, inches	14	14
Weight, per doz.	138	156

Packed in bundles of two

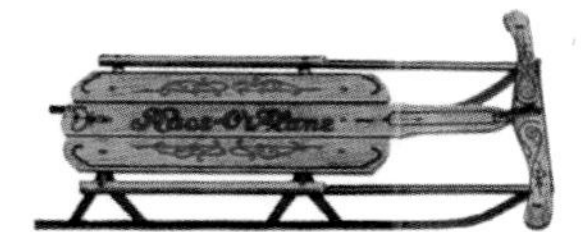

	No. 320	No. 360	No. 400
Length, inches	32	36	40
Height, inches	5½	5½	5½
Width, inches	11¼	12¾	12¾
Weight, per doz.	76	84	96

Packed in bundles of two

AT NO EXTRA COST, we offer you the following special, patented features, which are found exclusively in our line of Sleds:

A ONE PIECE BENCH, which supports the top boards and side-rails of the sled.

An ALL RIVITED CONSTRUCTION, which includes the top boards of the sled, thus preventing them from loosening or coming off, as is the case in the ordinary sled where the tops are nailed.

The STEEL BUMPER FRONT, a proven safeguard, formed for added strength, and in appearance is actually a bumper in place of an ordinary piece of iron as used on other sleds.

The lumber parts are of well seasoned, tough textured white ash, attractively decorated in three colors, and finished with two coats of water resisting varnish. The high carbon spring steel grooved runners, and sled benches, are finished with a bright red enamel.

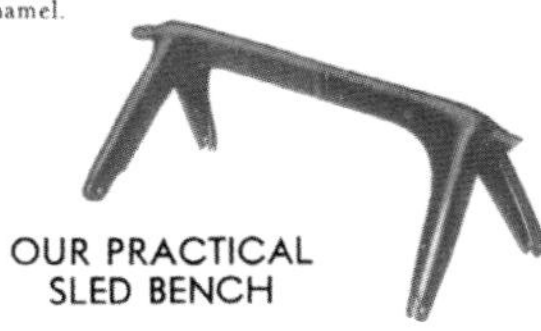

OUR PRACTICAL SLED BENCH

Our patented Sled Bench, a solid stamping of wrought steel, is lighter, neater in appearance, and has a strength far beyond any other form of construction. It is only half as heavy as a common sled bench, yet it carried 2200 pounds at the Toronto Exhibition.

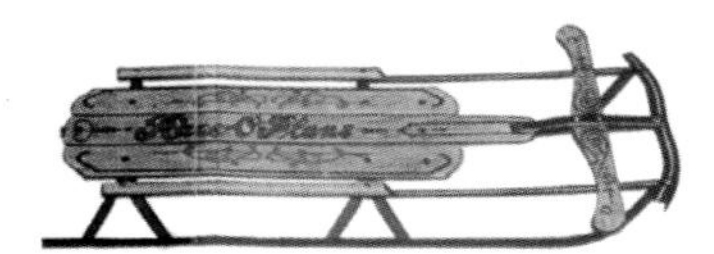

	No. 370	No. 410	No. 450
Length, inches	36	40	45
Height, inches	5½	5½	6¾
Width, inches	12¾	12¾	14
Weight, per doz.	85	96	115

Packed in bundles of two

Safety

	No. 33	No. 37	No. 41	No. 45
Length, inches	32	36	40	45
Height, inches	5½	5½	5½	5½
Width, inches	11¼	12¾	12¾	12¾
Weight, per doz.	70	78	84	90

Packed in bundles of two.

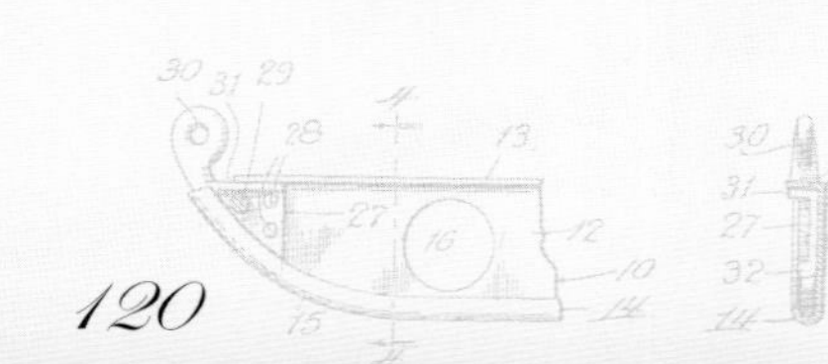

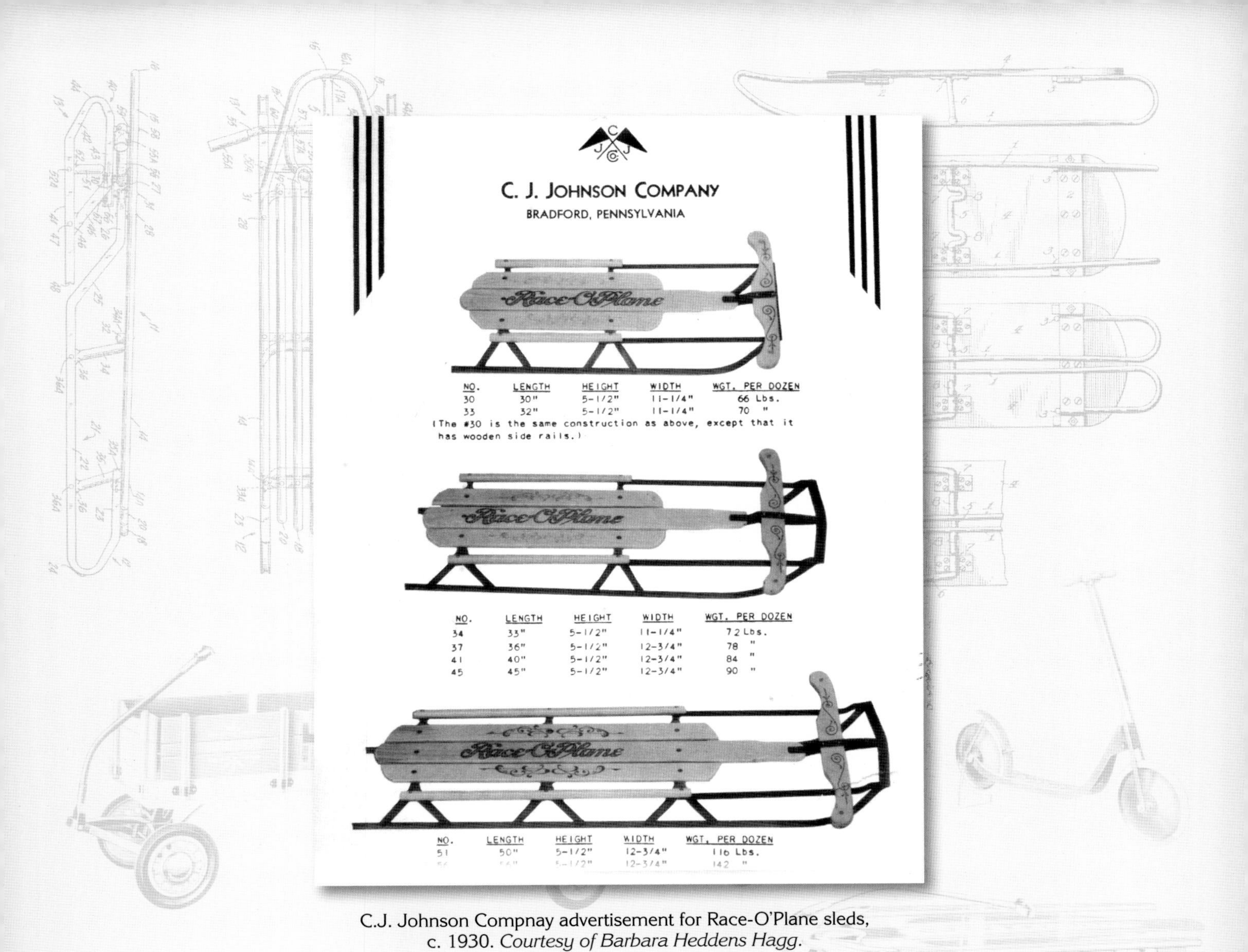

C.J. Johnson Compnay advertisement for Race-O'Plane sleds, c. 1930. *Courtesy of Barbara Heddens Hagg.*

Race-O'Plane

— THE EASIEST STEERING SLED —

OUR PATENTED NON-SKID RUNNERS, SLED BENCH, STEEL BUMPER FRONT AND EASY STEERING ARRANGEMENT MAKES THE RACE-O'PLANE,

THE PREMIER SLED LINE

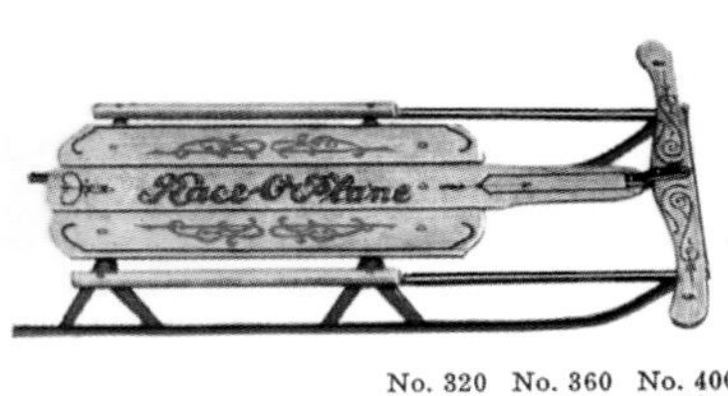

	No. 320	No. 360	No. 400
Length, inches	32	36	40
Height, inches	5½	5½	5½
Width, inches	11¼	12¾	12¾
Weight, pounds	6½	7½	8

Packed in bundles of two.

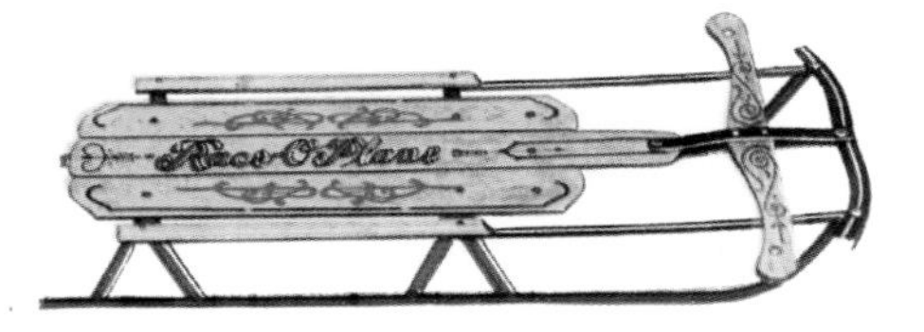

	No. 410	No. 450
Length, inches	41	45
Height, inches	5½	6¾
Width, inches	12¾	14¼
Weight, pounds	8½	11

Packed in bundles of two.

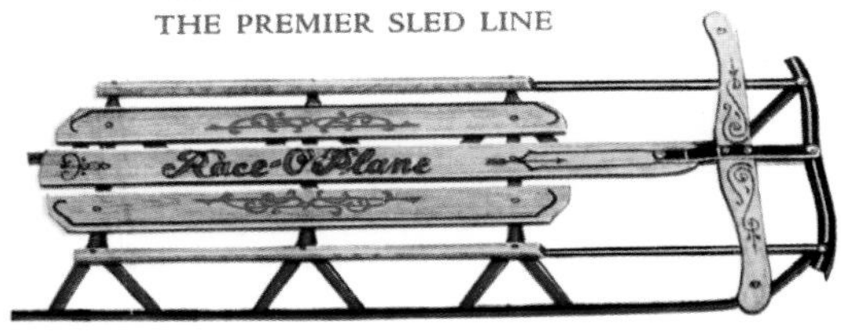

	No. 500	No. 510	No. 560
Length, inches	50	51	56
Height, inches	6¾	5	6¾
Width, inches	14¼	12¾	14¼
Weight, pounds	12	11½	14

Packed in bundles of two.

BUILT TO MEET THE DEMAND

POPULAR IN PRICE

C. J. JOHNSON COMPANY, BRADFORD, PENNSYLVANIA

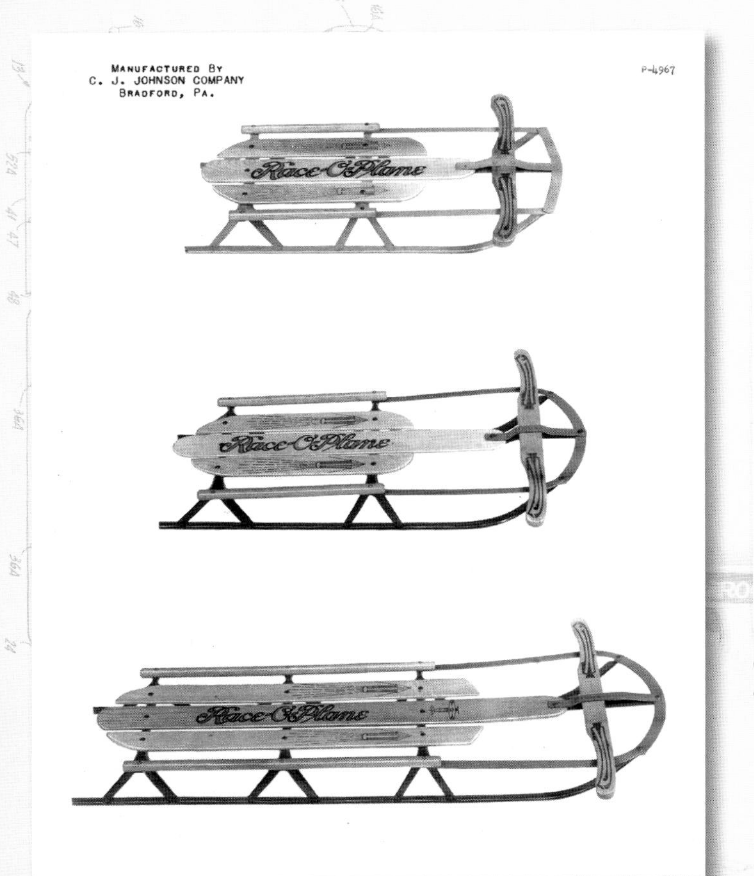

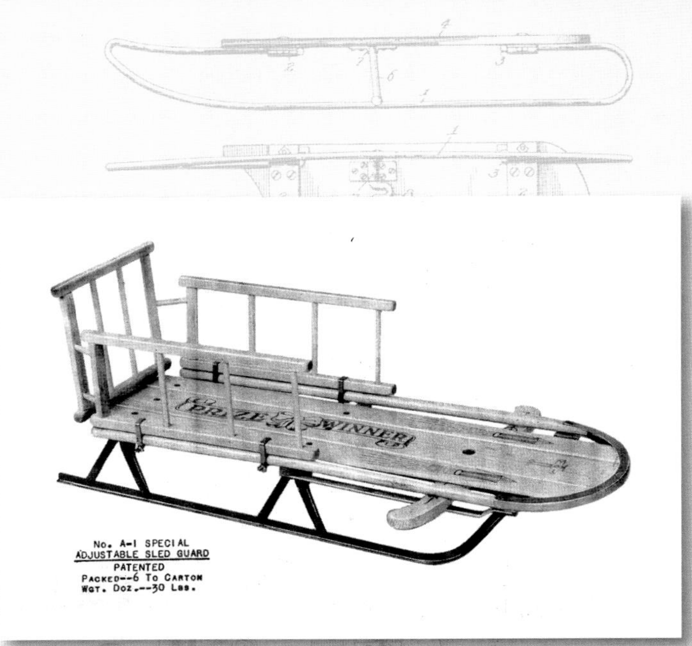

C.J. Johnson Compnay advertisement for Prize Winner Sled, c. 1935. *Courtesy of Barbara Heddens Hagg.*

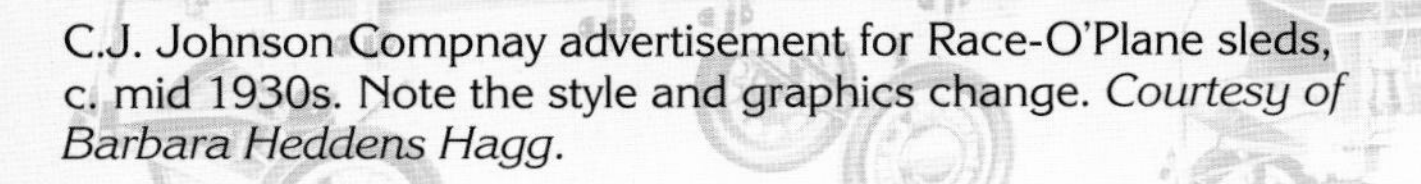

C.J. Johnson Compnay advertisement for Race-O'Plane sleds, c. mid 1930s. Note the style and graphics change. *Courtesy of Barbara Heddens Hagg.*

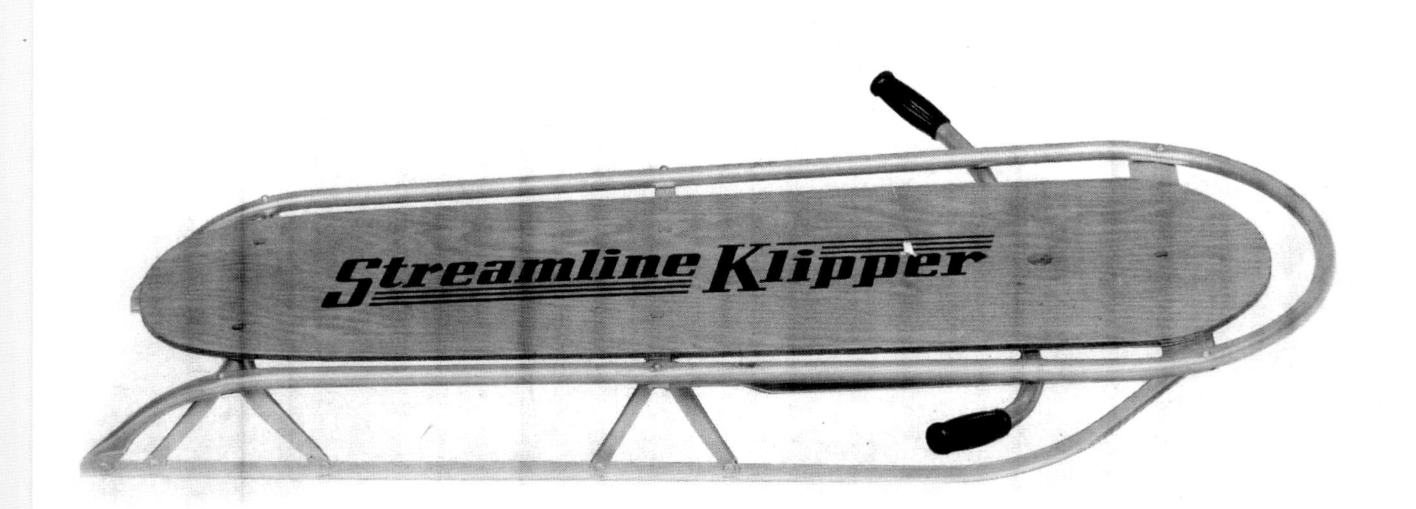

C.J. Johnson Compnay advertisement for Streamline Klipper Sled, c. 1935. *Courtesy of Barbara Heddens Hagg.*

Sherwood Brothers Mfg. Company

A patent was granted in 1914 for a new, ingenious invention called an "Auto Bob," in which an ordinary coaster could be changed from a Bob by adjusting a nut allowing a spring to loosen. This engaged the front set of runners creating a Bob Sled in less than one minute. In May 1914 William E. and John E. Sherwood founded the Sherwood Brothers Mfg. Company, taking over the holdings of the Tuttle Motor Company, under the control of the Watson Wagon Corporation, whose plant was located on the banks of the Erie Canal in Canastota, N.Y. By 1915, production was in full swing, one and a half years since the company was founded, 60,000 Auto Bobs had been sold. In October of that year William perfected a child's Coaster Wagon, predicting it would outsell the popular Auto Bob. Production for the unique "Spring Coaster Wagon" would begin in January 1916. The new coaster featured ball bearing wheels, a steering gear for easy operation, and the unique feature of a release mechanism for removing the box from the frame by pushing down on a spring, leaving the floor, so the wagon could be used as a coaster. The new wagon retailed for $3.75.

The year 1918 brought many changes due to World War I, however by 1919 Sherwood's business doubled and the demand for his Spring Coaster Wagons and Sleds exceeded expectations. Sherwood Brothers Mfg Company separated its offices from Watson Products Corporation, formerly known as Watson Wagon Corporation. Children's automobiles and wheelbarrows were added to the line. Later that year William succeeded in perfecting a great new improvement in steering sleds. His new invention was called the "Sherwood Steeroplane." By 1921, the popularity of the Sherwood Coasters gained national notoriety, when child star, Jackie Coogan, appeared in the *Syracuse Sunday Herald* taking a joy ride in his Sherwood Spring Coaster Wagon. F.C. Sylcox, Sales Manager of the Sherwood Brothers Mfg. Company immediately saw the opportunity to use star power to sell the Coasters, and he secured the consent to use the photo as an advertising tool. Despite management changes and rumors of impending plant shut downs, the Sherwood Company continued to survive. On February 1, 1924 the Sherwood Mfg. Company opened a new plant to manufacture roller skates, skis, and sleds. The roller skates and skis were new to the Sherwood Line. Samples were made and orders received well before displaying the complete line at the 9th annual New York Toy Show, February 4 to March 9, 1924. The famous Sherwood Spring Coaster Wagon continued to be produced exclusively at the original plant. A new model was introduced that same year, featuring rubber tires, single disc wheels, and finished in a rich cream color. Unfortunately two years later, on February 12, 1926, the Sherwood Brothers Mfg. Company met its demise, when De Ruyter creditors, petitioned them into bankruptcy, claiming they had never been paid for lumber sold to them. Canastotians can be proud of the rich history Sherwood Brothers Mfg. Company has given them. The very mention of the Sherwood Auto Bob or Spring Coaster Wagon brings a smile of delight to anyone who can still remember the thrill of the ride, and to the collector the high experienced at the mere thought of finding one in good to mint condition

Chronology of Sherwood Brothers Mfg. Co.

1914 • William E. Sherwood, granted patent for Auto Bob. Sherwood Brothers Mfg. Company, founded in Canastota, NY, to manufacture the Auto Bob and the Sherwood Junior (non Steerable sled)

1915 • William perfects child's Coaster Wagon & introduces the Sherwood Speeder

1916 • Production begins on "Spring Coaster Wagons"

1919 • "Sherwood Steeroplane" introduced, along with children's autos and wheelbarrows Sherwood Products are sold in Iowa, Illinois, Indiana, and Michigan under the name of "O.V.B." Representatives Hibbard, Spencer, Bartlett & Co. Chicago Ill.

1924 • New Coaster Wagons introduced finished in Cream color and equipped with rubber tires

1926 • Company closes due to bankruptcy

Sherwood Auto Bob #48, c.1914, $200-300.

MR. DEALER— Do You Want More Sales at Bigger Profits?

That is what the **Sherwood Auto Bob** offers. Twenty thousand were manufactured and sold during the last three months of 1914. One hundred thousand will be sold in 1915. The phenominal sale and our necessarily limited output is proof of our urging you to get our proposition at once. Our 1915 catalog is in the hands of the printer. Write for catalog and prices now.

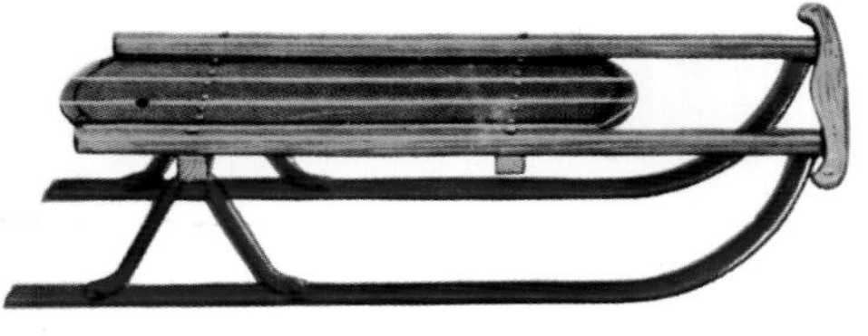

Sherwood Jr.
(Sled) **65c**

Length .... 28 in.
Height .... 6 in.
Width .... 11 in.
Weight .... 5 lbs.

No. 36
$2.00

Length .... 36 in.
Height .... 6 in.
Width .... 12 in.
Weight .... 9 lbs.

No. 42
$2.50

Length .... 42 in.
Height .... $6\frac{3}{8}$ in
Width .... $12\frac{1}{8}$ in.
Weight .... $10\frac{1}{4}$ lbs.

No. 48

No. 54

No. 48 — $3.25

Length .... 48 in. Width .... 14 in.
Height .... $7\frac{3}{4}$ in. Weight .... $17\frac{1}{2}$ lbs.

No. 54 — $4.00

Length .... 54 in. Width .... 14 in.
Height .... $7\frac{3}{4}$ in. Weight .... 19 lbs.

SHERWOOD BROS. MFG. CO., Inc. :: :: :: Canastota, N. Y.

MADE IN U. S. A.

Sherwood Auto Bob pinback button, c. 1915. Diameter: 1 ¼". $50-100.

Sherwood Brothers Mfg. Co. catalog, 1914. *Courtesy of Jim Pauzé.*

Experience has proven and it has been acknowledged by well known manufacturers that a small sled can not be operated with efficiency or safety. Sherwood Bobs overcome all this difficulty, as the steering device on a small size operates as easily as on the large size. That overcomes the steering problem which is the feature of the Sherwood Bobs.

MR. BUYER—place your order early, as the capacity of the season's output is limited and this is going to be the biggest selling proposition of the season. Prices and terms very interesting and quoted upon request.

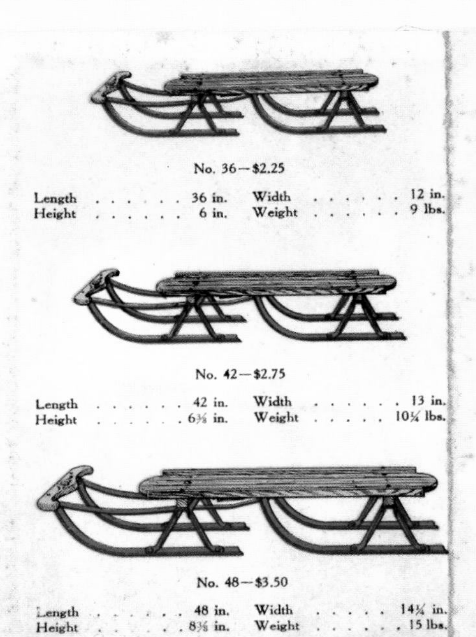

No. 36—$2.25

Length . . . . 36 in. Width . . . . . . 12 in.
Height . . . . 6 in. Weight . . . . . . 9 lbs.

No. 42—$2.75

Length . . . . 42 in. Width . . . . . . 13 in.
Height . . . . 6½ in. Weight . . . . . 10¼ lbs.

No. 48—$3.50

Length . . . . 48 in. Width . . . . . . 14¼ in.
Height . . . . 8½ in. Weight . . . . . . 15 lbs.

No. 54—$4.25

Length . . . . 54 in. Width . . . . . 14¼ in.
Height . . . . 8½ in. Weight . . . . . 17 lbs.

A successful buyer will investigate before he places his order this season

SHERWOOD AUTO BOB
TRADE MARK

Frank Wissig
737 West Lombard Street
Baltimore, Md.

Sherwood Auto Bob tri-fold, c. 1916. *Courtesy of Jim Pauzé.*

Photos, c. 1915, of Sherwood children riding Sherwood Spring Coaster Wagons, *Courtesy of the Canal Town Museum.*

Sherwood Spring Coaster Wagon #32. Body size: 14" x 32"; 8" artillery wheels. $500-750.

The Ladies' Home Journal for November, 1919 215

Built for Safety, Speed and Long, Rough Usage

SHERWOOD

SPRING COASTERS AND SLEDS

A *real* coaster wagon that children will enjoy for a long time. Pulls and glides so easily!—because of roller bearings. Rides *comfortably* because of 4 shock absorbing springs. Its fine, substantial construction—pressed steel frame and wheels, seasoned white ash body (removable)—mean long life in spite of rough usage.

Sherwood Sleds, too, provide unusual joy to the hearts of children. From the Junior model for the littlest tots to the clever foot-steering Auto-Bob model, they are all strong, rigidly braced, *quality*-made products, built for lasting service.

When children see a Sherwood Coaster or Sled their enthusiasm breaks loose immediately. When they actually own and use one their joy is good to see. Remember, both coaster and sleds keep children *outdoors* in healthful, wholesome, gladsome exercise.

On display by most department, hardware and toy stores. If not at your store, we will provide you direct. Write for Descriptive Matter.

SHERWOOD BROS. MFG. CO.

CANASTOTA, N. Y.

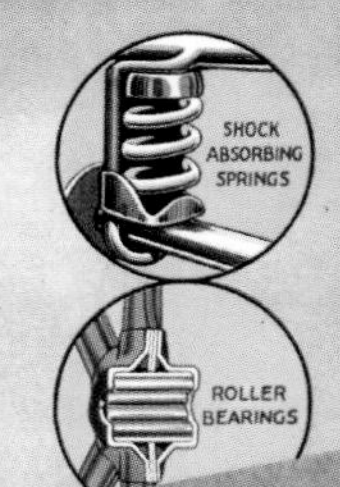

ROLLER BEARINGS

Sherwood Products are sold in Iowa, Illinois, Wisconsin, Indiana and Michigan under name of "O. V. B." Hibbard, Spencer, Bartlett & Co., Chicago, Ill., representatives.

Sherwood Spring Coaster advertisement in the *Ladies Home Journal*, 1919. $10-25.

The American Magazine 1919 285

SHERWOOD

SPRING COASTER

Great Outdoor Fun for the Children

Children have a world of fun with a Sherwood because it brings them what they most desire in a coaster wagon.

Roller Bearings—there's easy gliding speed. ***Steel shock absorbing springs***—there's comfortable, smooth riding, and protection from jar and jolt. ***All-steel wheels and frame***—there's safety and long wear under hard, rough usage.

The Sherwood is also a practical and handy wagon for moving heavy objects. Many a time mother or dad will be thankful for that easy-pulling wagon to lighten the day's work. Ends and sides are removable, making a platform wagon that will support a *ton*.

You can find them at most department, toy or hardware stores. If your dealer hasn't them, we will provide you direct. Write for Descriptive Circular.

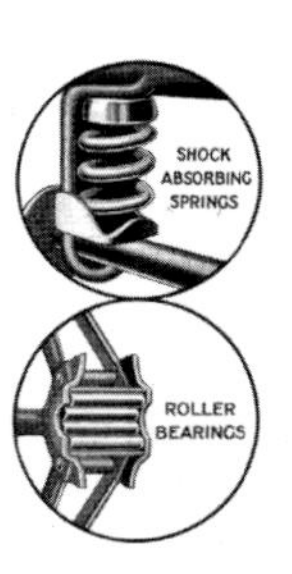

Sherwood Bros. Mfg. Co. **Canastota, N. Y.**

Sherwood Spring Coaster advertisement in the *American Magazine*, 1919. $5-10.

SHERWOOD SPRING COASTER

SHERWOOD SPRING COASTER

"See, These Springs Make My Coaster Ride So Smoothly"

FREE WHISTLE!

Sherwood Spring Coaster premium giveaway, with an advertisement for a free whistle, c. 1916. 5 ½" x 5". $25-50.

126 The American Magazine

SHERWOOD

SPRING COASTER

Speed, Comfort, Long Wear and Usefulness

Your boy or girl will have a world of fun with a Sherwood Spring Coaster.

Roller bearings for *speed*. Steel shock absorbing springs for *comfortable riding*, free from jar and jolt. All-steel wheels and frame for *safety* and *long wear*, even if it does receive rough usage.

Children find uses for it almost without end. It is at once a speedy coaster, a winning auto racer, a fire engine, a police patrol, a delivery wagon and what not. It carries heavy things from the grocer, and will help dad and mother with many a heavy load, around the house. Ends and sides are easily removable. Its steel-strength will support a *ton*.

There's a great deal of usefulness and no end of healthful outdoor fun for the children in a Sherwood Spring Coaster.

Sold by most hardware, department and toy stores. If not in your vicinity, we will supply you direct. Send for Circular.

SHOCK ABSORBING SPRINGS

ROLLER BEARINGS

SHERWOOD

Steeroplane Sleds

Exceptionally Easy to Steer

Your boy or girl will get a great deal more fun out of winter coasting with a Sherwood Steeroplane, because it steers so easily. The Steeroplane's improved arrangement makes it possible to draw the steering bar back all the way, with a light, easy pull.

A wide curve is thus obtainable on the runners which ensures real, effective steering, in a short radius.

This short turning is further ensured by the Steeroplane's *grooved runners* which grip the snow and hold the sled on its course.

Sherwood Bros. Mfg. Co. Canastota, N. Y.

Sherwood Steeroplane Sled advertisement in the *American Magazine*, 1919. $20-30.

Sherwood Auto Bob paperweight reproduction. $7-10.

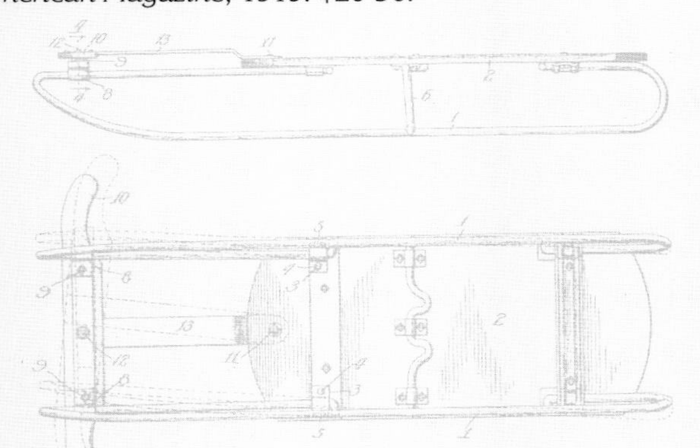

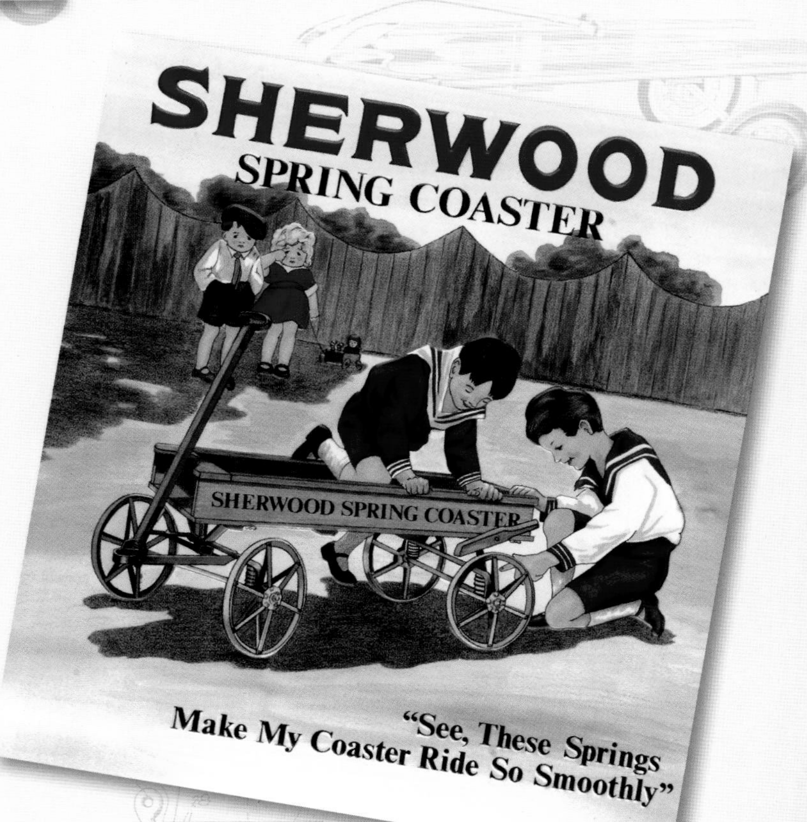

Sherwood metal sign, reproduction. $7-10.

Hibbard, Spencer, Bartlett & Company

It is no wonder that Sherwood Brothers and Garton Toys, sold their products through H.S.B.& Co. one of the largest Merchantile Houses in the country. Supplying products in Illinois, as well as Iowa, Indiana, and Michigan, under the company trademarks "O.V.B." ("Our Very Best!"), Rev-O-Noc, Cruso, & Hibbard. A large rooster accompanied the Cruso trademark. Rumor has it that the rooster was inspired by William Hibbard's son, who had a pet rooster by the name of Cruso

Sherwood's Spring Coaster wagon along with the Speeder were sold under the O.V.B. trademark. The "Rev-O-Noc" Coaster Wagon is none other than Garton's Badger Coaster Wagon, Cruso Express Wagon, is the Garton Bulldog and the Hibbard Ball Bearing Coaster is Garton's Dreadnaught Ball Bearing Coaster. H.S.B. & Companies largest competitor was Richards-Conover Hardware Company of Kansas City, Missouri, known to the trade by its nickname Rich-Con. Rich-Con supplied the Southwest Territory which included Missouri, Kansas, Oklahoma, Iowa, Nebraska, Colorado, Arkansas, Texas, New Mexico, Wyoming, and Arizona. It made perfect business sense to sell products under the "Rich-Con" name with such a vast territory. Without company catalogs, or advertisements, I have only been able to validate that Garton did sell one wagon, the "Iron Clad" under the Rich-Con trademark. The company closed its doors, in 1958, leaving behind a hardware empire that spanned eighty three years.

HIBBARD, SPENCER, BARTLETT & CO. STATE STREET BRIDGE

NEW OVB SPEEDER STEERING SLEDS

WITH ALL STEEL FRONT

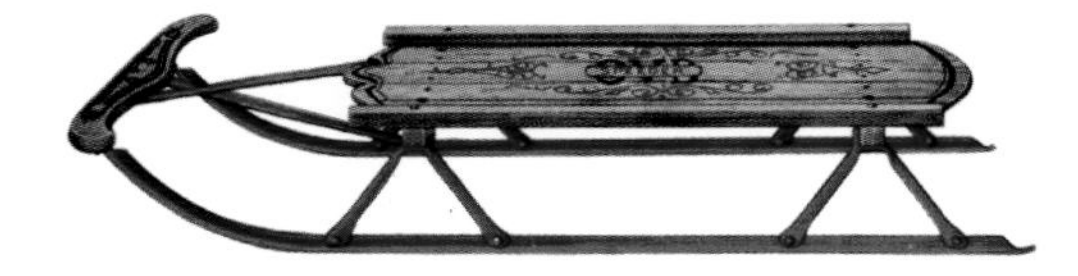

WE were first to introduce the idea of removing the side rail to the front, giving more strength and foot room, and this is now recognized by other manufacturers.

Year by year our sales have increased, and with our New Steering Device added, the OVB Speeder will lead all others.

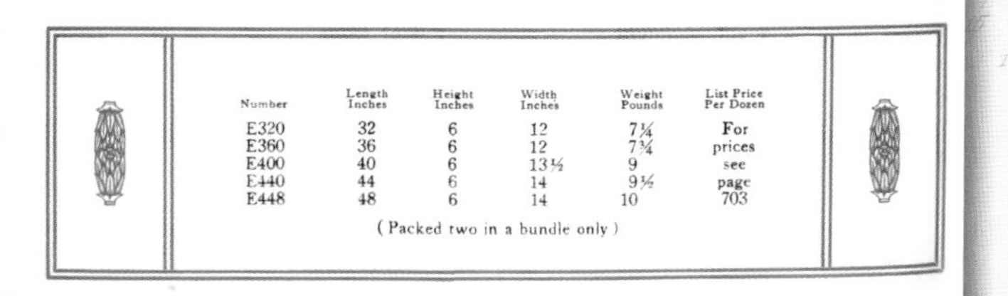

Number	Length Inches	Height Inches	Width Inches	Weight Pounds	List Price Per Dozen
E320	32	6	12	7¼	For
E360	36	6	12	7¾	prices
E400	40	6	13½	9	see
E440	44	6	14	9½	page
E448	48	6	14	10	703

(Packed two in a bundle only)

Hibbard, Spencer, Bartlett & Co., catalog, 1919. O.V.B. Speeder Sleds (Sherwood Brothers). *Courtesy of Jim Escher.*

HIBBARD, SPENCER, BARTLETT & CO. STATE STREET BRIDGE

OVB SPRING COASTER

The World's Best Coaster Wagon

OVB SPRING COASTERS are recognized by the Jobbers and Dealers to be the leader for sales and profits.

WHY? Because OVB was built for SERVICE, STRENGTH and DURABILITY. For easy riding the OVB is in a class by itself.

Follow the black arrows and read the reasons why OVB excels.

For Five Years the boys' most famous coaster.

Order now!

Increased manufacturing facilities enable us to make early and prompt shipments.

SIZES AND LIST PRICES

Number	Body Size Inches	Stamped Steel Wheel Wheel Size, Inches	Weight Pounds	List Price Per Dozen
N232	14 x 32	8	31	For
N234	14 x 34	10	34	prices
N238	16 x 38	10	38	see page
N240	18 x 40	10	43	701

The OVB Wheel

Everybody knows the O V B wheel to be the best and strongest wheel made.

It's of pressed steel as we said above. The spokes are hollow—practically two strong spokes in one. The tire, too, is of strong steel.

Then see how the roller bearings are concealed so that grit can't get in. That's why O V B wagons run easily *all the time.*

Sectional View of the OVB Wheel

Hibbard, Spencer, Bartlett & Co., catalog, 1919. O.V.B. Spring Coaster Wagon (Sherwood Brothers). *Courtesy of Jim Escher.*

Hibbard, Spencer, Bartlett & Co., Rev-O-Noc trademark photo, c. 1919. *Courtesy of Jim Escher.*

HIBBARD, SPENCER, BARTLETT & CO. — STATE STREET BRIDGE — 2055

REVONOC EXPRESS WAGONS

olid body of 24 ga. steel, outside painted vermilion, ornamented with gold stripes, inside green. Solid substantial wheels and handle, painted vermilion. Steel loop grip.

IRON HUB WHEELS, ¾ IN. SQUARE HANDLE

Nos.	N70	N71
Height of body, in.	4½	4½
Body, in.	12×24	13×26
Wheels, in.	8 & 12	9 & 13
W't doz., crated, lbs.	264	282
Per dozen	$71.10	78.60

Nos.	N72	N73
Height of body, in.	5	5
Body, in.	14×28	15×30
Wheels, in.	11 & 14	12 & 15
W't doz., crated, lbs.	300	336
Per dozen	$86.10	93.70

WAGONS SIXTH DOZEN IN A CRATE. WE DO NOT BREAK CRATES.

EXTRA HANDLES

No. N071—Per dozen $3.30

EXTRA WHEELS

Nos.	N078	N079	N0711	N0712
Inches	8	9	11	12
Per dozen	$10.00	10.50	10.50	10.50

Nos.	N0713	N0714	N0715
Inches	13	14	15
Per dozen	$10.50	10.50	10.50

CRUSO EXPRESS WAGONS

olid body of 25 ga. steel, painted red, stenciled in gold. Extra heavy steel wheels painted black, nickel hub caps, wrought gear with extra brace on rear axle, finished in black coke enamel, 7/16 in. round steel axles, bent hande, grip, painted red, four cleats across bottom.

Handle ⅞ in. Square.

	N81	N83
ght of body, in.	5⅛	5⅛
, in.	12×26	14×30
els, in.	10	11
doz., crated, lbs.	204	252
dozen	$47.40	56.15

O IN A CRATE. WE DO NOT BREAK CRATES

EXTRA AXLES

	N704	N714	N724	N734
agons No.	N70	N71	N72	N73
zen	$6.60	6.60	8.10	8.10

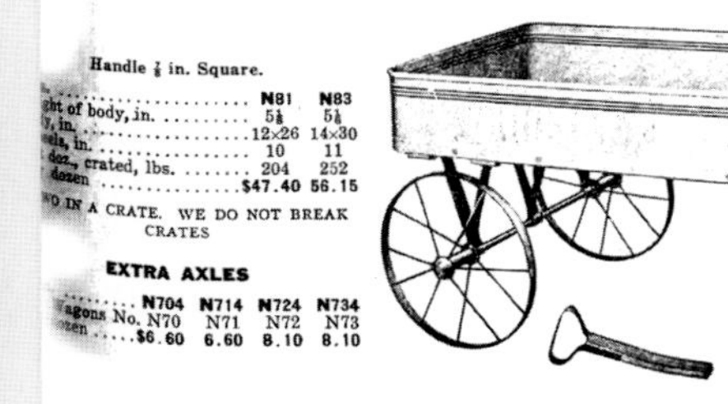

Hibbard, Spencer, Bartlett & Co., catalog, 1923. Rev-O-Noc Coaster Wagon (Garton's Badger Coaster), Cruso Express Wagon (Garton's Bulldog Wagon). *Courtesy of Jim Escher.*

2054 — HIBBARD, SPENCER, BARTLETT & CO. — STATE STREET BRIDGE

Hibbard BALL BEARING COASTER WAGONS

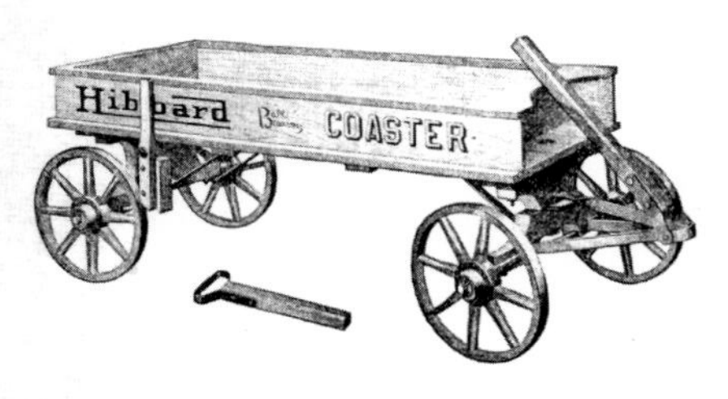

Removable body. Body and handle of well seasoned hardwood, varnished, body decorated in red and black. Extra heavy wood wheels painted red and fitted with serviceable brakes, iron hubs and frame, rims of bent rock elm, wrought steel tires welded and shrunk on, gear of selected stock securely braced to body, steel loop grip on handle.

1½ in. Square Handle.

Nos. Hibbard	N89	N90	N91	N92
Height of body, in.	4½	4½	4½	4½
Size body, in.	13×28	14×32	16×36	18×40
Wheels, in.	8	10	10	10
Shipping w't, lbs.	224	432	480	552
Per dozen	$83.20	106.35	117.95	131.90

ONE IN A CARTON

EXTRA WHEELS FOR NOS. N90, N91, N92

No. N090—Per dozen $14.40

BADGER COASTER WAGONS

WITH HEAVY METAL BEARING WHEELS

Body, handle and brace of well seasoned hardwood, varnished, extra heavy metal bearing wheels with round spokes, ¾×½ in. bent rock elm rims, ¾×½ in. wrought steel tires welded and shrunk on, malleable iron hub caps and bands. Al selected stock gears, securely braced to body with 1/32×¾ in. round edge iron and 9/16 in. round steel axles. Extra heavy malleable iron and hardwood steering gears

STATIONARY BOX, 3 INCHES HIGH

No. N32—Body 13×28 in., wheels 8 in., w't per doz. crated 245 lbs., per dozen $67.00

REMOVABLE BOX

Nos.	N33	N34	N35
Height of body, in.	4½	4½	4½
Body, in.	14×32	16×36	18×40
Wheels, in.	10	10	10
Shipping w't, lbs.	390	420	450
Per dozen	$92.50	104.10	117.95

EXTRA WHEELS

Nos.	N032	N033
Size, in.	8	10
Per dozen	$10.50	10.50

WITH MALLEABLE IRON WHEELS. PLAIN BEARING.

Body, handle and brace of well seasoned hardwood, varnished, Al selected stock gears, securely braced to body with 1/32×¾ in. round edge iron and 9/16 in. round steel axles. Extra heavy malleable iron and hardwood steering gears.

STATIONARY BOX, 3 INCHES HIGH

No. N532—Body 13×28 in. wheels 8 in., w't per dozen crated 240 lbs. .... per dozen $63.00

REMOVABLE BOX

Nos.	N533	N534
Height of body, in.	4½	4½
Body, in.	14×32	16×36
Wheels, in.	10	10
Shipping w't, lbs.	444	474
Per dozen	$90.00	99.00

NOS. N32 AND N532, ONE IN A CRATE; OTHERS SIXTH DOZEN IN A CRATE. WE DO NOT BREAK CRATES

Hibbard, Spencer, Bartlett & Co., catalog, 1923. Hibbard Ballbearing Coaster (Garton's Dreadnaught Ballbearing Coaster). *Courtesy of Jim Escher.*

2056 HIBBARD, SPENCER, BARTLETT & CO. STATE STREET BRIDGE

STEEL EXPRESS WAGONS

Steel body and wheels, body painted bright vermilion, striped and varnished, wheels painted black, inside painted green. Top of body bent over wrought iron rod, making it very firm, rounded corners, steel gears. Wood handle with cross bar grip.

STEEL BODY AND WHEELS

Handle ¾×½ in.

Nos.	N04	N02
Height of body, in.	2¾	[illegible]
Body, in.	8×18	[illegible]
Wheels, in.	6	8
W't doz. crated, lbs.	70	90
Per dozen	$21.70	25.[illegible]

Handle ⅞ in. Square

Nos.	N1	N3	N4
Height of body, in.	5½	5½	[illegible]
Body, in.	12×26	14×30	[illegible]
Wheels, in.	10	11	11
W't doz., crated, lbs.	160	234	245
Per dozen	$39.15	47.40	55.[illegible]

No. N04 ONE THIRD DOZEN; BALANCE ONE SIXTH DOZEN IN A BOX

FARM WAGONS

Body 18×36×7 in., bent hounds, adjustable reach, all parts strongly ironed and braced, 14 and 20 in. wheels, shaved spokes, heavy welded tires, plate wrought iron hub boxes and hub caps 9/16 in. round steel axles, complete with seat and tongue.

No. N3550—W't per doz. 660 lbs., per dozen $185.40

Extra hardwood shafts for dog or goats, w't per doz. 60 lbs., per dozen $21.00

TWO IN A CRATE WE DO NOT BREAK CRATES

This resembles a farm wagon in every respect. The sides and ends can be taken off, leavng bed with stakes.

SEASONED HARDWOOD FRAME AND WHEELS

Hibbard, Spencer, Bartlett & Co catalog, 1923, Garton's Iron Clad Wagon. *Courtesy of Jim Escher.*

Richards-Conover Hardware trademark, Rich-Con, photo c. 1919 (Garton's Iron Clad Wagon.) *Courtesy of Jim Escher.*

The Wilkinson Manufacturing Company:

Prior to 1900 Binghamton, New York, was a hub for the manufacturing of sleighs, wagons, children's carriages, and sleds. In 1860, The Binghamton Sled Company was formed. In 1863, The Winton Manufacturing Company was founded for the purposes of manufacturing children's carriages by Winton & Doolittle. The business continued until 1875, when it was organized into a stock company; for the next seven years they made Sleds exclusively.

Unfortunately there are no known catalogs, but only an advertisement from 1869 in Boyd's Binghamton Directory (courtesy Broome County Historical Society). It is not clear when Davis, Wilkinson & Co., manufactures of children's carriages, sleighs, & velocipedes, was established, however, by 1879 Wilkinson & Eastwood were manufacturing boy's express wagons and sleds, according to the Binghamton City Directory. After thirty years and several name changes The Wilkinson Manufacturing Company emerged in 1890, producing juvenile autos, police & fire wagons, express, delivery, & farm wagons, Daisy Wagons, Safety Coaster Wagons and Columbian Coaster Wagons. One can only surmise that the Columbian was dedicated or named after the 1893 Columbian Exposition (Chicago's World Fair) In addition to clippers and bow runner sleds, they produced baby sleighs, The top of the line was "The Daisy," offering upholstery in plush or satin russé, and finished in white enamel for an extra charge, "The Excelsior" was a step down from "The Daisy," upholstered in velour or other desirable fabrics, and also offered in white enamel for an extra charge. With the advent of the Flexible Flyer, the demand for steerable sleds was increasing and The Wilkinson Line introduced their most successful sled, "The Storm King," available in seven sizes. (No. 0 – 37inches long to No. 6 – 101 inches long). The National Coaster was introduced as their inexpensive line in five sizes (No. 10 – 33inches long to No 14 – 46 inches long). Does the numbering look familiar? Remember every company copied the famous Flexible Flyer right down to the advertising in some cases. Referred to as Coasters (actually Bobs) the Double Ripper No. 1, advertised as a suitable size for young men, had a deck ten feet long with steering and braking attachments. The Boys' "Double Ripper" No. 3, had a deck six feet long with steering and brakes, and the No. 4 was same as No. 3 without brakes. All models were painted vermillion and ornamentally decorated. It is not clear as to when their Bobs were introduced, but the Company went out of business in the 1930s.

Chronology of the Wilkinson Manufacturing Company

1860 • Binghamton Sled Company formed
1863-1874 • Winton Manufacturing Co. founded by Winton & Doolittle to manufacturer children's carriages
1875 • Organized into a Stock Co. for manufacturing Sleds exclusively. Davis, Wilkinson & Company established
1879 • Wilkinson & Eastwood manufactures of boys Express Wagons and Sleds established
1890 • Wilkinson Manufacturing Co. emerged after thirty years of name changing
1900 • Columbian Coaster Wagons introduced
1912. The Storm King and National Coaster Sleds manufactured
1920s • Double Ripper Bob Sleds appear
1930s • Company went out of business

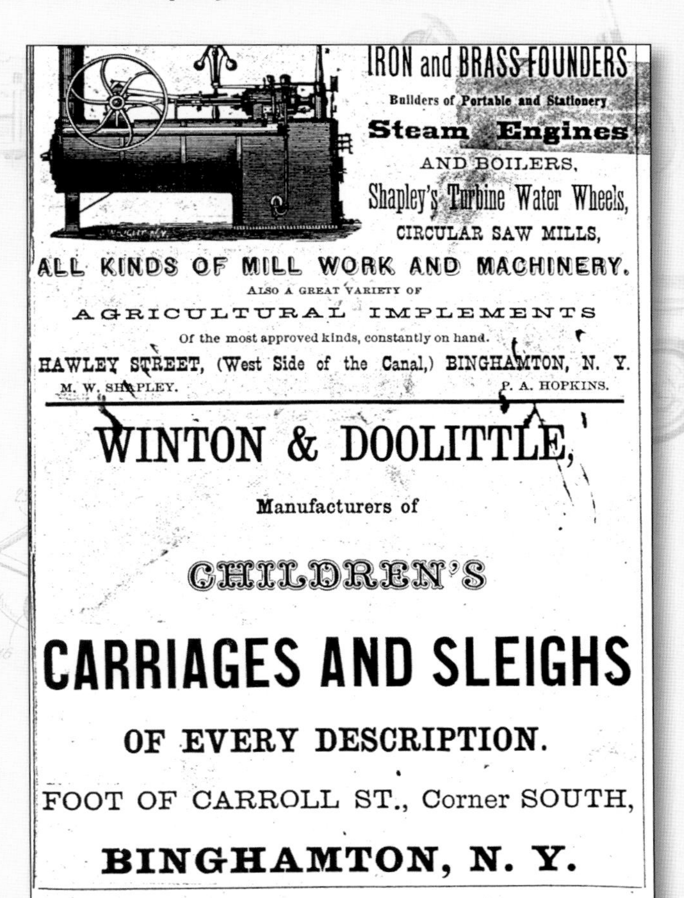

Winton & Doolittle advertisement from Boyd's Binghamton Directory, 1869-1870. Courtesy of Broome County Library, Binghamton, New York.

Wilkinson Line model, c. 1880. "The Gypsy," #36. Cala lily. L 36", W 14", H 6 ½". $750-1850. *Courtesy of Carole and Lou Scudillo.*

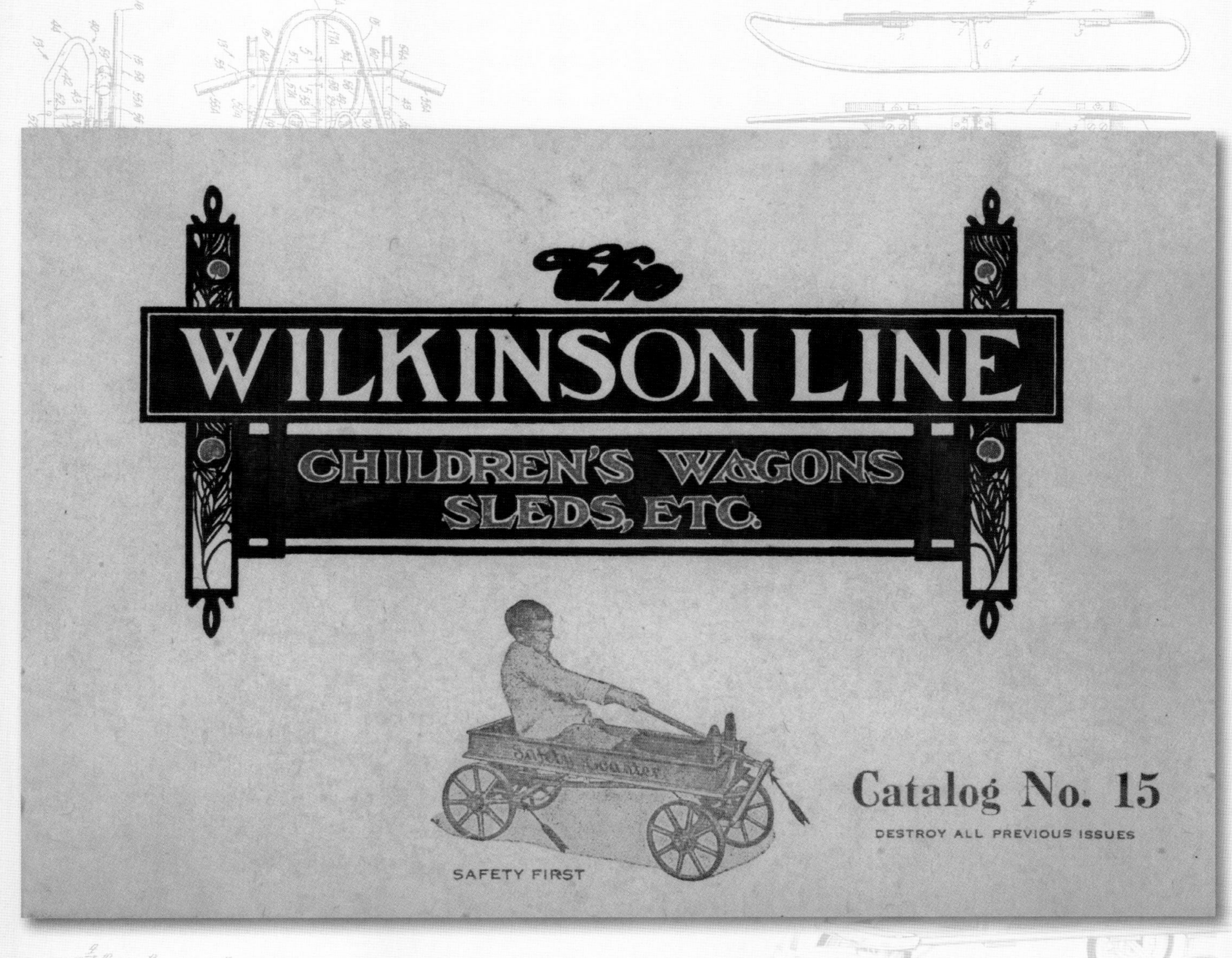

Wilkinson Line catalog #15, c. early1910. $50-100. *(Continued on the next four pages)*

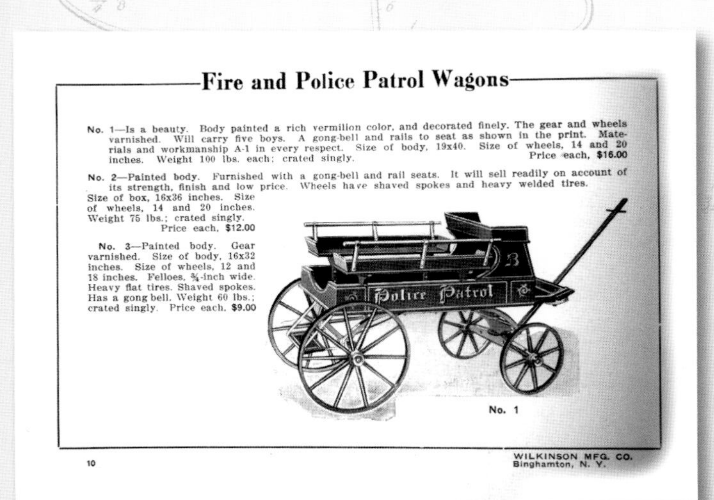

Fire and Police Patrol Wagons

No. 1—Is a beauty. Body painted a rich vermilion color, and decorated finely. The gear and wheels varnished. Will carry five boys. A gong-bell and rails to seat as shown in the print. Materials and workmanship A-1 in every respect. Size of body, 19x40. Size of wheels, 14 and 20 inches. Weight 100 lbs. each; crated singly. Price each, **$16.00**

No. 2—Painted body. Furnished with a gong-bell and rail seats. It will sell readily on account of its strength, finish and low price. Wheels have shaved spokes and heavy welded tires. Size of box, 16x36 inches. Size of wheels, 14 and 20 inches. Weight 75 lbs.; crated singly. Price each, **$12.00**

No. 3—Painted body. Gear varnished. Size of body, 16x32 inches. Size of wheels, 12 and 18 inches. Felloes, ¾-inch wide. Heavy flat tires. Shaved spokes. Has a gong bell. Weight 60 lbs.; crated singly. Price each, **$9.00**

No. 1

10

WILKINSON MFG. CO.
Binghamton, N. Y.

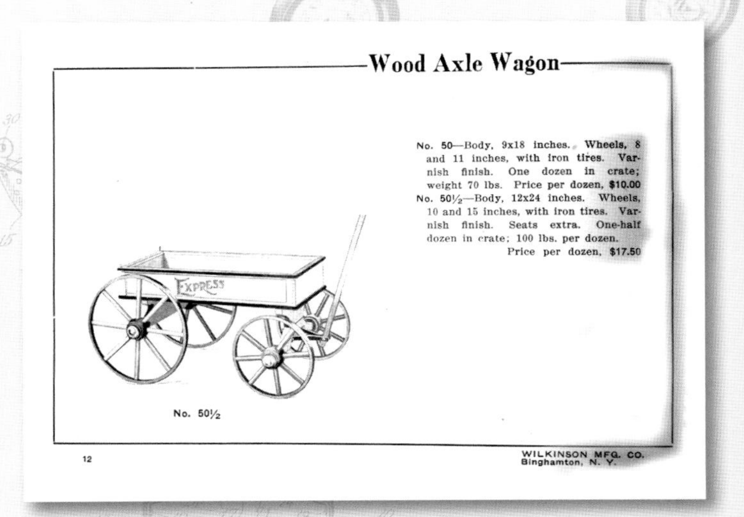

Wood Axle Wagon

No. 50—Body, 9x18 inches. Wheels, 8 and 11 inches, with iron tires. Varnish finish. One dozen in crate; weight 70 lbs. Price per dozen, **$10.00**

No. 50½—Body, 12x24 inches. Wheels, 10 and 15 inches, with iron tires. Varnish finish. Seats extra. One-half dozen in crate; 100 lbs. per dozen. Price per dozen, **$17.50**

No. 50½

12

WILKINSON MFG. CO.
Binghamton, N. Y.

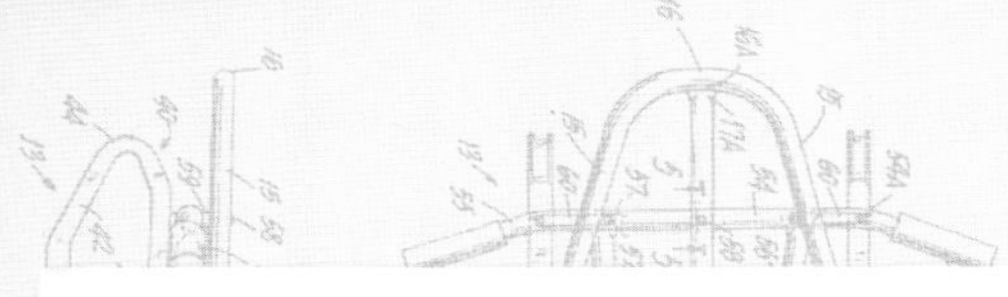

Steel Axle Wagons

No. 51—Body, 12x24 inches. Wheels 8 and 12 inches, with steel tires. Varnish finish. One-half dozen in crate; 140 lbs. per dozen.
Price per dozen, $18.00

No. 54X

No. 52—Body, 13x26 inches. Wheels, 10 and 15 inches, with steel tires. Varnish finish. One-half dozen in crate; 190 lbs. per dozen.
Price per dozen, $20.00

No. 52X—Same as No. 52, except they have half-oval tires and hub caps on the wheels. One-half dozen in crate; 240 lbs. per doz.
Price per dozen, $24.00

No. 54X—Body, 14x28 inches. Wheels, 10 and 15 inches with half-oval tires and hub caps. One-half dozen in crate; 250 lbs. per dozen.
Price per dozen, $32.00

All steel axle wagons have right and left hand nuts on axles, and steel boxes in hubs.

The above line of wagons have a steel brace running from the front axle to the under side of the box. Seats extra.

Steel Axle Wagons

These are wagons of high grade. They are finely finished with two coats of varnish, and nicely ornamented. Are made from the best materials, and put together with great care. We furnish a steel brace to run from the lower end of the king-bolt to the under side of the box. Shaved spokes welded tires, and hub caps; also extra rear axle brace.

No. 62—Box, 14x30 inches. Wheels, 11 and 16 inches. One-half dozen in crate; 275 lbs. per dozen.
Price per dozen, $40.00

No. 63—Box, 15½x31 inches. Wheels, 12 and 18 inches. One-half dozen in crate; 300 lbs. per dozen.
Price per dozen, $48.00

No. 63½—Box, 16x32 inches. Oak wheels, 12 and 18 inches. Heavy rims and tires, ¾ inches wide. One-half dozen in crate; 325 lbs. per doz.
Price per dozen, $53.33

Seats extra.

Nos. 62, 63 and 63½

Boys' Delivery Wagon

The most desirable we have ever made for boys to draw loads with. We furnish steel braces to run from king-bolt and rear axles to under side of box.

No. 64X—Body, 14x30 inches. Wheels, 12 and 18 inches, with heavy, welded, half oval tires. Shaved spokes tenoned into rims. Tin caps on hubs. Removable sideboards and dash as shown in cut. One-third dozen in crate; 300 lbs. per dozen.
Price per dozen, $48.00

Nos. 64X, 65 and 65X

No. 65—Body ,16x32 inches. Oak wheels, 12 and 18 inches. Heavy rims and tires ¾ inches wide. Removable sideboards and dash; one-third dozen in crate; 360 lbs. per dozen.
Price per dozen, $62.66

No. 65X—Box, 16x32 inches. Has a frame body, hardwood sills. Very strong oak wheels, 14 and 20 inches, with shaved spokes. Removable sideboards and dash; one-third dozen in crate; 450 lbs. per dozen.
Price per dozen, $70.00

Seats extra.

Boys' Delivery Wagon

No. 66—Has a frame bottom, hardwood sills, locked corners to the box, also corner irons and thoroughly braced. ⅝ square axles. Shaved spokes, heavy welded tires, iron bands on hubs. Spokes and rim to wheels made from oak and are heavy. Side and dash boards as shown in the cut. Size of body 19x36½ inches. Wheels, 14 and 20 inches. One-third dozen in crate; 680 lbs. per dozen.
Price per dozen, $83.00

No. 68—A large, very heavy and well made delivery wagon with sideboards, also wood reach running from front to rear axles to strengthen the running parts. Size of box, 22x42 inches. Has heavy shaved spoke wheels, 14 and 20 inches. A wagon made for service. One-sixth dozen in crate; 900 lbs. per dozen.
Price per dozen, $136.00

No. 66

No. 67—Same as No. 66, except has no side or dash boards. One-third dozen in crate; 600 lbs. per dozen.
Price per dozen, $74.00

Farm Wagons

No. 74—Sides and ends are easily removed. It can be used as a platform wagon. Size of body, 18x36 inches. Heavy oak wheels with shaved spokes, hub and caps and welded tires, 14 and 20 inches. Body painted a rich vermilion, or any other color if so ordered. Has a seat and stakes as represented in cut; packed one in a crate; weight 50 lbs. each.

Price each, $10.00

No. 74

Cab Wagons

For small children, or a suitable substitute for a baby carriage.

No. 77—Body, 14x28 inches, with flaring sides. Wood wheels, 11 and 16 inches. Shaved spokes, welded tires, and hub caps. Varnish nicely decorated; one-third dozen in a crate; weight 250 lbs. per dozen.
Price per dozen, $40.00

No. 78—Body painted vermilion, tinned-steel wheels. Otherwise like No. 77. One-third dozen in a crate; 270 lbs. per dozen.
Price per dozen, $44.00

No. 79—Wood wheels, with shaved spokes, welded tires, and caps on hubs. Otherwise like No. 78. One-third dozen in a crate; 250 lbs. per dozen.
Price per dozen, $44.00

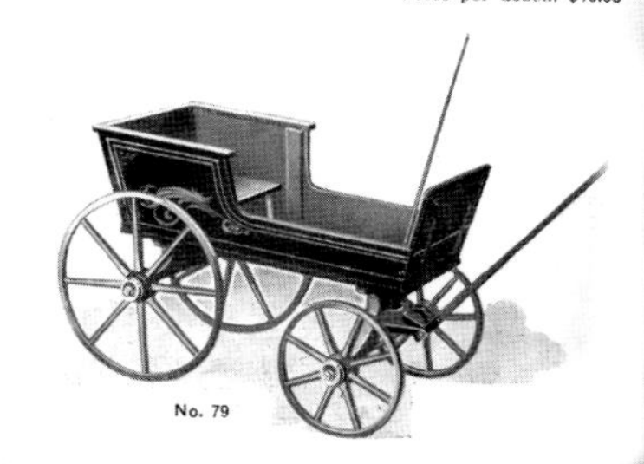

No. 79

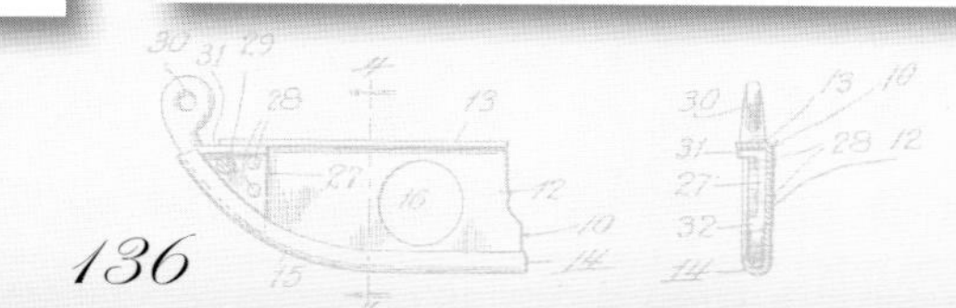

Daisy Wagons

This style of wagon has proved to be a great seller. We make it in two sizes. You will notice that the wagon is handsomely designed, and has a graceful high-back. The workmanship is first-class.

No. 100—Box, 13x26 inches. Wheels, 11 and 16 inches, shaved spokes, hub capped, half oval welded tires. With seat and whip. One-half dozen in a crate; 250 lbs. per dozen. Price per dozen, **$35.00**

No. 101—Box, 15½x31 inches. Oak wheels, 12 and 18 inches, shaved spokes, welded tires and hub caps. With seat and whip. One-half dozen in a crate; 325 lbs. per dozen. Price per dozen, **$53.34**

No. 102—Box, 13x26 inches. Wheels 10 and 15 inches, round spokes, hub capped, half oval welded tires. With seat and whip. One-half dozen in a crate; 240 lbs. per dozen. Price per dozen, **$32.00**

No. 101

WILKINSON MFG. CO.
Binghamton, N. Y.

19

Safety Coaster Wagons

ROLLER BEARING

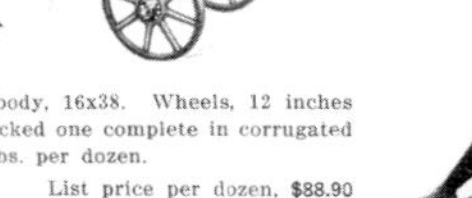

Simple manner of detaching Box and Hinge in Tongue are features peculiar to these Coasters.

No. 210—Size of body, 14x36. Wheels, 12 inches in diameter. Packed one complete in corrugated container; 480 lbs. per dozen. List price per dozen, **$80.00**

No. 211—Size of body, 16x38. Wheels, 12 inches in diameter. Packed one complete in corrugated container; 500 lbs. per dozen. List price per dozen, **$88.90**

No. 212—Size of body, 18x40. Wheels, 12 inches in diameter. Packed one complete in corrugated container; 570 lbs. per dozen. List price per dozen, **$100.00**

20

WILKINSON MFG. CO.
Binghamton, N. Y.

Columbian Coaster Wagons

ROLLER BEARING

Made in five sizes; No. 0 has a rigid body while on all larger sizes the body can be instantaneously removed. All packed in Dust Proof Corrugated containers. Rear axle-risers doweled to insure strength and safety; hubs, bushings, axles and roller bearings all made from cold rolled steel and spokes tenoned into felloe insures greatest possible strength and efficiency.

No. 0—12x28	8 in.	2 pieces	40 lbs.	**$50.00**
No. 1—14x32	8 in.	2 pieces	45 lbs.	**60.00**
No. 2—14x34	10 in.	1 piece	30 lbs.	**66.00**
No. 3—16x38	10 in.	1 piece	35 lbs.	**74.00**
No. 4—18x40	12 in.	1 piece	40 lbs.	**84.00**

WILKINSON MFG. CO.
Binghamton, N. Y.

21

Standard Oak Sleighs

Standard medium-priced oak or rock elm frame sled. Finished on the natural wood. Braces brightly tinned; half-oval steel shoes; packed six in a crate.

No. 14—Size, 13x38 inches; 80 lbs. per dozen.................................Price per dozen, **$25.00**
No. 16—Size, 12x34 inches; 75 lbs. per dozen.................................Price per dozen, **21.00**
No. 18—Size, 11x32 inches, knee braces only; 60 lbs. per dozen.............Price per dozen, **19.00**

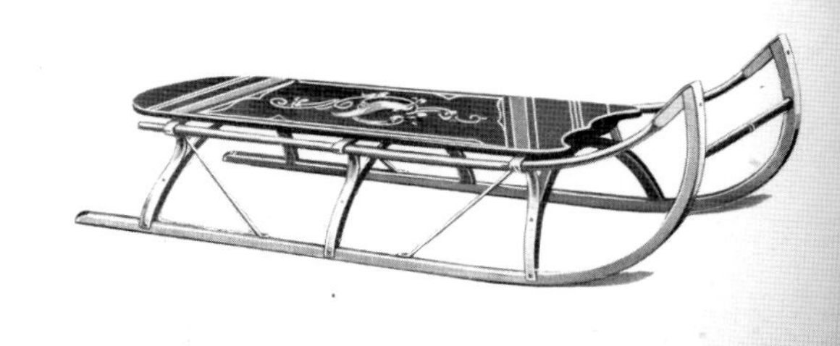

26

WILKINSON MFG. CO.
Binghamton, N. Y.

Red Oak Frame

No. 24—This sled has a wide rave, bent knees, strongly braced with half-oval steel from the runners to the beams; thoroughly riveted. Braces and malleable iron swan necks, brightly tinned. Half-oval steel shoes. Top board handsomely painted and ornamented. Size, 13x36 inches; weight 75 lbs. per dozen. Price per dozen, **$22.00**

No. 26—Like No. 24, except it has no swan neck. Half oval shoes; weight 70 lbs. per dozen. Price per dozen, **$20.00**

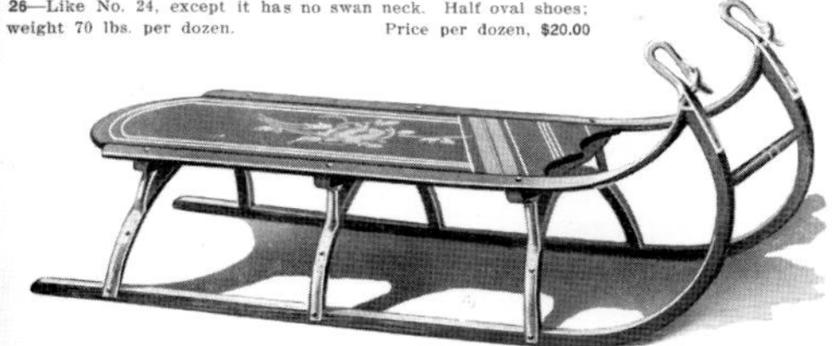

Packed one-half dozen in a crate.

WILKINSON MFG. CO.
Binghamton, N. Y.

27

Bow Runner Sleigh

No. 32—A bow runner, oak or ash sled, with artistically painted top board. Runners and knees finished on the natural wood. Bent knees, half-oval steel braces from the runners to the beams, thoroughly riveted. Braces brightly tinned. Half-oval steel shoes. Striped and well ornamented. Size, 13x36 inches; 85 lbs. per dozen. **Price per dozen, $22,00**

No. 34—Made and finished like No. 32, except that it has two beams instead of three. **Size, 12½x32** inches; 70 lbs. per dozen. **Price per dozen, $18.00**

Packed one-half dozen in a crate.

28

WILKINSON MFG. CO.
Binghamton, N. Y.

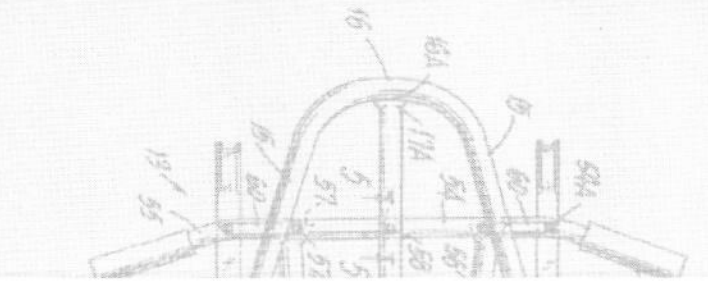

The Gipsy

Made from good oak or ash lumber. Top board painted; other parts finished on the natural wood; nicely striped and ornamented. Packed one half dozen in a crate.

No. 36—Has four beams. Half-oval shoes. Size, 14x36 inches; 75 lbs. per dozen.
Price per dozen, **$18.30**

No. 38—Has three beams. Flat shoes. Size, 13x32 inches; 60 lbs. per dozen.
Price per dozen, **$16.10**

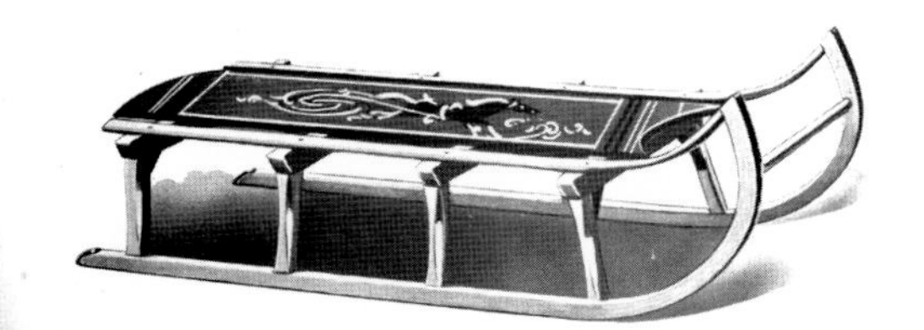

WILKINSON MFG. CO.
Binghamton, N. Y.

29

Curved-Side Maple Clipper

WITH HANDHOLES

No. 70—Size, 10½x40 inches; 115 lbs. per dozen .... Price per dozen, **$17.00**

No. 72—Size, 10x36 inches; 90 lbs. per dozen .... Price per dozen, **$14.00**

Packed one-half dozen in a crate.

30

WILKINSON MFG. CO.
Binghamton, N. Y.

Maple Clipper

With hand holes, top board painted and ornamented, sides varnished on the wood.

No. 90—Size 10x33 inches, round spring shoes; 70 lbs. per dozen.......... Price per dozen, $10.00
No. 92—Size, 10x36 inches, round spring shoes; 90 lbs. per dozen.......... Price per dozen, 12.00
No. 93—Size, 10½x40 inches, round spring shoes; 100 lbs. per dozen.......... Price per dozen, 14.00
No. 94—Size, 11x44 inches, round spring shoes; 115 lbs. per dozen.......... Price per dozen, 16.00

Nos. 92, 93 and 94 have larger shoes and are better finished than No. 90.

Packed one-half dozen in a crate.

WILKINSON MFG. CO.
Binghamton, N. Y.

31

The Daisy

The Daisy is what we all call this Baby Sleigh. It is roomy and comfortable, graceful and stylish in form and finish. The following numbers are all furnished with a push handle as in cut. We also have a draw handle or tongue, as shown, which has a very convenient connection. This we furnish whenever desired for 50 cents net.

No. 96—Upholstered in plush, with side rails as shown in cut. Weight 35 lbs. each. Price each, **$13.34**

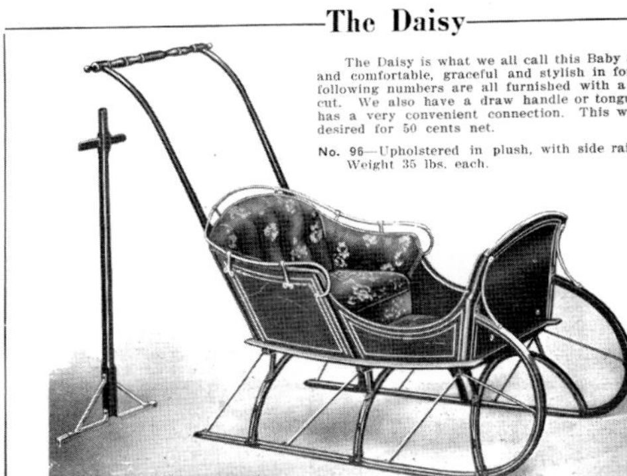

No. 97—Upholstered with tapestry, or goods of similar quality, with side rails as shown in cut. Weight 35 lbs. each. Price each, **$11.12**

No. 98—Upholstered in satin russe; this number has no side rails. Weight 30 lbs. each. Price each, **$9.50** Packed one in a crate.

For white enamel finish add $2.00 to above.

32

WILKINSON MFG. CO.
Binghamton, N. Y.

Excelsior Cutter

No. 104—A desirable and quick-selling Cutter, well made and nicely finished. Upholstered with velour or other desirable coverings. Weight 30 lbs. each Price each, **$7.80**
For white enamel finish add $1.50 to above.
Packed one in a crate.

WILKINSON MFG. CO.
Binghamton, N. Y.

33

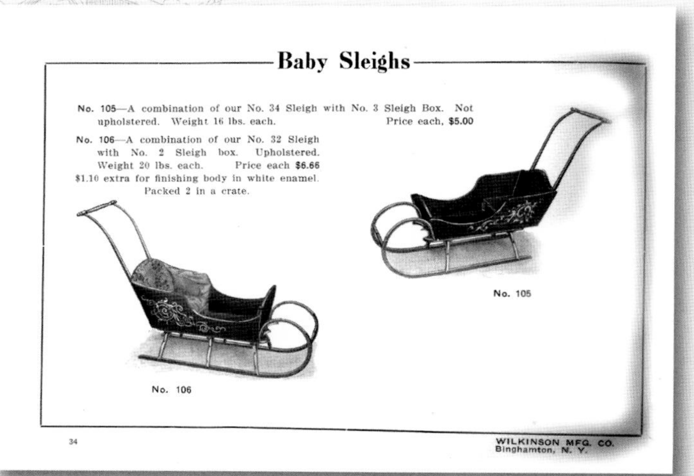

Baby Sleighs

No. 105—A combination of our No. 34 Sleigh with No. 3 Sleigh Box. Not upholstered. Weight 16 lbs. each. Price each, **$5.00**

No. 106—A combination of our No. 32 Sleigh with No. 2 Sleigh box. Upholstered. Weight 20 lbs. each. Price each **$6.66**
$1.10 extra for finishing body in white enamel.
Packed 2 in a crate.

No. 105

No. 106

34

WILKINSON MFG. CO.
Binghamton, N. Y.

Ideal Cutter

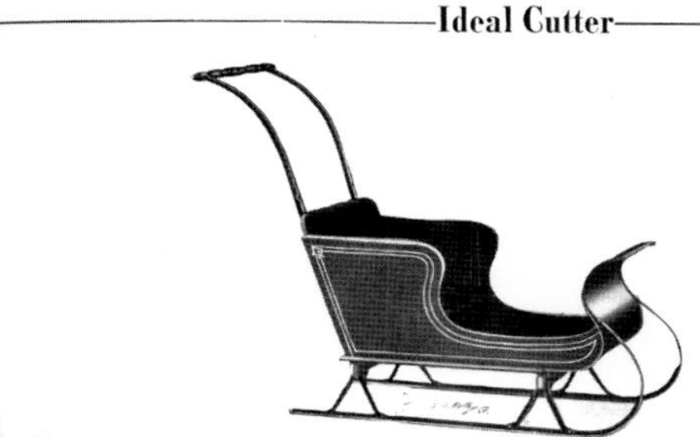

No. 110—A high grade Baby Cutter at popular price, upholstered with velour. Weight 35 lbs. each
List price, $7.80 each

No. 110X—Same as above with silk upholstering.
List price, $10.00 each

Packed one in a crate.

For white enamel finish add $2.00.

WILKINSON MFG. CO.
Binghamton, N. Y.

35

The Storm King

This is the sled that the boys and girls are bound to have. It is made to steer with the feet or hands by pressing the cross bar in front, either to the right or left. By this arrangement the feet do not come in contact with the ground and retard the speed of the sled, also does not wear out shoe leather.

The runners are made from crucible spring steel rolled T shape. They are of high temper, yet flexible and easily deflected right or left and then back to a straight position.

The knees are pressed into shape from sheet steel and are very strong. This line of sleds is finely finished and ornamented.

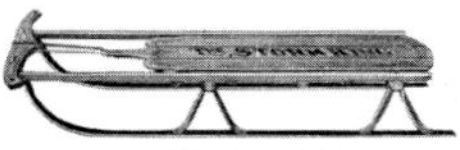

Nos. 0, 1 and 2

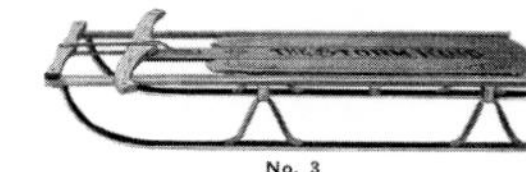

No. 3

Packed two each size in a bundle.

No. 0—33 inches long, 11½ inches wide, 4 knees, 2 on a side. Weight 80 lbs. per dozen.
Price per dozen, $15.56

No. 1—37 inches long, 12 inches wide, 4 knees, 2 on a side. Weight 96 lbs. per dozen.
Price per dozen, $17.78

No. 2—41 inches long, 13 inches wide, 4 knees, 2 on a side. Weight 108 lbs. per dozen.
Price per dozen, $22.22

No. 3—45 inches long, 14 inches wide, 4 knees, 2 on a side. Weight 132 lbs. per dozen.
Price per dozen, $28.88

Racer—59 inches long, 14 inches wide, 6 knees, 3 on a side. Weight 168 lbs. per dozen.
Price per dozen, $33.34

38

WILKINSON MFG. CO.
Binghamton, N. Y.

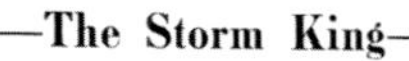

The Storm King

Nos. 4, 5 and 6

No. 4—50 inches long, 16 inches wide, 4 knees, 2 on a side, 2 iron foot rests. Weight 180 lbs. per dozen.
Price per dozen, $36.66

No. 5—62 inches long, 16 inches wide, 6 knees, 3 on a side, 4 iron foot rests. Weight 222 lbs. per dozen.
Price per dozen, $48.88

No. 6—101 inches long, 16 inches wide, 10 knees, 5 on a side, 8 iron foot rests. Weight 410 lbs. per dozen.
Price per dozen, $88.88

Packed two in a bundle

WILKINSON MFG. CO.
Binghamton, N. Y.

39

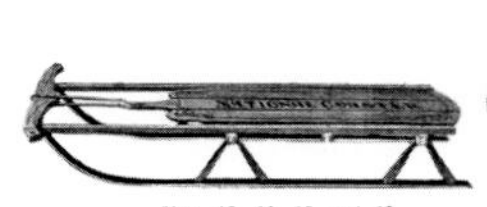

Nos. 10, 11, 12 and 13

No. 14

Steering Sled

Pressed steel knees. T shaped runners, designed especially for these sleds, made from our own rolls.

No. 10—Length 33 inches. Width 11½ inches. 4 knees. 80 lbs. per dozen. Price per doz..$14.44

No. 11—Length 37 inches. Width 12 inches. 4 knees. 90 lbs. per dozen. Price per doz.. 16.66

No. 12—Length 41 inches. Width 13 inches. 4 knees. 102 lbs. per dozen. Price per doz.. 20.00

No. 13—Length 46 inches. Width 13½ inches. 4 knees. 108 lbs. per dozen. Price per doz.. 23.34

No. 14—National Coaster, length 46 inches. Width 13½ inches. Flexible carbon spring tee steel runners. 115 lbs. per dozen. List price ....$25.55 per doz.

Packed two each in a bundle.

40

WILKINSON MFG. CO.
Binghamton, N. Y.

Double Ripper

These Coasters are made very strong, and have steering and brake attachments, which are completely under the control of the rider. Suitable in size for young men.

No. 1—Top board, 10 feet long, painted, not upholstered; ½-inch steel shoes on runners. Weight 80 lbs. each.
Price each, $16.00

Boys' "Double Ripper"

No. 3—Similar to No. 1, with steering and brake attachments. Top board, 6 feet long. Top board painted vermilion. Weight 30 lbs. each.
Price each, $6.66

No. 4—Same as No. 3, without brake. Weight 28 lbs. each.
Price each, $5.56

Packed 1 in a crate.

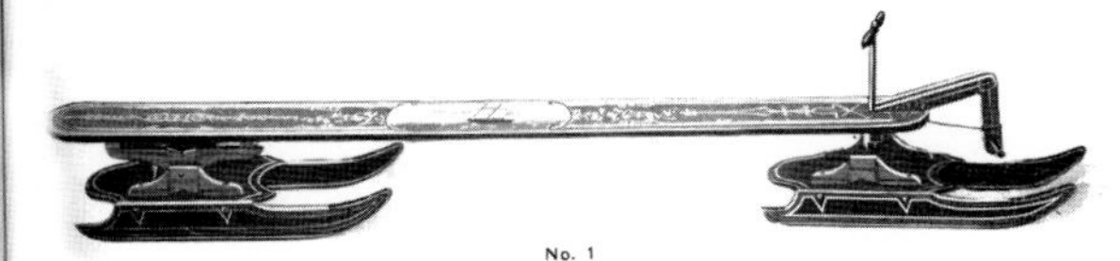

No. 1

WILKINSON MFG. CO.
Binghamton, N. Y.

41

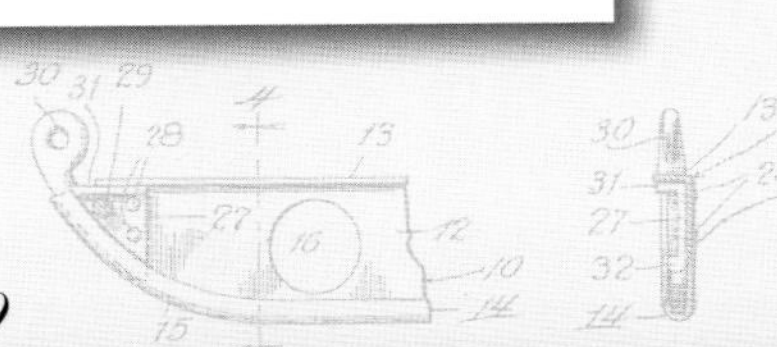

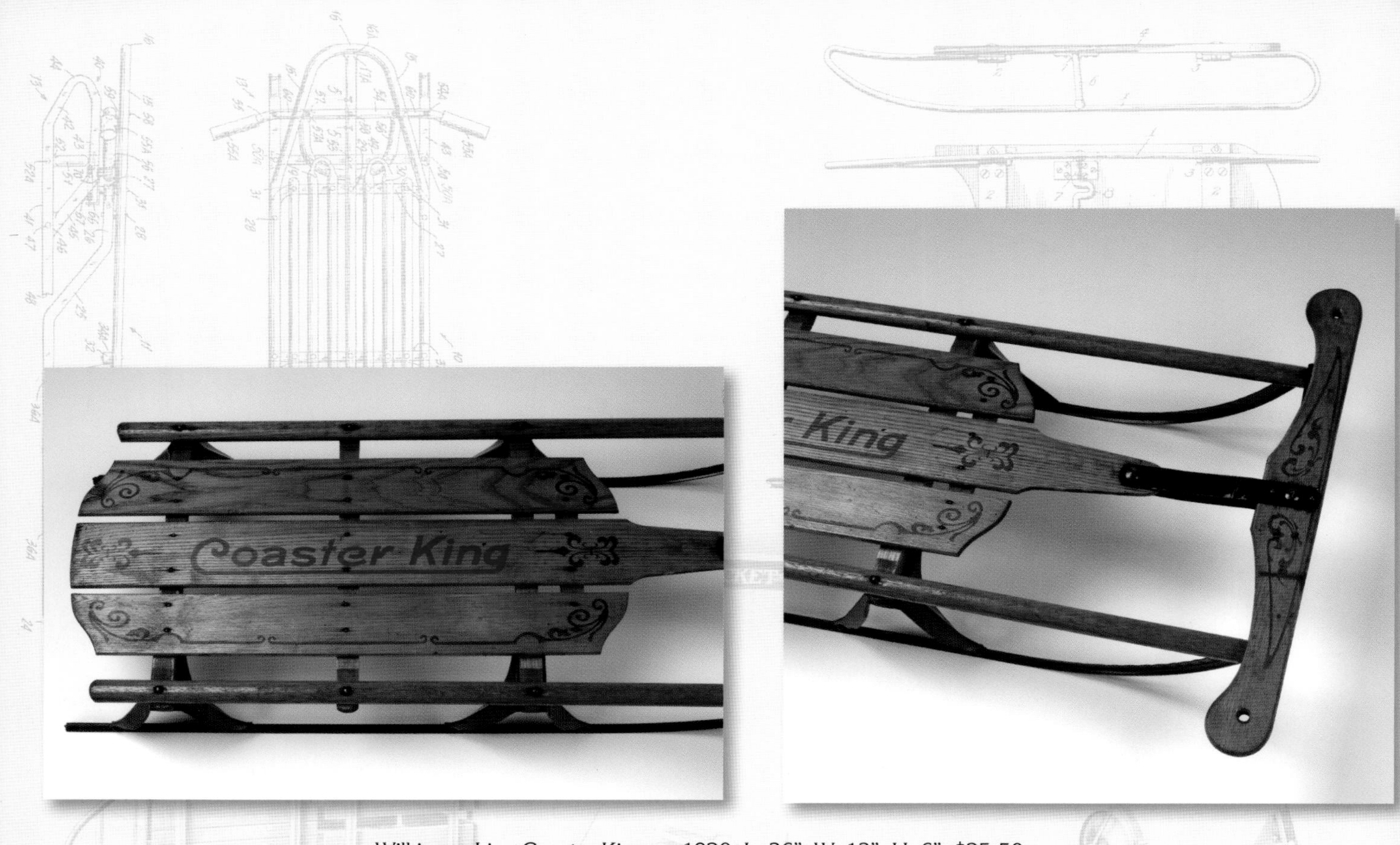

Wilkinson Line Coaster King, c. 1920. L: 36"; W: 13"; H: 6". $25-50.

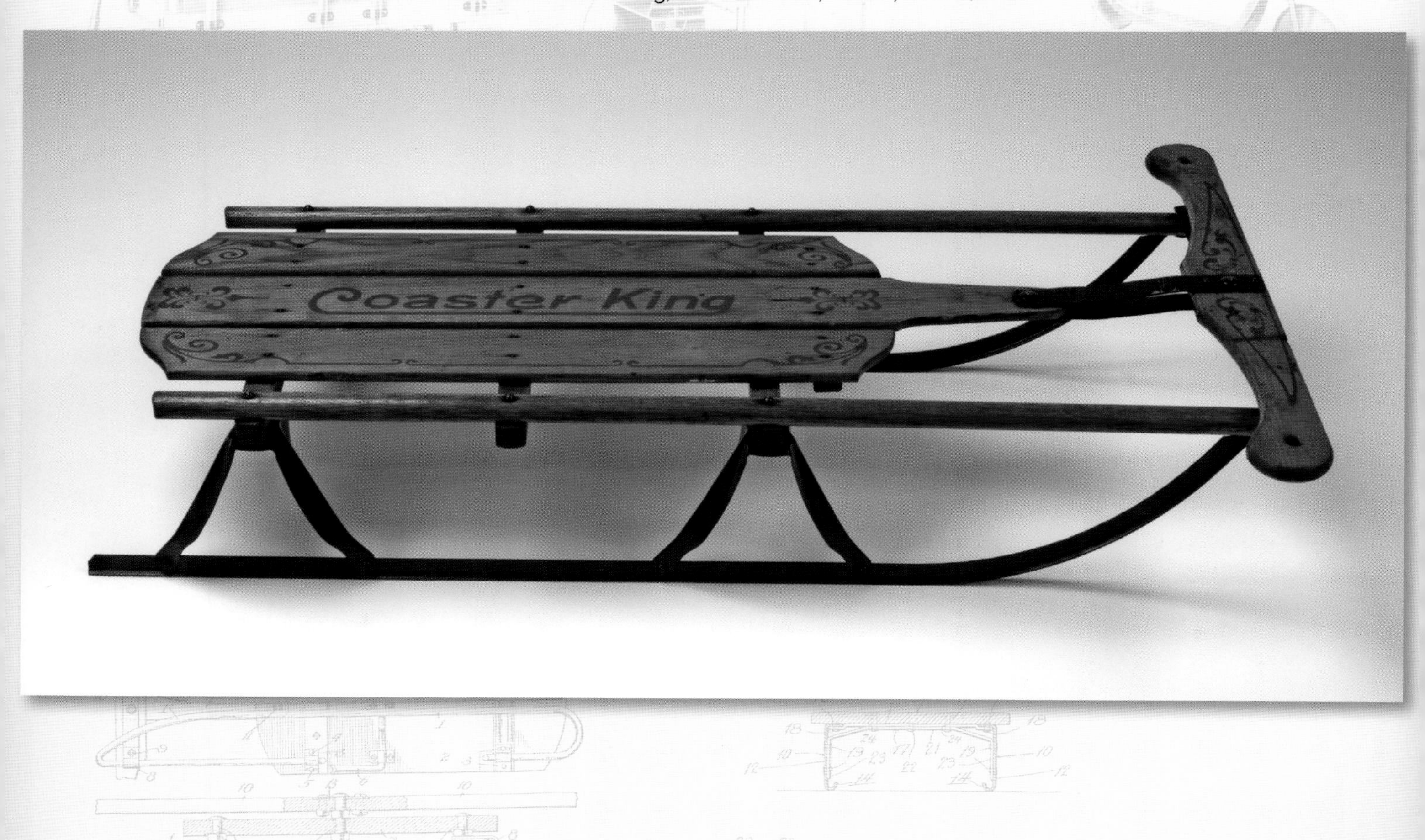

Withington

Wayne O. Stevens, was the man behind Withington. No stranger to winter sports, he retired from competitive skiing and purchased a brush business from a man named Withington. The company was founded in the forties to manufacture archery equipment, adding skis, toboggans, ski skates and bow guns later on. Although not a major player in the manufacture of sleds, I would be remiss not to include Withington. The forty-something generation is now trying to identify their childhood icons. For instance, the Space Sled, introduced in 1956-57, could be used standing, sitting, or prone. The Sno-Scoota, a new and imaginative glider scooter, was added to their line in 1958. Perhaps their most famous sled was the "Bob-O-Link" bobsled, the sled that skis, introduced in 1963, in three different sizes. The largest, 72", was the only one equipped with brakes. Much like Flexible Flyer, a Club Membership was offered along with a certificate, a colorful embroidered arm patch, and a newsletter. The design of the Bob-O-Link changed slightly in 1965. Hood sizes were increased 2" in length and height. The handle bar size was also enlarged to withstand rough usage. Brakes were also added as standard equipment on all models. Wheel conversion kits were offered sometime during 1965. "The unique wheel kit transforms your Bob-O-Link from a free sliding bobsled into a rugged, racy, wheeling, dealing year round luxury vehicle," as stated in the newsletter. In addition to the Bob-O-Link they manufactured toboggans, tow sleds, and trailer-type tow sleds. Withington disappeared from the market place sometime in the seventies. It is now up to the next generation of sled collectors to guarantee that, although gone from the market place, Withington will not be forgotten.

Withington 1958 catalog. $25-30.

Chronology of Withington

1940s • Withington of West Minot Maine, was founded by Wayne O. Stevens

1950s •Skiis, Skee Skates and Bow Guns were manufactured

1956-1957 • The Space Sled introduced

1958 • The Algonquin Toboggan introduced in four sizes along with the Yukon Trailmaster Dog Sled and Sno Scoota

1963 • Bob-O- Link, bobsled introduced. Club membership certificate, newsletter and arm patch offered. Toboggans added to their line: The Continental, Trailmaster, and Pursuit

1965 • Wheel Conversion Kit offered; Skee Mee #40 introduced

1970s • Operations ceased

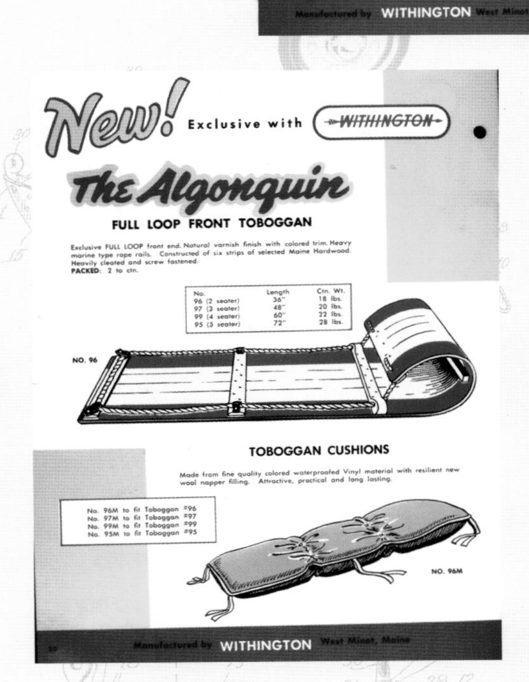

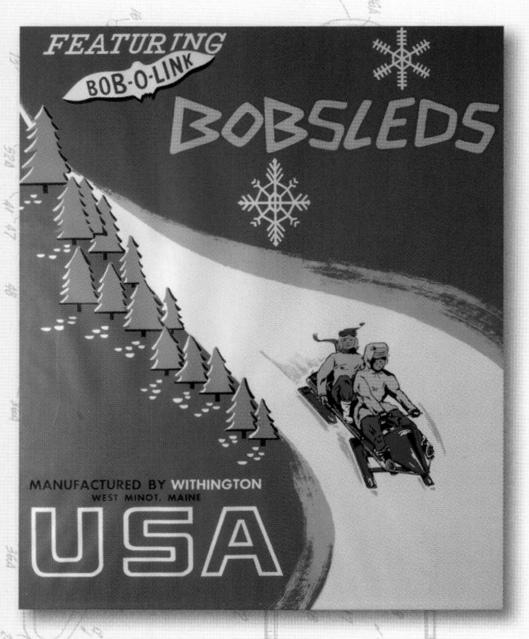

Withington poster featuring Bob-O-Link bobsleds. $50-100.

Withington Bob-O-Link Junior, c. 1965. $150-200. . *Courtesy of Paul Cote.*

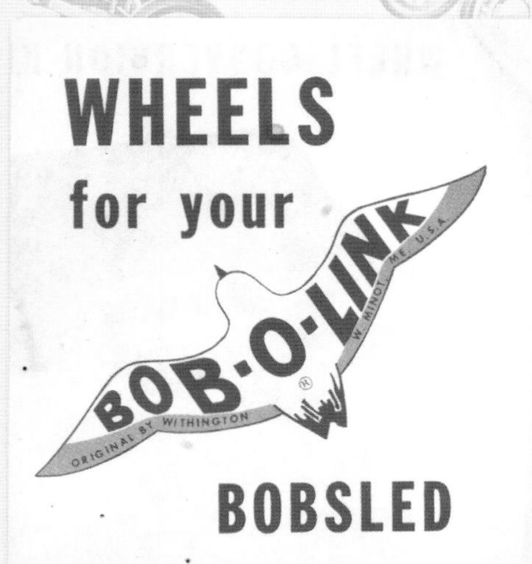

WHEELS for your Bobsled

Wheels on your Bobsled are optional at extra cost. If your sled does not have the wheels, and they are not available at your dealer, the conversion may be obtained direct from the factory. Simply state the overall length of your Bobsled or Model Number from the decal on the hood.

Red Wheels 8¼" diameter with Semi-pneumatic 1¾" Tire Tread Width, with ½" steel Axles slung under the Bobsleds through holes already provided, as shown on illustration. (Bobsleds without the hole provision can easily be drilled for this wheel assembly.)

This conversion to Wheels makes your BOB-O-LINK a Year-round Sports Car, a 45 pound, wide steering piece of luxurious machinery.

Besides this, of course, those Bobsleds with Brakes still have the safety factor of being able to stop in emergencies.

You should not be without this added feature of Bobsledding. Write directly to us at the factory, enclosing check or money order for $9.95 which includes postage charges, and we will rush this Wheel Conversion Kit to you.

Bob-O-Link wheel conversion kit, 1965 brochure. $5-10.

WHEEL CONVERSION KIT .. $9.95 delivered to you.

THE WHEEL ASSEMBLY INCLUDES:

4 — Rubber Tired Wheels
2 — Wheel Axles
4 — Wheel Spacers
4 — Cotter Pins and Screws
2 — Sled Stabilizers
(See diagram at left for easy attachment.)

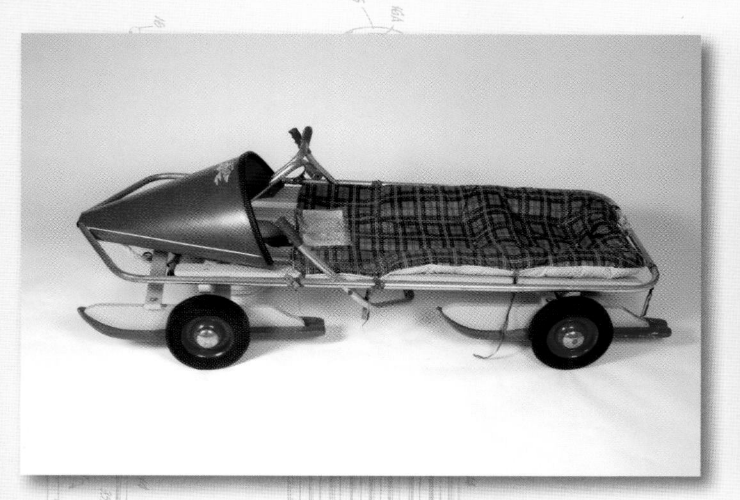

Withington Bob-O-Link, c. 1965, featuring a wheel conversion kit. $150-200. . *Courtesy of Paul Cote.*

Bob-O-Link company advertisement, c. 1965.
Note the brakes on all models.

Bob-O-Link Club certificate and newsletter, c. 1965 brochure. $15-25.

Vol. 1 No. 2

Withington
WEST MINOT, MAINE

BOB-O-LINK BOBSLED NEWSLETTER

Now, my friends, is the time to wax up the runners on your Bob-O-Link Bobsleds in happy anticipation of a cold, hard winter's fun. Be on the lookout for a long, steep slope loaded with bumps, swerves and jumps. Your bobsleds will perform like champs on this kind of rugged terrain. Because the bobsled steers, it does not require a fixed run like a toboggan does. It does require a hard packed snow surface for top performance. As a matter of fact, the bobsled is so named because it can bob-up, bob-down, bob-right, bob-left and even all the way around! And still, the passengers stay firmly aboard saddled in the low-slung bucket seats.

We're sure that you will be interested to learn that Senator Robert Kennedy's children are enjoying a Bob-O-Link Bobsled along with all of you. Another enthusiast is Jack Paar of television.

The 1965 Bob-O-Links have a 2" longer and higher hood. The steering handle bars have been increased in size to withstand rough usuage. In most cases, the brakes have become standard equipment.

A wheel conversion kit is offered this year as optional equipment for your bobsled. The kits can be used on all sleds made in 1963, 1964 and 1965 simply by boring a 1/2" hole in each riser. This unique wheel kit transforms your Bob-O-Link from a free sliding bobsled into a rugged, racy, wheeling dealing year round luxury vehicle. The conversion kit costs $14.25 delivered to you. You will find a sheet enclosed giving you further details.

Another new product flash - Withington is manufacturing snap-on skis for coaster wagons. Its as simple as it sounds, just snap the skis on the wheels.

Withington also manufactures deluxe toboggans and dogsleds of the finest quality. Catalogs are free for the asking.

If you should misplace your parts booklet just send in a sketch of the part involved with a short description. Individual parts are always promptly available. Breakage may occurr in this rugged sport and we feel it is our obligation to keep all the bobsleds bobbing.

A **Withington** ORIGINAL

#3035C (6 ft. 2 inch) Eskimo Type Tow Sled for Cargo and Passengers

22½ inch wide – constructed of Ash and Oak with Plated Screws and Bolts. Interchangeable Parts in stock and available at all times.

Steel Tow Bar with Swivel at connecting point to snowmobile.

Three Red Runners with balance in Natural. Weight 75 lbs.

#3035C
HEAVY DUTY ESKIMO TYPE TOW SLED WITH SEAT AND CUSHION REMOVED

UNIVERSAL HITCH

#3035C
(Same as above)
HEAVY DUTY PASSENGER TOW SLED COMPLETE

Withington Winter Sports catalog, c. 1963, with advertisement for a Bob-O-Link arm patch. $15-25.

Withington TRAILER-TYPE TOW SLEDS

DESIGNED FOR SNOW VEHICLES . . . CARGO OR PASSENGERS

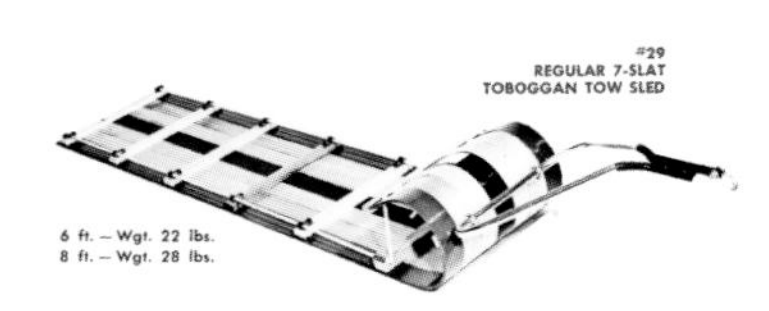

#29
REGULAR 7-SLAT
TOBOGGAN TOW SLED

6 ft. – Wgt. 22 lbs.
8 ft. – Wgt. 28 lbs.

#30
OVERSIZED 9-SLAT
TOBOGGAN TOW SLED

6 ft. – Wgt. 27 lbs.
8 ft. – Wgt. 36 lbs.

#29 (6 ft. or 8 ft.) Regular Width Toboggan Tow Sled

17 inch wide – constructed like regular quality Toboggans with Swivel Tow Bar at connecting point with snowmobile. Three Runners are Red with the balance in Natural Color.

#30 (6 ft. or 8 ft.) Wide Toboggan Tow Sled

Extra wide (22 inch) gives this Trailing Toboggan needed stability, buoyancy and carrying capacity. It is also wide enough to straddle the snowmobile track. Two Iron Strips are attached to the underside for strength and wearing qualities. Steel Tow Bar with Swivel at connecting point to snowmobile.
Three Runners are Red with the balance in Natural Color.

#29M or #30M Extra – Cushion

Made of Fabric-backed Vinyl filled with new wool. Blue Color.

Withington TRAILER-TYPE TOW SLEDS, TOBOGGANS and BOBSLEDS (ORIGINALS)

PULLED BEHIND ANY SNO-TRAVELLER FOR:
Winter Sports Fun, Cruising through the Woods, Winter Camping, Utility and Lumber Companies, Winter Resorts, Ice Fishing and Hunting.

#40B Box Type Passenger and Cargo Carrier Tow Sled (illustrated in color on front cover)

The Trailer Body is made of 1/2 inch Plywood with Plastic overlay.
A Steel Rail serves as a protective bushwhacker.
The Cushion is made of Fabric-backed Vinyl filled with new wool napper.

#40T Tarpaulin

A Tarpaulin with the ties attached is available as an extra.

INSIDE DIMENSIONS: 48 inches long x 24 inches wide x 12 inches deep.
OVERALL DIMENSIONS: 74 inches long x 29 inches wide x 25 inches high. WEIGHT: 96 lbs.

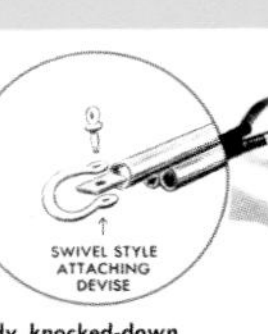

Shipped body knocked-down
Carton size: 59 x 28 x 8 inches

- TWO FLEXIBLE BOBBING SLEDS SYNCHRONIZED TO TURN TOGETHER
- BUILT WELL OFF THE GROUND FOR ADEQUATE CLEARANCE
- SKIS ARE EQUIPPED WITH STEEL UNDER RUNNERS

TOBOGGANS FOR TOWING

IN FULL REGULATION SIZES
made from selected Maine Hardwood

FRONT LOOPS BENT FULL THICKNESS THROUGHOUT

#6T

#8T

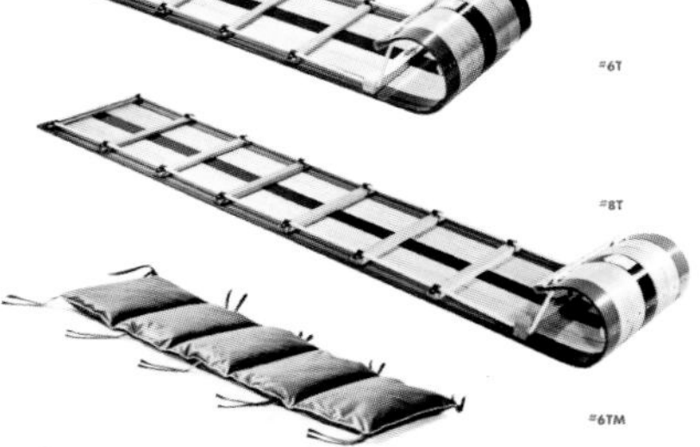

#6TM

The Trailmaster Toboggans are Full Regulation Size – constructed throughout of selected Maine Hardwood. The slats are heavily cleated and screwed. They are bent full thickness throughout the front loop. These quality Toboggans are beautifully finished in Red, Black and Natural Colors with Marine Cotton Hand Ropes.

#5T – 5 ft. seats 4, pkd. 3/ctn., wgt. 48 lbs.
#6T – 6 ft. seats 5, pkd. 3/ctn., wgt. 57 lbs.
#7T – 7 ft. seats 6, pkd. 3/ctn., wgt. 60 lbs.
#8T – 8 ft. seats 7, pkd. 3/ctn., wgt. 68 lbs.

Blue Matching Vinyl Mattresses with New Wool Filling

#5TM – pkd. 4/ctn., wgt. 16 lbs.
#6TM – pkd. 4/ctn., wgt. 20 lbs.
#7TM – pkd. 4/ctn., wgt. 22 lbs.
#8TM – pkd. 4/ctn., wgt. 24 lbs.

THE BOBSLEDS FOR TOWING

EXCLUSIVE! with Brakes and Tow Ring

#60
FOR 3 CHILDREN OR 2 ADULTS

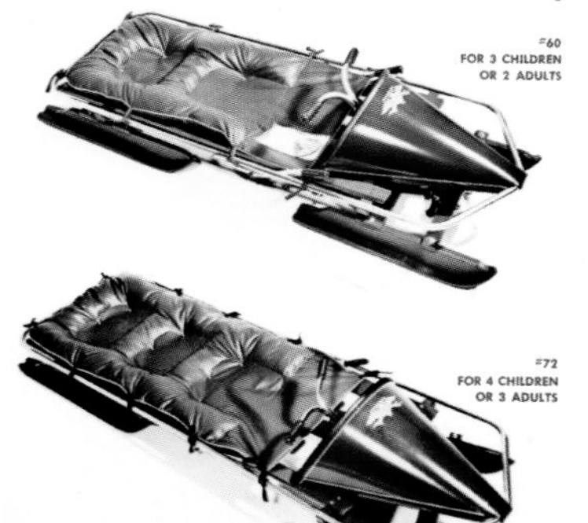

#72
FOR 4 CHILDREN OR 3 ADULTS

COLORFUL EMBROIDERED ARM PATCH WITH EACH SLED

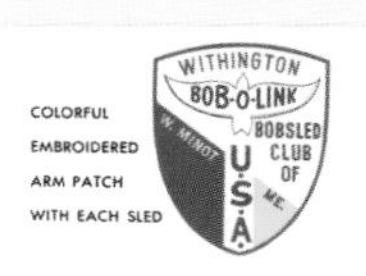

These Sturdy Bobsleds are fabricated of Hardwood – Heavy Gauge Steel Hood and Steel Frame – Automotive Vinyl Fabric. Finished in Solid Red with Blue Bucket Type Padded Cushion with shock absorbing qualities. Full Floating Bobs with 3 inch wide Skis; having Steel Runners attached to undersides.

#60
59 in. long, 22 in. wide, pkd. 1/ctn., wgt. 40 lbs.

#72
72 in. long, 23 in. wide, pkd. 1/ctn., wgt. 48 lbs.

SLIDE THE SLOPES
with a SLED
that SKIS!

MADE IN U.S.A.
BY
WITHINGTON MFG. INC., W. MINOT, MAINE
TEL. 207 – 966 – 2600
ORDER NOW!

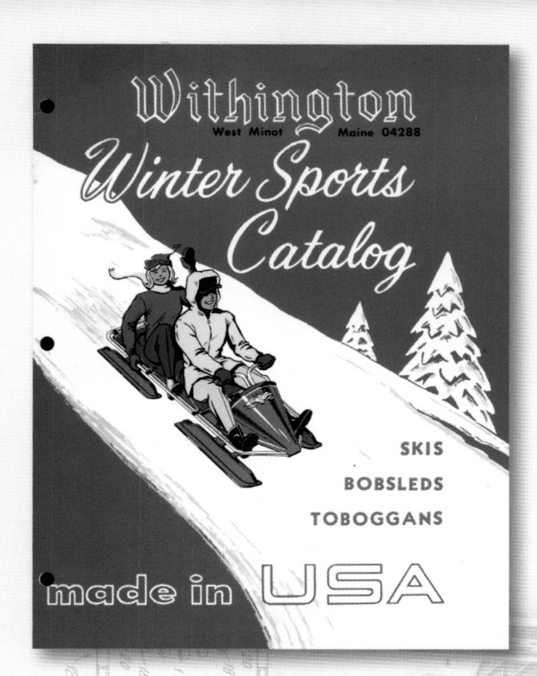

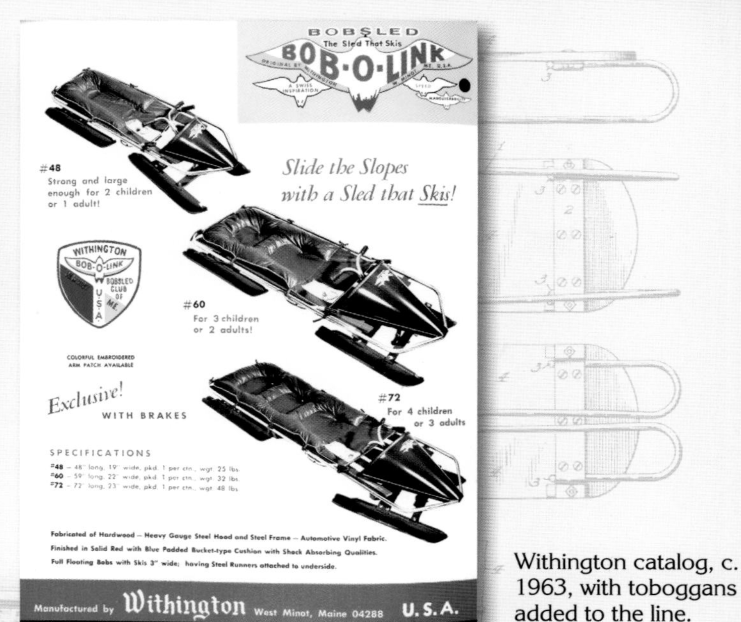

Withington catalog, c. 1963, with toboggans added to the line.

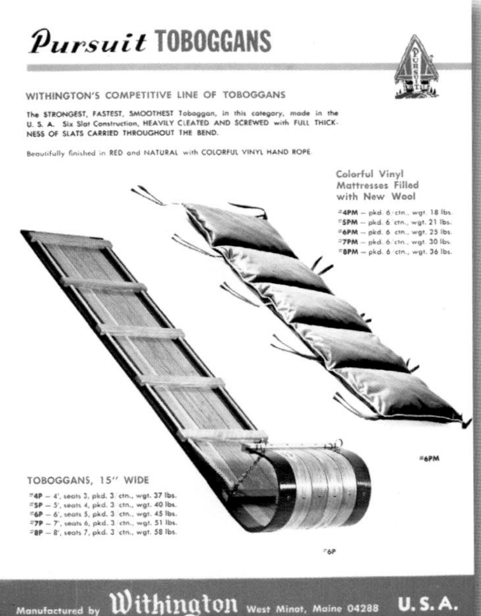

Withington Skee-Mee, c. 1965. $3-5.

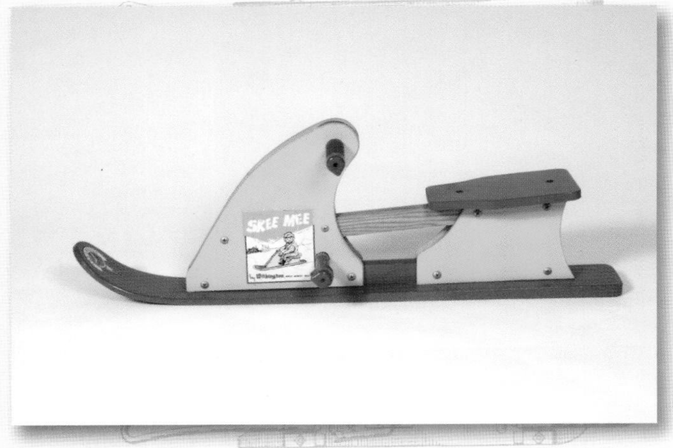

Withington Skee-Mee, c. 1965. $50-100. . *Courtesy of Paul Cote.*

Miscellaneous Companies & Unknown Manufacturers

Chicago Roller Skate Co. advertisement, c. 1920. $3-5.

Unusual Snow Hawk bike sled, distributed by American Golf Co. of Boston, manufacturer unknown. $125-250.

Postcard, 1882, Maine Manufacturing Company. There are no records available about this company, but note the diamond logo, perhaps indicating an early Ellingwood Turning Company connection. $30-50.

Columbia
Cycles

Model W-28

"COLUMBIA" COASTERS appeal to every youngster. Sturdy in construction. Glide along speedily and noiselessly.

Have regulation size Bodies, 16 x 36 Inches. Wheels have 1¼ Inch OVERSIZE RUBBER TIRES. Convenient Hand Brake on Rear Wheel

There are four models.

Models	Wheels	Body	Trimming	Price Each
D-26	11" Steel Disc	Natural Var.	Redient	$9.75
W-26	11½" Bicycle Spokes	" "	"	9.75
D-28	11" Steel Disc	Berkshire Blue	"	10.50
W-28	11½" Bicycle Spokes	" "	"	10.50

SUGGESTION:—"Columbia" Cycles last a long time, therefore it is advisable to let your Child grow into his or her "Columbia." Select a size which is large enough and be guided by the SIZE OF YOUR CHILD, rather than by age.

[12]

Columbia Cycles catalog, Coaster Wagon.

Special Terms on HOLIDAY GOODS See Pink Insert 313

"KING-OF-THE-HILL" STEERING SLEDS

Sold exclusively by Butler Brothers

$10.75 | $14.00 | $16.75

$20.00 | $25.00

"FLEXIBLE FLYER" STEERING SLEDS

$2.50 | 3.17 | 4.00

$4.33 | 5.83

"Flexible Flyer Racers"

Extra Long—Extra Strong

The Slim, Trim Racer That Every Boy Wants

$3.50 | $4.33

HIGH SLEDS WITH GUARDS

$21.00 | $3.00

SLED GUARDS

Adjustable

$7.50 | $10.75

"GO DINK" SCOOTER SLED

$16.00

"SKEEBOGGAN"

$1.70

MENGEL'S NOVELTY "SEE-SAW" CIRCLE

$6.00

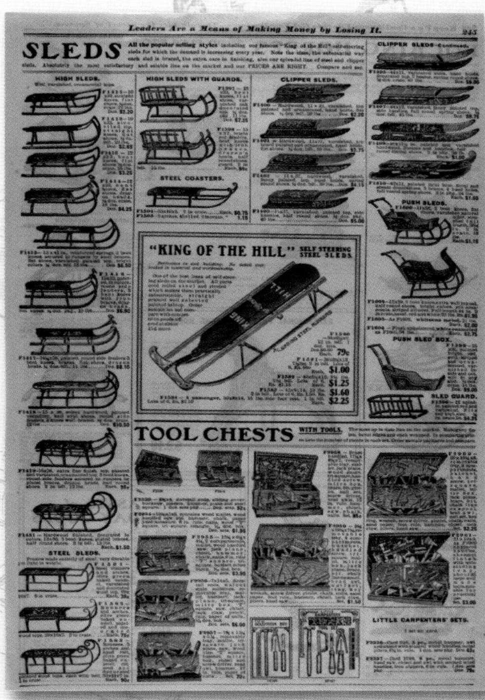

Leaders Are a Means of Making Money by Losing It.

SLEDS

HIGH SLEDS.

HIGH SLEDS WITH GUARDS.

CLIPPER SLEDS.

STEEL COASTERS.

"KING OF THE HILL" SELF STEERING STEEL SLEDS

STEEL SLEDS.

PUSH SLEDS.

PUSH SLED BOX.

SLED GUARD.

TOOL CHESTS WITH TOOLS.

LITTLE CARPENTERS' SETS.

King of the Hill Steering Sleds advertisements, c. 1915. Manufacturer unknown. $5-7.

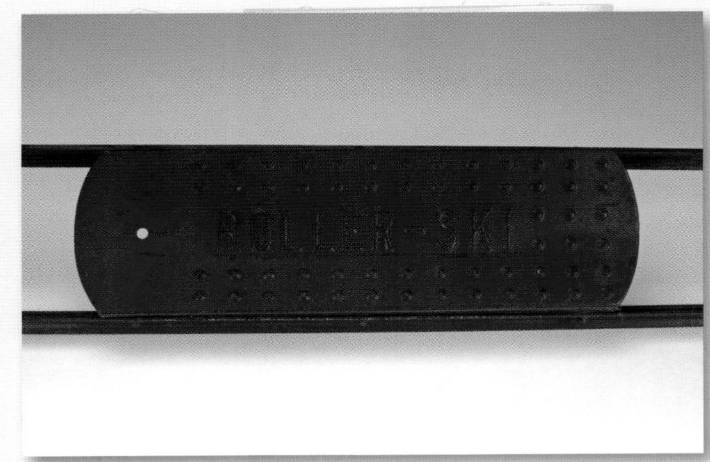

Pollock Roller Ski, c. 1928. L: 28"; W: 4"; H: 24". $150-200.

Rockaway Coaster Company advertisement, *Saturday Evening Post*, December 5, 1908. $5-10.

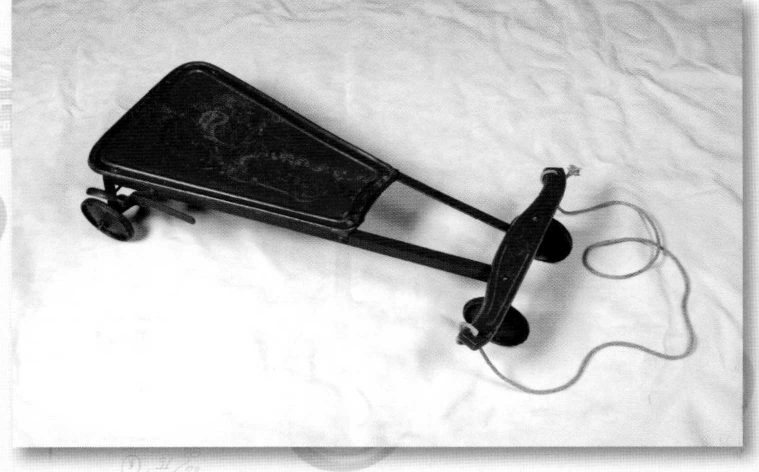

Rockaway "Snowless Coasting" sled. Note the hand brake and wheels. *Courtesy of Carroll "Skip" Palmer.* $350-500.

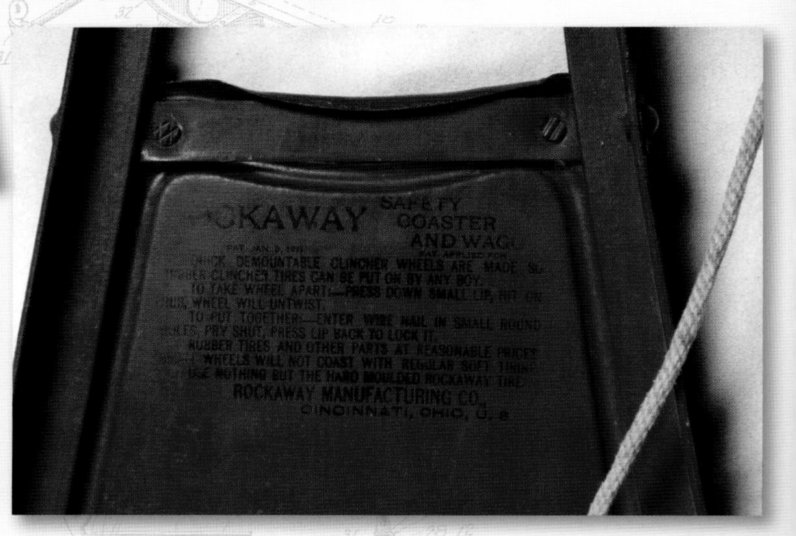

Skeeter, maker unknown, c. 1910. L: 34" overall, 21" deck; W: 5 1/2"; H: 23 1/2". $150-200.

WABASH STEEL HAND SLEDS

PATENTED APRIL 10, 1906

EXTRA FINE FINISH

Wabash Sled No. 136.

An extra long and stout 3-knee sled, with runners of 5/8 inch tee steel. Bright red and green, with hand decorations and striping. Assorted colors in crate.

Weight 120 lbs. per dozen.

Packed —— doz. K. D. in a crate.

No. 136—36 in. long, 12 in. wide, 6 in. high .... Price ——

Wabash Coaster No. 242.

Runners 5/8 inch tee steel. Extra long coaster, with three knees, finished and decorated same as No. 136. Assorted colors in crate.

Weight 120 lbs. per dozen.

Packed —— doz. K. D. in a crate.

No. 242—42 in. long, 12 in. wide, 4½ in. high .... *Price* ——

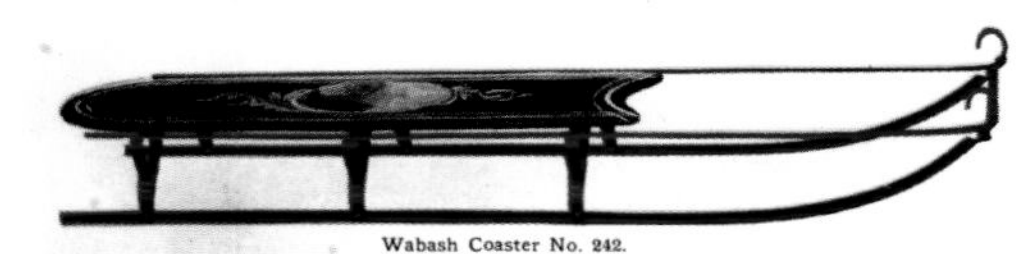

Wabash Sled No. 130.

Runners are 5/8 inch tee steel; beautifully finished in red and green; hand decorated and striped, assorted colors in crate.

Weight 100 lbs. per dozen.

Packed —— doz. K. D. in a crate.

No. 130—30 in. long, 12 in. wide, 6 in. high. Price ——

No. 133—33 in. long, 12 in. wide, 6 in. high. Price ——

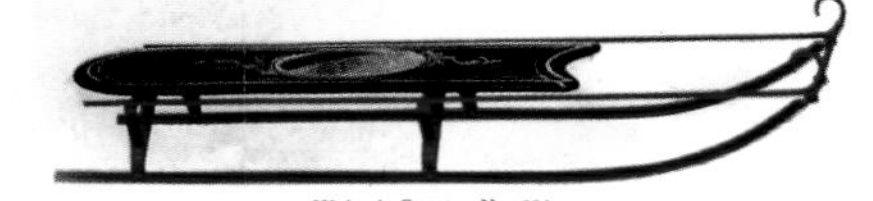

Wabash Coaster No. 234.

Runners are 5/8 inch tee steel; finished same as No. 130.

Weight 100 lbs. per dozen.

Packed —— doz. K. D. in a crate.

No. 234—34 in. long, 12 in. wide, 4½ in. high .... Price ——

No. 238—38 in. long, 12 in. wide, 4½ in. high .... Price ——

Wabash Steel Hand Sleds, manufacturer unknown. Advertising sheet, April 10, 1906. *Courtesy of Jim Pauzé.*

RACE-O-PLANE SLEDS

Race-o-Plane Sleds have always held the preference of youngsters because of their appealing snap in line and color. And the parents will appreciate the low prices at which these popular sleds are available. We suggest that you order your supply now to meet the first heavy snowfall demand.

A popular, low priced sled, sturdily constructed throughout to withstand any amount of rough usage. Runners are made of high carbon spring steel and finished in red. Top is made of clear grain hardwood, attractively decorated in bright red and protected with two coats of best spar varnish. Responds instantly to slightest pressure on steering bar. Size, 32x11¼ in.
No. 7301. List, $1.60. Each, 94c. Lots of 3, each, 89c, less 2%, net........................ **87¢**

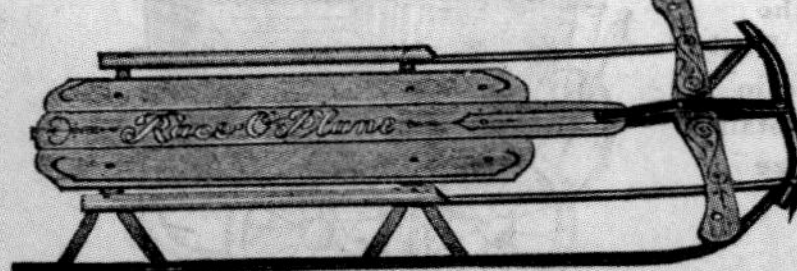

This swanky looking sled will strike joy to the heart of any red-blooded youngster. Designed for class, durability and speed. It is well made of heavy selected top boards and spring steel runners. Steering bars make it possible to have full control of direction. It is beautifully decorated and finished. Runners in red. Size, 45x12¾ in.
No. 7302. List, $3.00. Each, $1.80. Lots of 3, each, $1.65, less 2%, net........................ **1.62**

De Luxe Double Runner ICE SKEETER

A Real Toy for the Real Boy!

Any boy's or girl's heart will be thrilled with pride and joy as an owner of one of these new Ice Skeeters. It is a toy that will furnish them much pleasure and joy and at the same time afford them an opportunity to indulge in healthful outdoor exercise. It is new—it is different. Every boy and girl will want one. Sturdily made of steel throughout and fitted with bright polished steel double runners. Equipped with an anti-skid device which prevents the foot from slipping under any snow and ice conditions. This is a very important feature and will mean many more sales for you if properly stressed. The platform is 14x9 in. Size over all is 30 in. high by 42½ in. long by 5¾ in. wide. Finished in deep maroon. Case packed one to a carton—unassembled. Easy to assemble. Priced unusually low for high grade toy of this type. Weight, 9¼ lbs.

No. 7300. List, $1.75. Dealer's price, each, $1.20. Lots of 6, each, $1.10, less 2%.......... **1.08** Net

130

Ice Skeeter and Race-O-Plane sled advertisement, c. 1918. $5-10.

Maker unknown, attributed to the Goodyear Estate, c. 1880. Blue deck, horse head and flower. L: 38", W: 12" H: 7". $1500-3200. *Courtesy of Carole and Lou Scudillo.*

Cutter, maker unknown, c. 1880. Blue deck with lake and mountain scene. L: 36", W: 15" H: 9". $950-2400. *Courtesy of Carole and Lou Scudillo.*

Unidentified bob sled on the cover of *American Boy* magazine, January 1934.

Scene 3

Childs Toy, Salesman's Sample or U.S. Patent Model?

Let's begin with a child's toy. Toys are made for the purpose of bringing pleasure to any child regardless of age. They are magical and allow us to role play and perform the function of the original by using our imaginations. Toys are small versions, of the originals, not made to scale, and for the most part lack the detail, and in some cases lack the primary function of the original.

In contrast salesman's samples are exact replicas, made to scale and able to perform the same functions as the originals. Many companies made these samples to give their salesman a marketing tool to entice the perspective buyer into purchasing their product. In the case of sleds, one could have a hands-on demonstration of the unique features of the salesman's product. Salesman's samples vary in size but must be more than twelve inches square; most are between eighteen and twenty nine inches in length. As collectors we know how difficult they are to come by and how expensive they have gotten in the last decade. Unlike the sleds themselves, these miniatures are very rare and remain elusive. Remember rarity, supply and demand drives the prices up.

U.S. patent models are by far the ultimate "high" for any collector when they stumble upon one. These miniatures are as scarce as "snowballs in"... you guessed it... Hawaii. Companies produced different models of salesman's samples, to coincide with each model they sold, but U.S. patent models were not reproduced. In 1790 the U.S. Patent Office was created and a three member Patent Commission was established. The first Commission consisted of Thomas Jefferson, Secretary of State, Henry Knox, Secretary of War, and Edmond Randolph, Attorney General. They established the requirement that a working model of each invention be produced in miniature. Anyone applying for a patent from 1790 to 1880 was required to submit paperwork with diagrams explaining the items purpose, construction and operation, along with a working model not to exceed twelve inches square.

Each model was to be neatly made, with the name of the inventor printed or engraved on it. Although the requirement for models was discontinued in 1890, models were still being made into the early 1900s. If you are fortunate to come across a patent model, be prepared to pay anywhere from $500 up, the limit being what someone is willing to pay.

Once again, let me caution you about price. The prices that appear in this book are to be used only as a guide. I do not claim or intend to set a fair market price. The prices that appear in this book are based on my own experiences in the years that I have been collecting. Ultimately a fair market price is determined solely by the buyer and seller.

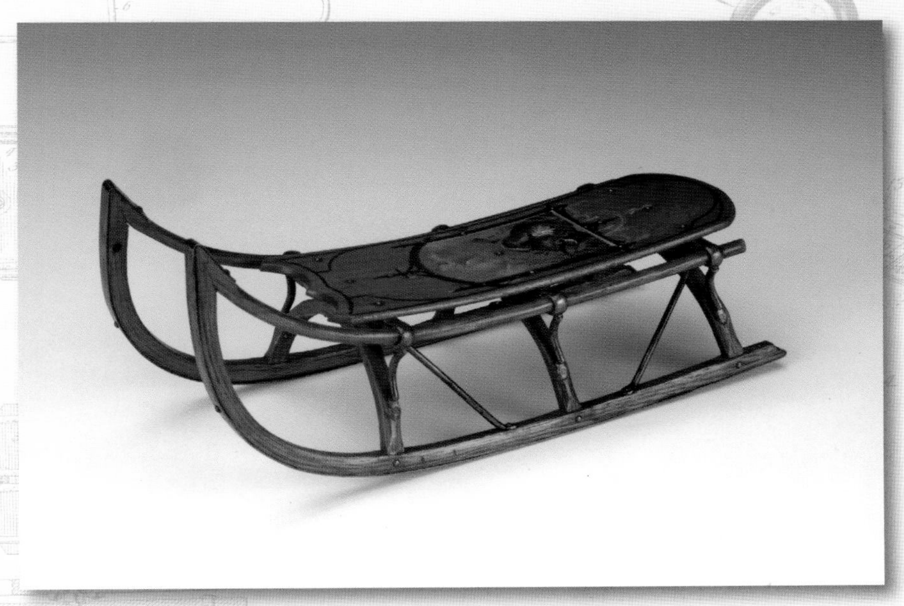

Paris Mfg. Co., U.S. Patent model, c. mid-1860s. L: 9 ½" W 3 ¾" at runners, H 2". $5,500-6,500. *Courtesy of Sam Tressler.* *(continued on following page)*

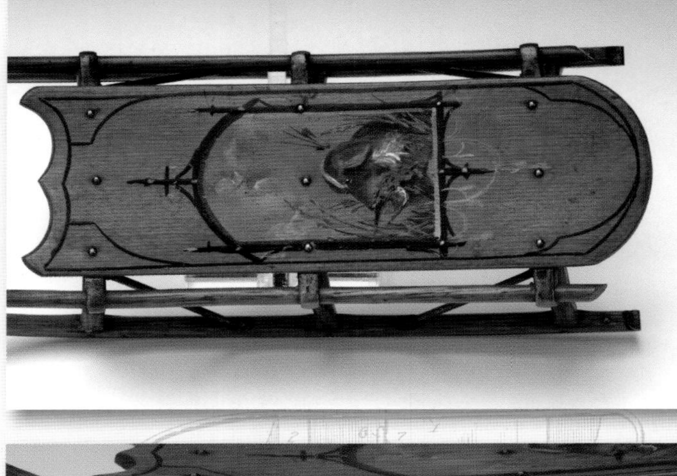

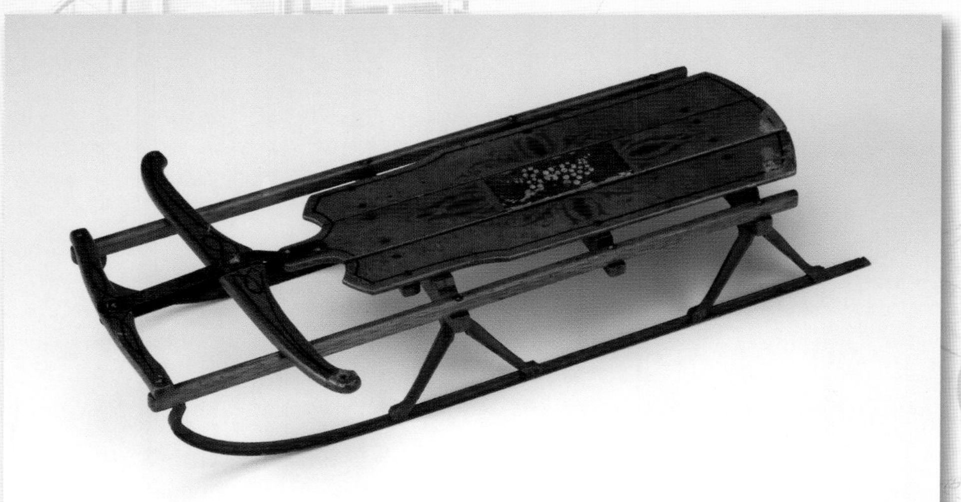

Flexible Flyer, c. 1888. U.S. patent model. L: 11"; W: 3-1/4"; H: 1-1/2". *Courtesy Sam L. Allen III.*

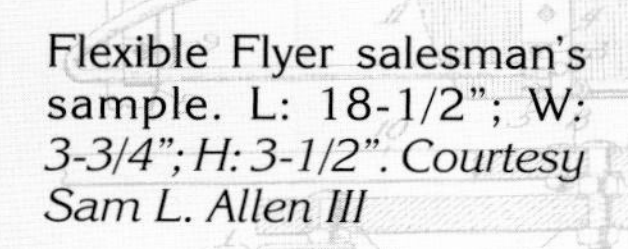

Flexible Flyer salesman's sample. L: 18-1/2"; W: *3-3/4"; H: 3-1/2". Courtesy Sam L. Allen III*

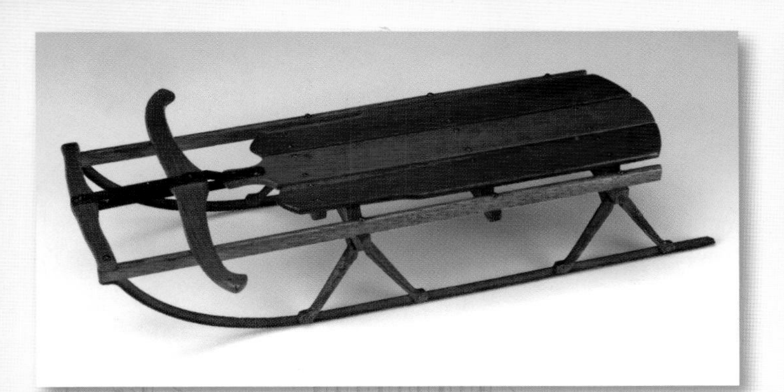

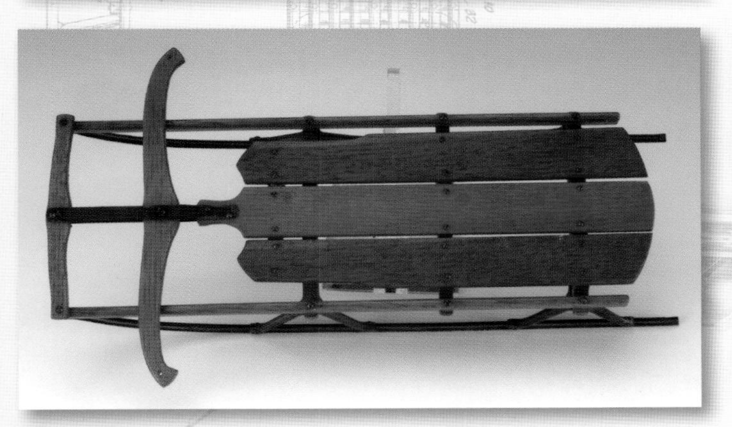

Flexible Flyer, U.S. Patent model, c. late 1890s. L: 10 7/8"; W. 3 ¾" at runners; H 2". $900-1,000. *Courtesy of Sam Tressler.*

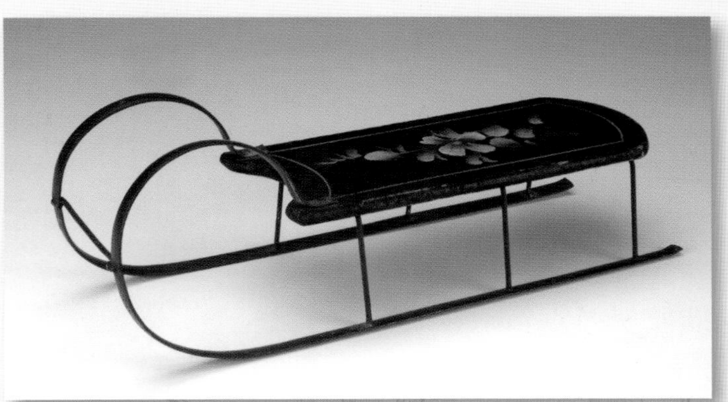

Painted doll sled, c. 1900. L: 15 ½"; W: 5 ½"; H: 3 ¼". $1200. *Courtesy of Sam Tressler.*

Flex-O-Fold U.S. patent model, c. 1901. L: 12"; W: 4"; H: 2". $500-750.

Raithel Cart "Baby Carriage" salesman's sample, c. 1904. L: 21 1/2"; W 9 1/2"; H: 17 1/2". $2,500-3,000. *Courtesy of Sam Tressler.*

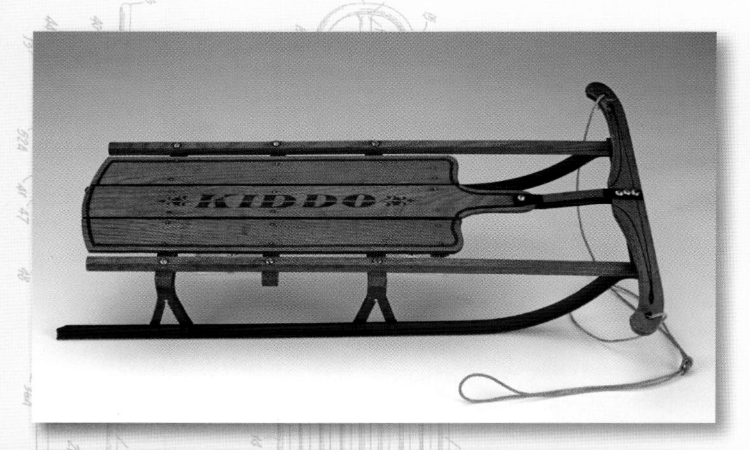

Champion Racer salesman's sample, c. 1910. L: 18 ¾"; W: 5 7/8"; H: 3 1/8". $2,000-$2,800. *Courtesy of Sam Tressler.*

Kiddo salesman's sample, c. 1908. L: 18 ¾"; w: 6 ¾"; H: 3 ¼". $1,800-2,500. *Courtesy of Sam Tressler.*

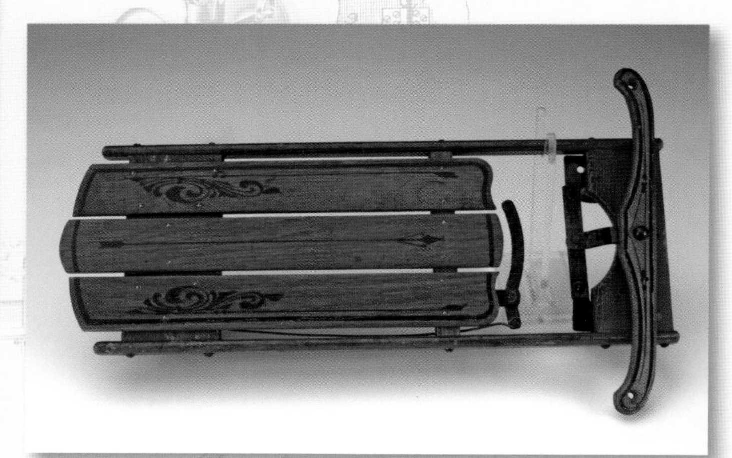

Buffalo Sled Co. Glider Roll, wheeled coaster, salesman's sample, August 16, 1910. L: 17"; W 6 ¼"; H: 3". $2,500-3,000. Note the handbrake assembly. *Courtesy of Sam Tressler.*

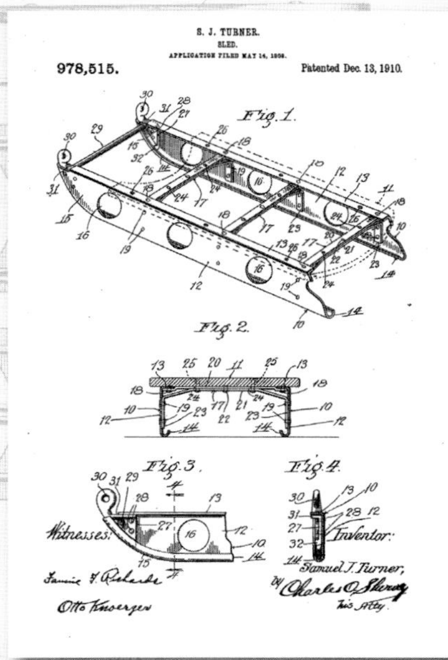

Turner & Thomas, Chicago, Illinois, Steel Flyer #10, U.S. Patent model, December 13, 1910. L: 8 ½"; W: 3"; H: 1 ¼". There are no company records available. Very rare, $1,800-$2,500. *Courtesy of Sam Tressler.*

Patent drawing, 1910, for the Turner & Thomas Steel Flyer.

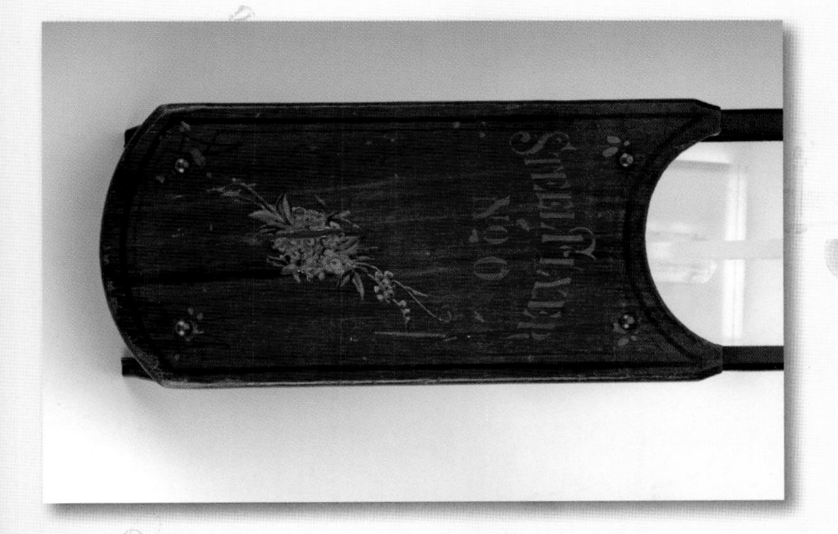

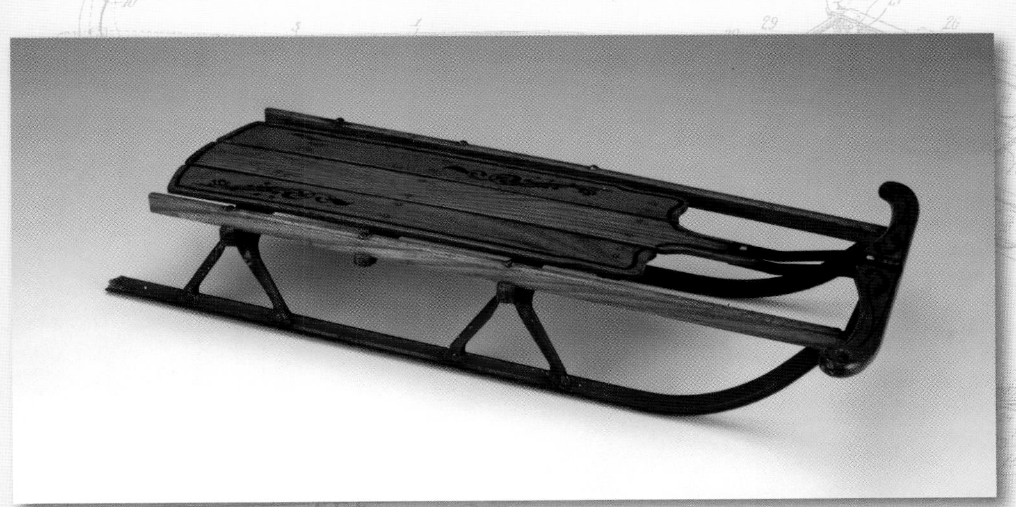

Buffalo Sled Co., unmarked Fleetwing salesman's sample, c. 1915. L: 20"; W 6 3/8"; H: 3 ¼". $1,500-2,000. *Courtesy of Sam Tressler.*

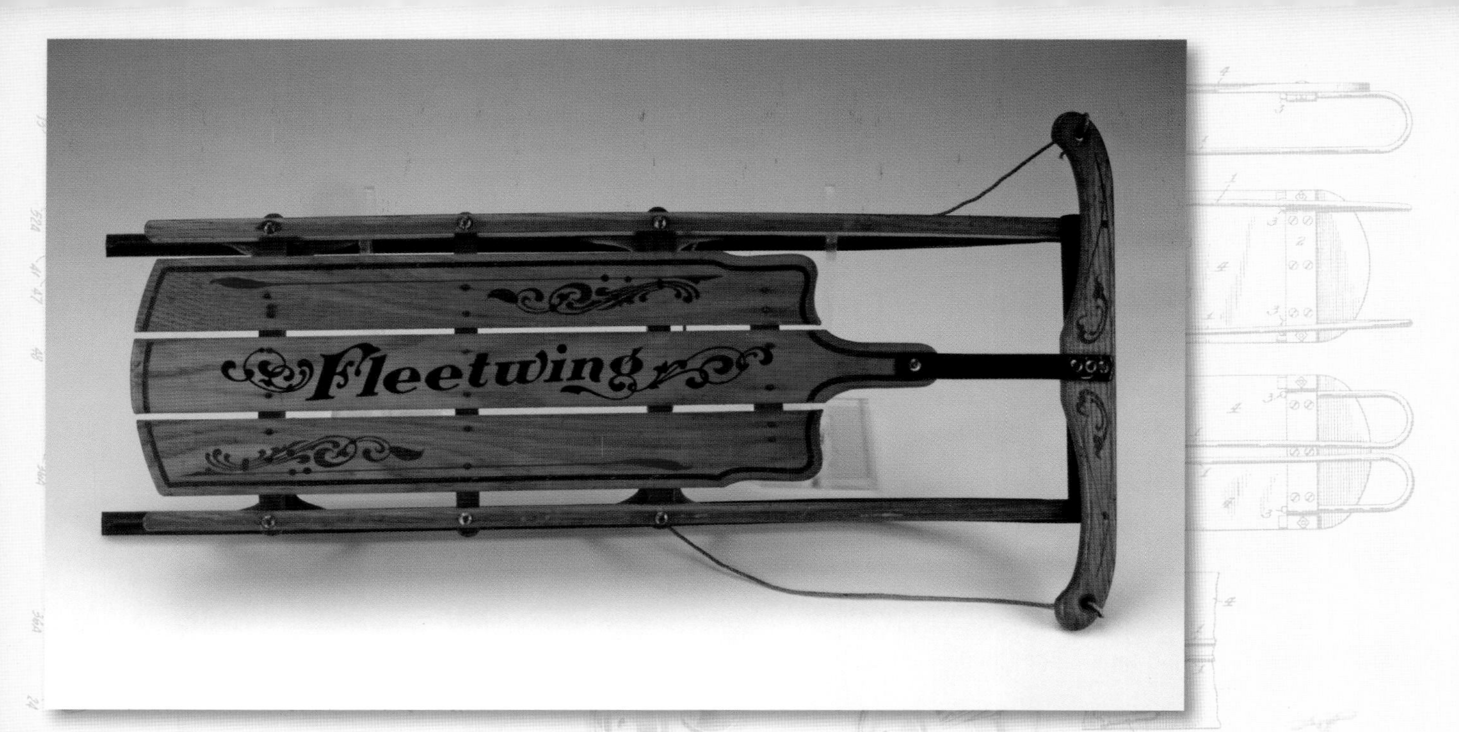

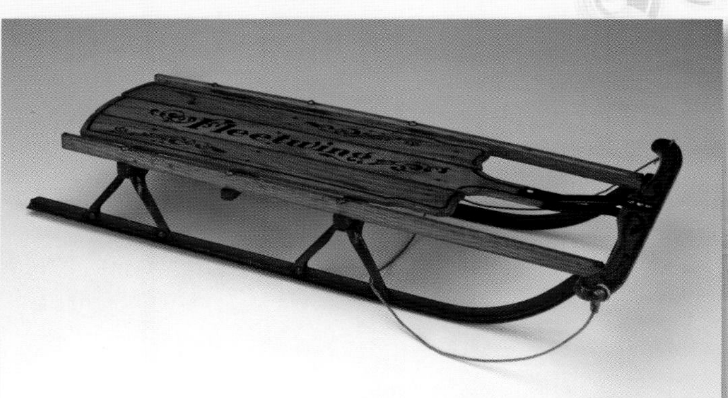

Buffalo Sled Co., Fleetwing (black) salesman's sample, c. 1915. L: 20"; W 6 3/8"; H: 3 ¼". $2,500-3,000. *Courtesy of Sam Tressler.*

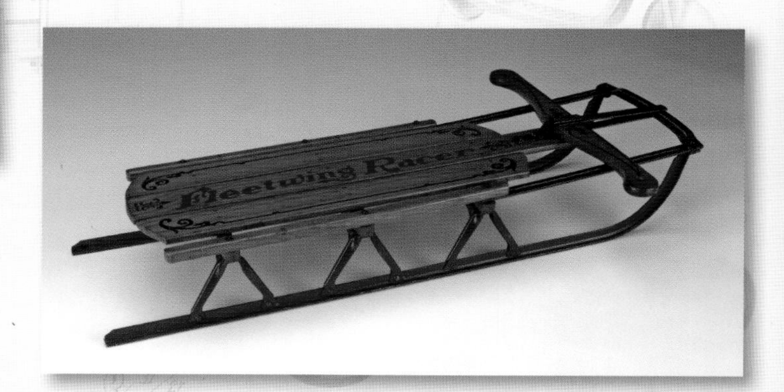

Auto Wheel Coaster Co., Fleetwing Racer salesman's sample, c. 1921. L: 22 1/2"; W 6 1/2"; H: 3 ¼". $2,500-3,000. *Courtesy of Sam Tressler.*

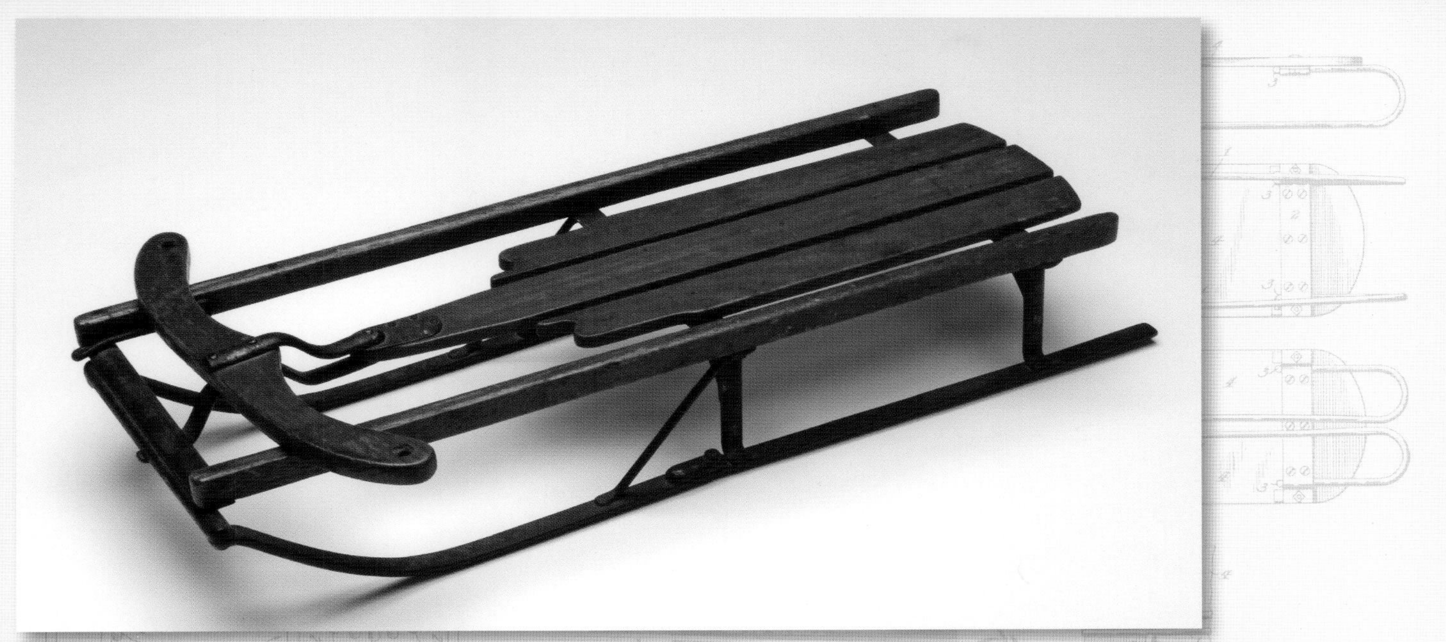

Safety Sled Co., salesman's sample. C. 1915, unmarked. L: 15"; W: 5-1-1/2"; H: 2-3/4". $250-500. *Courtesy of Randy Hagg.*

Safety Sled Co., Joy Ride, c. 1916. U.S. patent model. L: 9"; W: 3"; H: 1-3/4". $500-1000. *Courtesy of Randy Hagg.*

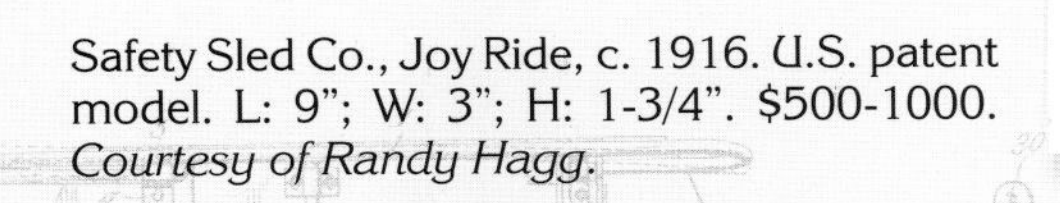

Safety Sled Co., Race-O-Plane, c. 1919. U.S. patent model. L: 10"; W: 3-1/2"; H: 1-3/4". $500-1000. . *Courtesy of Randy Hagg.*

Auto Wheel Coaster Co., Imperial Racer salesman's sample, c. 1930. L: 27"; W: 7"; H: 3 1/2". $750-1000.

16 AUTO-WHEEL COASTER COMPANY, Inc.

King of 'Em All
Exclusive Design
Superior Construction
Attractively Finished

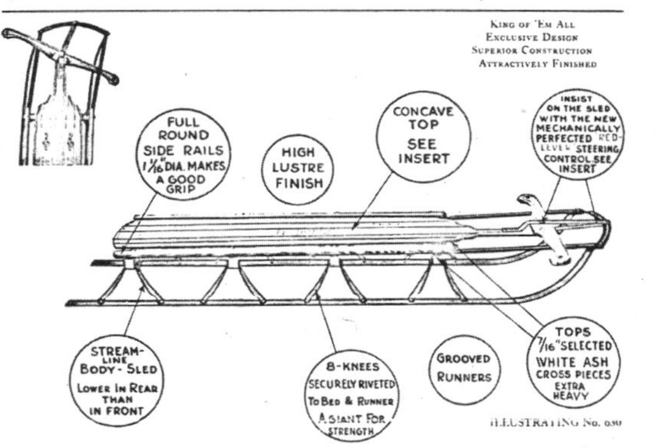

Concave Top

THE IMPERIAL RACER

BUILT FOR SPEED AND SERVICE

STREAM line Body, made from five pieces of selected kiln-dried Northern White Ash. Center and side tops are handsomely shaped, securely nailed to heavy curved supports, making these models distinctive in design and construction. Side rails also made from Northern White Ash perfectly rounded and free from sharp edges.

Steering bar is made from selected kiln-dried Ash, handsomely shaped, set back and operated through a mechanically perfect RED LEVER arrangement, which insures immediate response with less strain.

Steel front formed with an upward sweep supported by two heavy steel side rails, while the runners are manufactured from high grade special analysis grooved steel, securely riveted to body and so designed that the runners remain parallel.

Finish—Body and side rails attractive green stain finish, ornamented in red, lettered in red. Steel front enameled in red and attractively shaped. Runners enameled in bright red.

SPECIFICATIONS

MODEL NO.	SIZE	HEIGHT FRONT	HEIGHT REAR	KNEES	WEIGHT PER BUNDLE	PACKED
370	37" x 13½"	6¾"	6½"	4	20½ lbs.	2 per bundle—set up
420	42" x 13½"	6¾"	6½"	4	23 lbs.	2 per bundle—set up
470	47" x 13¾"	7½"	7"	6	28½ lbs.	2 per bundle—set up
630	63" x 14¾"	7½"	6¾"	8	38 lbs.	2 per bundle—set up

Imperial Racer, Auto-Wheel Coaster Company advertisement, c. 1930. $5-10.

Ellingwood, Oxford salesman's sample sled, L: 21"; W: 5-1/2"; H: 2-1/2". $1200-1500. . *Courtesy of Paul Cote.*

Scene 4

Coming Attractions:

Every year the market is flooded with new Snow Vehicles. Names like Snow Tracker, Blizzard, and Tornado are beckoning to today's youth "flop on" and enjoy the thrill of a life time! Sheer speed is what sells. The trend for fast sleds has sparked a new industry aimed at adults. Be on the lookout for the new kids on the block, Captain Avalanche, The Hammerhead. and Mad River Rocket to name a few. These are not your parents, "Flexible Flyers" Will they stand the test of time? Only you can be the judge. Today's Sleds are the Collectibles that will be tomorrows Antiques!

References

Broome County Library, Binghamton, NY. Mr. Charles Browne (Wilkerson Mfg. Co.)

History of Broome County, Syracuse, NY. D. Mason & Co. Publishers 1885 (Wilkerson Mfg. Co.)

Canal Town Museum, Canastota, NY. Mr. Joseph Di Giorgio, President & Village Historian, David Sadler (Sherwood Brothers Mfg. Co.)

Coldwater Public Library, Coldwater, Mich. Ms. Nola Baker (Pratt Mfg. Co.)

Greater Harvard Area Historical Society, Harvard, Ill. Ms. Elaine Fiducci Curator (Hunt, Helm, & Ferris)

Historical Society of Aurora, Ill. Mr. Dennis Buck, Historian & Ms. Barbara Peck, Researcher. (Richards-WilcoxMfg. Co)

Historical Society of Seneca Falls, NY. Ms. Kathy Jans-Duffy (Maynard Miller)

Historical Society of the Tonawanda's, NY Mr. Richard Dutton, Historian & Ms. Jane Penrose past Curator

Klyne Esopus Museum, Hamlet of Ulster Park, NY. Mr.Alexander Contini, President of the Klyne Esopus Museum and Kingston Historian Mr. Edwin Ford (Crosby-Gilzinger &Co.)

Maplewood Craft Academy, Hutchinson, Minn. Mr. Kenneth Ellstrom, Vice President for Finance (Steermaster BobSki)

Mt. Jewett Public Library, Mt. Jewett, PA. Ms. Edith Raught (Safety Sled Co.)